Methods in Molecular Biology

Series Editor
John M. Walker
School of Life and Medical Sciences
University of Hertfordshire
Hatfield, Hertfordshire, UK

For further volumes:
http://www.springer.com/series/7651

For over 35 years, biological scientists have come to rely on the research protocols and methodologies in the critically acclaimed Methods in Molecular Biology series. The series was the first to introduce the step-by-step protocols approach that has become the standard in all biomedical protocol publishing. Each protocol is provided in readily-reproducible step-by step fashion, opening with an introductory overview, a list of the materials and reagents needed to complete the experiment, and followed by a detailed procedure that is supported with a helpful notes section offering tips and tricks of the trade as well as troubleshooting advice. These hallmark features were introduced by series editor Dr. John Walker and constitute the key ingredient in each and every volume of the Methods in Molecular Biology series. Tested and trusted, comprehensive and reliable, all protocols from the series are indexed in PubMed.

Manuscript Instructions https://drive.google.com/file/d/1K4Az5vpAFVzYadJ3jKx2-QSo7Dqa_SNA/view?usp=sharing

Biofuels from Algae

Methods and Protocols

Second Edition

Edited by

Kristian Spilling

Marine and Freshwater Solutions Unit, Finnish Environment Institute, Helsinki, Finland

Humana Press

Editor
Kristian Spilling
Marine and Freshwater Solutions Unit
Finnish Environment Institute
Helsinki, Finland

ISSN 1064-3745 ISSN 1940-6029 (electronic)
Methods in Molecular Biology
ISBN 978-1-0716-5030-1 ISBN 978-1-0716-5031-8 (eBook)
https://doi.org/10.1007/978-1-0716-5031-8

This Humana imprint is published by the registered company Springer Science+Business Media, LLC, part of Springer Nature
The registered company address is: 1 New York Plaza, New York, NY 10004, U.S.A.

Preface

Algae are a diverse group of photosynthetic organisms, ranging from microscopic microalgae to large, multicellular macroalgae (commonly known as seaweeds) that, with some exceptions, live in aquatic environments. Unlike terrestrial crops, algae do not require arable land or freshwater to thrive, and many species grow rapidly, converting carbon dioxide and sunlight into biomass with remarkable efficiency. This unique set of characteristics has made algae the focus of sustained interest as a renewable source of raw material for biofuel production and other biotechnological applications.

In recent decades, global challenges such as climate change, energy security, and the need for sustainable resource utilization have spurred extensive research into alternative fuel sources. Among these, algal biofuels stand out for their high productivity potential and the versatility of the biomass they produce. Algal species can be cultivated in controlled systems or harvested from natural waters, and depending on their composition—rich in lipids, carbohydrates, or proteins—can be processed into biodiesel, bioethanol, biogas, or even jet fuel. Still, this is done only at a miniscule scale, and a major challenge has been to develop cost-effective cultivation and harvesting technologies.

This second edition of *Biofuels from Algae: Methods and Protocols* reflects the rapid advances in algal biotechnology over the past years and incorporates both foundational and newly developed protocols for working with algal cultivation systems and biomass.

This volume is structured to guide the reader from initial cultivation to advanced biochemical analysis, with a particular focus on reproducibility and accessibility:

- The first part of this book provides a broad overview of algal biology, the rationale for using algae in biofuel production, and considerations for selecting appropriate strains and cultivation systems. These chapters are particularly suitable for newcomers to the field and for researchers transitioning from other areas of biotechnology.
- This is followed by chapters providing advanced laboratory protocols for quantifying key biochemical components of algal biomass, including carbohydrates, lipids, and proteins. Although these methods require specialized equipment and expertise, they are essential for evaluating both the cultivation environment and the algal biomass produced.
- The final chapters focus on what to consider when developing an algal industry, such as its environmental impact, and what tools and methods can be used in the continued development of this field.

Our goal with this book is to make algal biofuel research more accessible, supporting a wide audience that includes students, early-career researchers, and experienced practitioners. All chapters from the first edition that made it into the second edition have been updated to reflect current best practices, emerging technologies, and recent insights in the field. In addition, this second edition features ten new chapters covering previously unaddressed topics.

We hope that this second edition will serve as a practical resource for all those interested in algal biofuel research, enabling continued innovation in the cultivation and utilization of algae.

Helsinki, Finland *Kristian Spilling*

Contents

Contributors

ABBAS GHASSEMI • *Department of Civil and Environmental Engineering, University of California Merced, Merced, USA*

ALAWI SULAIMAN • *Faculty of Plantation and Agrotechnology, Universiti Teknologi MARA (UiTM), Shah Alam, Selangor, Malaysia*

ALEXANDER GOLBERG • *School of Mechanical Engineering, Faculty of Engineering, Tel Aviv University, Tel Aviv, Israel*

DANIEL A. WHITE • *Plymouth Marine Laboratory, Plymouth, Devon, UK*

DANIELLE BANSFIELD • *Department of Built Environment, Aalto University, Espoo, Finland; Marine and Freshwater Solutions Unit, Finnish Environment Institute, Helsinki, Finland*

DAVID J. COOK • *The University of Nottingham, Loughborough, Leicestershire, UK*

DAVID W. TEMPLETON • *BioEconomy and Sustainable Transportation Directorate, National Renewable Energy Laboratory, Golden, CO, USA*

ELIF DEMIRKAN • *Department of Biology, Faculty of Arts and Sciences, Bursa Uludag University, Bursa, Turkey*

EMILY T. KOSTAS • *UCL Manufacturing Futures Lab, Department of Biochemical Engineering, University College London, London, UK*

FAKHRA LIAQAT • *Department of Biology, Faculty of Arts and Sciences, Bursa Uludag University, Bursa, Turkey*

HEIKO RISCHER • *VTT Technical Research Centre of Finland Ltd, Espoo, Finland*

HELI NYGREN • *VTT Technical Research Centre of Finland Ltd, Espoo, Finland*

HOMA HOSSEINZADEH-BANDBAFHA • *Environmental Research and Sustainability Division, Malus Ecolife Inc., Halifax, NS, Canada*

INGRID MARIBU • *Department of Marine Biotechnology, Nofima – Norwegian Institute of Food, Fisheries and Aquaculture Research, Tromsø, Norway; Norwegian College of Fishery Science, Faculty of Biosciences, Fisheries and Economics, UiT – The Arctic University of Norway,, Tromsø, Norway*

IRFANA IQBAL • *Department of Biotechnology, Lahore College for Women University, Lahore, Pakistan*

JAANA KOISTINEN • *Tvärminne Zoological Station, University of Helsinki, Hanko, Finland*

JACQUELINE JERNEY • *eb&p Umweltbüro GmbH, Klagenfurt, Austria*

JAN ARNE ARNESEN • *Department of Marine Biotechnology, Nofima – Norwegian Institute of Food, Fisheries and Aquaculture Research, Tromsø, Norway*

JEAN-BAPTISTE THOMAS • *Department of Sustainable Development, Environmental Science and Engineering, KTH Royal Institute of Technology, Stockholm, Sweden*

JESSICA L. OLSTAD • *BioEconomy and Sustainable Transportation Directorate, National Renewable Energy Laboratory, Golden, CO, USA*

JUKKA SEPPÄLÄ • *Finnish Environment Institute, Helsinki, Finland*

KATARIINA NATUNEN • *Finnish Environment Institute, Helsinki, Finland*

KIIA SILVENNOINEN • *Circular Economy Solutions Unit, Finnish Environment Institute, Helsinki, Finland*

KRISTIAN SPILLING • *Centre for Coastal Research, University of Agder, Kristiansand, Norway; Marine and Freshwater Solutions Unit, Finnish Environment Institute, Helsinki, Finland*

LARA D. MATEOS • *Aalto University, Espoo, Finland; Marine and Freshwater Solutions Unit, Finnish Environment Institute, Helsinki, Finland*

LÉA BRAUD • *Department of Sustainable Development, Environmental Science and Engineering, KTH Royal Institute of Technology, Stockholm, Sweden*

LIEVE M.L. LAURENS • *BioEconomy and Sustainable Transportation Directorate, National Renewable Energy Laboratory, Golden, CO, USA*

MAHAMMED ILYAS KHAZI • *Department of Biology, Faculty of Arts and Sciences, Bursa Uludag University, Bursa, Turkey; Department of Biotechnology, Faculty of Natural and Applied Sciences, Ege University, Izmir, Turkey*

MARI VANHARANTA • *Research Infrastructure, Finnish Environment Institute, Helsinki, Finland*

MEISAM TABATABAEI • *Higher Institution Centre of Excellence (HICoE), Institute of Tropical Aquaculture and Fisheries (AKUATROP), Universiti Malaysia Terengganu, Kuala Nerus, Terengganu, Malaysia*

MELTEM CONK DALAY • *Department of Bioengineering, Faculty of Engineering, Ege University, Izmir, Turkey*

MERVI SJÖBLOM • *Tvärminne Zoological Station, University of Helsinki, Hanko, Finland*

MOHAMMADALI KIEHBADROUDINEZHAD • *Department of Process Engineering and Applied Science, Dalhousie University, Halifax, NS, Canada; Research and Innovation Division, Malus Ecolife Inc., Halifax, NS, Canada*

MOHAMMAD POOYA NAGHSHBANDI • *Department of Microbial Biotechnology, School of Biology, College of Science, University of Tehran, Tehran, Iran*

MORTAZA AGHBASHLO • *Department of Agricultural Machinery, Faculty of Agriculture, College of Agriculture and Natural Resources, University of Tehran, Karaj, Iran*

MUHAMMAD NAUMAN AFTAB • *Institute of Industrial Biotechnology, Government College University, Lahore, Pakistan*

NIELS VAN BOXTEL • *VTT Technical Research Centre of Finland Ltd, Espoo, Finland*

PASI YLÖSTALO • *Finnish Environment Institute, Helsinki, Finland*

SARA SAUKKONEN • *Circular Economy Solutions Unit, Finnish Environment Institute, Helsinki, Finland*

STEFANIE VAN WYCHEN • *BioEconomy and Sustainable Transportation Directorate, National Renewable Energy Laboratory, Golden, CO, USA*

STUART J. WILKINSON • *The University of Nottingham, Loughborough, Leicestershire, UK*

TANJA MYLLYVIITA • *Climate Solutions Unit, Finnish Environment Institute, Helsinki, Finland*

THOMAS ROHRLACK • *Faculty of Environmental Sciences and Natural Resource Management, Ås, Norway*

TIMO JENSEN • *Coastal Research & Management GbR, Kiel, Germany*

TOBIAS LIPSEWERS • *Independent Consultant, Rietberg, Germany*

TUULIKKI SEPPÄNEN-LAAKSO • *VTT Technical Research Centre of Finland Ltd, Espoo, Finland*

ZELIHA DEMIREL • *Department of Bioengineering, Faculty of Engineering, Ege University, Izmir, Turkey*

Chapter 1

Basic Methods for Isolating and Culturing Microalgae

Kristian Spilling

Abstract

This chapter presents basic methods for isolating and setting up a microalgae culture. Isolation can be done from local algal strains with a micropipette and microscope. Determining algal growth is important for optimizing growth and can be done by counting cells under a microscope or by determining the change in dry weight (DW).

Keywords Microalgae, Cultivation, Growth rate

1 Introduction

Microalgae are found all around us and exhibit great taxonomic and functional diversity. A single drop of water may contain thousands of algal cells and they can be found free living in aquatic habitats (lakes, streams, and ocean), in symbiosis with other organisms (lichen, corals), or forming biofilms on moist surfaces (forming mats on the shore line). Common for all microalgae is that they convert CO_2 to organic carbon during photosynthesis; in this process, they need light energy and inorganic nutrients, most notably nitrogen (N) and phosphorus (P) in addition to some trace elements like iron (Fe). With all the diversity, several microalgae species offer possibilities to grow and harvest valuable compounds like fatty acids (e.g., omega 3 fatty acids), carbohydrates proteins, or more complex secondary metabolites. For biofuel purposes, lipids in particular have attracted a lot of attention [1].

Although there have been a lot of studies over the past decades, e.g., on algal lipid production [1, 2], most of the scientific literature is concentrated on a handful of easily cultivated species, and new approaches are needed [3]. In order to expand cultivation efforts to taxa that have been less studied, there is a need for isolating, culturing, and studying more species of microalga. In this chapter I outline some basic methods for doing this: isolating new algal strains and following their growth.

Kristian Spilling (ed.), *Biofuels from Algae*, Methods in Molecular Biology, vol. 2996,
https://doi.org/10.1007/978-1-0716-5031-8_1,

2 Materials

2.1 Isolation of Microalgae

1. Inverted microscope
2. Glass slides
3. Growth medium (*see* **Note 1**)
4. Glass pipette
5. Rubber bulb for the glass pipettes
6. Bunsen burner
7. Cultivation well plates (5 ml)
8. Cultivation flasks (50 ml)
9. Cultivation light with timer

2.2 Determining Biomass Concentration

1. Inverted microscope
2. Sedimentation chamber with grid (e.g. Sedgwick-Rafter counting chamber)
3. High precision scale
4. Glass fiber filters (Ø 25 mm)
5. Filtration stand and vaccum pump
6. Oven(s) capable of keeping 60°C and 450°C

3 Methods

3.1 Establishing an Algal Culture

1. Collect water from a local environment that could contain species suitable for cultivation (*see* **Note 2**).
2. Make a micropipette from a normal glass pipette by heating it up over a Bunsen burner, remove it from the flame, and pull the glass tube to a very thin capillary glass tube (*see* **Note 3**).
3. Break or cut the capillary glass tube 1–2 cm from the base of the glass pipette.
4. Put a drop of the collected water on a glass slide and place it under the microscope.
5. Move and focus on a single algal cell/unit that is of interest.
6. Move the micropipette over the field of view above the drop (the shade of it can be seen) and dip it into the water. See that the cell of interest is sucked into the capillary glass tip of the micropipette (*see* **Note 4**).
7. Put a drop of clean medium on a glass slide.
8. Push the content in the micropipette into the clean medium drop (you can do this by either applying pressure with your thumb or attaching the rubber bulb).

9. Put the new glass slide under the microscope and search for the cell of interest.
10. Repeat Steps 4–9 until you are left with only the cell of interest.
11. Suck up the whole content and put it into a 5 ml well that you have pre-filled with growth medium.
12. Repeat Steps 4–12 until you have established several clones of the same species or several species that you are interested in getting in culture.
13. Place the well plate under the light which should be set to a 16 h light: 8 h dark cycle. The temperature should preferably be close to the temperature of the collected water (Step 1).
14. Inspect the well plate visually on a daily basis; when the culture turns visibly colored, place a small volume under the microscope in order to check for contamination by unwanted algal species or heterotrophic organisms.
15. If no contamination can be seen, take a sub-sample with a pipette and inoculate the 50 ml cultivation bottle that has been prefilled with cultivation medium. This is your base culture (*see* **Note 5**).
16. The base culture needs to be monitored regularly (weekly) and reinoculated into fresh medium when growth slows down. The base culture can also be used for setting up larger cultures (*see* **Note 6**).

3.2 Monitoring Growth

1. Take a small sub-sample from your cultivation and put it onto a sedimentation chamber with or without a grid (the optimal size of the sedimentation chamber will depend on the concentration of cells).
2. Count the cells in a grid or fields of view and enumerate >400 cells. If the cells are motile, you could try using heat shock or adding a fixative to make counting easier.
3. Calculate the number of cells per milliliter based on the grids counted; if you use fields of view you have to first calculate the area in the field of view based on the microscope objective that you are using (*see* **Note 7**).
4. After monitoring cell concentration over time, the growth rate per day (μ) can be calculated according to: $\mu = \ln(\text{cell}_t/\text{cell}_0)/dt$; where cell_0 and cell_t are the cell concentration at time 0 and t, respectively, and dt is the difference in time (in days) between time 0 and t.
5. For determining dry weigh (DW), prepare glass fiber filters by combusting them at 450°C for 4 h and weigh them. You need to keep track of individual filters as their weight may vary up to 20%.

6. Apply the filter in a filtration stand, apply suction, and filter a known volume of culture onto the filter.
7. Dry the filter over night at 60°C
8. Weight the filter again and calculate the DW from the difference in weight before and after filtration. Divide the DW by the filtered volume to get DW/L.
9. Combust the filter, 450°C for 4 h, and weigh it again.
10. Calculate the ash weight (AW) from the difference between the weight of the filter after combustion and before filtration.
11. Calculate the ash free dry weight (AFDW) from the difference between DW and AW (*see* **Note 8**).

4 Notes

1. A range of different cultivation medium exists; the main components are a source of nitrogen (normally ammonium or nitrate), phosphorus (orthophosphate), and trace elements like iron. See, e.g., Harrison and Berges [4] for an overview. A good overall fertilizer for agriculture can be used, but some species will require special medium recipes to grow well. Testing different growth media is important when setting up new cultures.
2. For example, extreme habitats such as rock pools along the coast will be inhabited by species adapted to large fluctuations in temperature and salinity, which are traits that could make them suitable for cultivation.
3. It is critical to remove the pipette from the flame before pulling; if not the tip of the micropipette will close and cannot be used. An example is shown in Fig. 1.

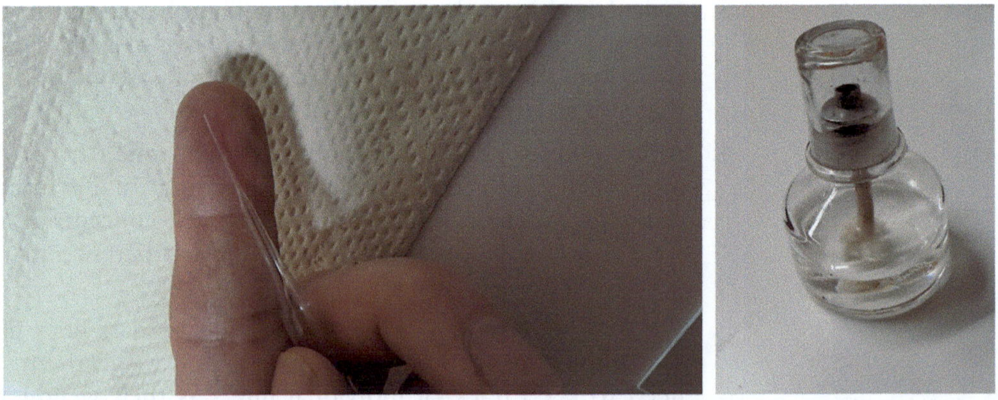

Fig. 1 An example of micropipette easily made from a standard glass pipette with the help of a burner. Photo by K. Spilling

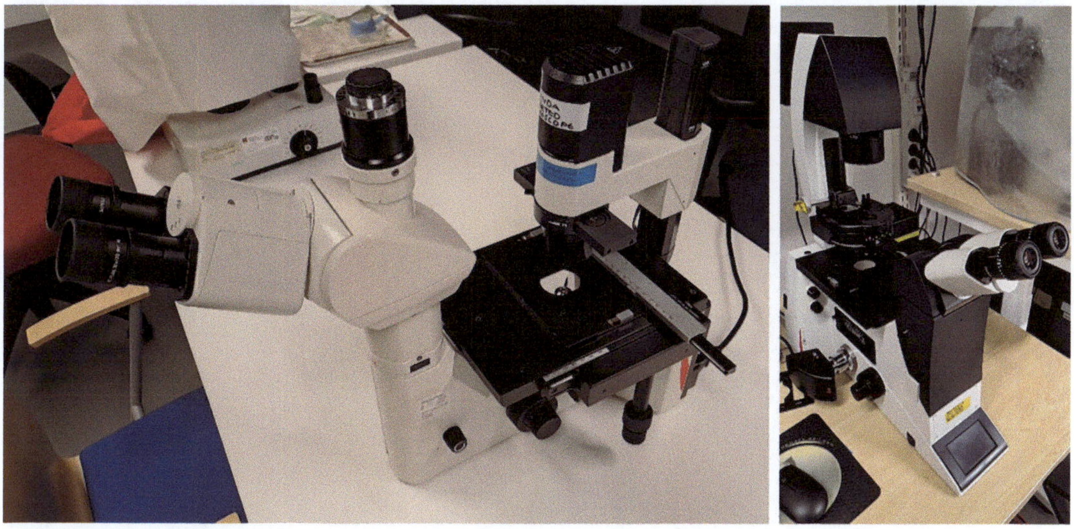

Fig. 2 Examples of different microscopes that could be used for isolation. The simpler version on the left is better for this purpose as it has more room on top of the microscopy slide than the more advanced microscope on the right. Photo by K. Spilling

4. When doing this, a simple inverted microscope with more room for maneuvering is better than a more advanced microscope that often has less room above the microscopy slide (Fig. 2). You would likely have sucked up other cells that you are not interested in and one or several washing steps might be needed in order to isolate only the cell of interest (Fig. 3).
5. Following the growth of the initial culture closely is recommended, then you will get a good overview of how long the culture grows under the cultivation conditions that you use. The main thing is to reinoculate and start the culture growing again before it dies.
6. Going from small-scale cultures to large-scale cultivation imposes challenges, for example, the surface to volume ratio of the culture typically decreases. With lower surface to volume ratio, it is important to keep track of how this affects the gas exchange. Providing sufficient CO_2 (and stable pH) and removing excess O_2 are vital for keeping optimal conditions for algal photosynthesis and growth. Also, the light penetration will most likely be different due to the different surface to volume ratio. In a dense culture, light will not penetrate far, which can be counteracted by increasing the surface area:volume ratio, or by increasing the turbulence in the cultivation. The latter will increase the frequency that the cells are mixed from high light at the surface of the cultivation unit to darkness in the middle of the culture. It has been known for a long time that increasing this mixing frequency can increase the productivity [5].

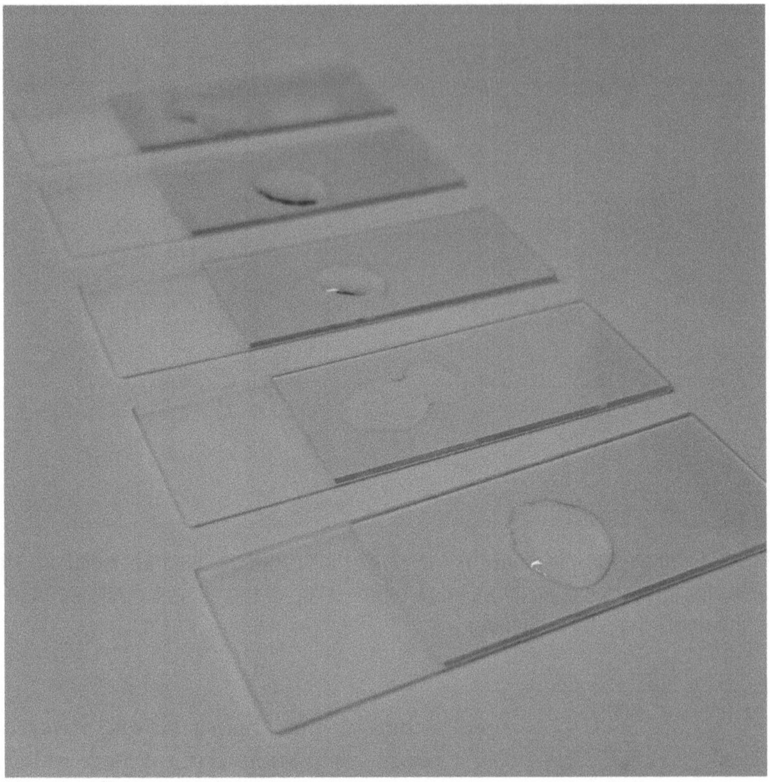

Fig. 3 Example of washing steps. Each slide has a drop of medium where the cell of interest is picked, ending up with only one cell in the final step that can be transferred to the initial cultivation chamber/flask. Photo by K. Spilling

7. Chambers with grid normally contains 1 ml and has 1000 grid areas; thus each grid area represents 1 μl, making the calculation of cells/ml = counted cells * 1000/grid areas. If you use fields of view you have to first establish the area of the field of view and the volume that it represents based on the volume in the counting chamber.
8. In most cases, DW is the most straightforward measure of biomass. However, depending on your purpose and/or species of algae you may want to calculate the AFDW. For example, when comparing lipid content as percent of biomass, using AFDW could be a better way of comparing different algal groups as, e.g., diatoms have silicate frustules, which remain in the ash part after combustion. Using lipids as percent of AFDW will give a more comparable data between species/groups, e.g., if you are interested in what fraction of the carbon is in the form of lipids. Be sure to mention specifically what you use to normalize the data in any reporting of the results.

Acknowledgments

This updated chapter was supported by the AlgaeProBANOS project, co-funded by the European Union through the Horizon Europe program (project number 101112943), and used equipment part of the Finnish marine research infrastructure (FINMARI) consortium.

References

1. Sheehan J, Dunahay T, Benemann J, Roessler P (1998) A look back at the US Department of Energy's aquatic species program: biodiesel from algae. Nation Renew Energ Lab Rep NREL/TP-580-24190
2. Williams PJ leB, Laurens LML (2010) Microalgae as biodiesel & biomass feedstocks: review & analysis of the biochemistry, energetics & economics. Energy Environ Sci 3:554–590
3. Nalley JO, Stockenreiter M, Litchman E (2014) Community ecology of algal biofuels: complementarity and trait-based approaches. Industr Biotechnol 10:191–201
4. Harrison PJ, Berges JA (2005) Marine culture media. In: Andersen RA (ed) Algal culturing techniques. Academic Press, Cambridge
5. Burlew JS (ed) (1953) Algal culture, from laboratory to pilot plant. Carnegie Institution of Washington Publication, Washington

Chapter 2

Measurement of Fluorescence for Monitoring Algal Growth and Health

Kristian Spilling, Pasi Ylöstalo and Jukka Seppälä

Abstract

Measuring fluorescence is a noninvasive, inexpensive, and quick way of estimating the biomass concentration and the health of the cultivated algae. Fluorescence is generally positively correlated with Chlorophyll *a* and can be used as a proxy for biomass. In addition, the proportion variable fluorescence of maximal fluorescence is a measure of photochemical efficiency, which is affected by stress in a very early stage and can be used as a proxy for algal health.

Keywords Variable fluorescence, Microalgae, DCMU, OJIP curves

1 Introduction

Light provides the energy needed for running photosynthesis and is an absolute requirement for autotrophic (photosynthetically driven) growth [1]. The main light harvesting pigment in algae, Chlorophyll *a* (Chl *a*), is fluorescent. When light energy is absorbed by Chl *a*, part of this energy re-emitted as a light photon, i.e., fluorescence, after an exited electron returns to the ground state. The Chl *a* fluorescence light is emitted with a peak at approximately 680 nm, and measurements of fluorescence can be used as a proxy for Chl *a*, photochemical efficiency, and electron transport rate. The measurement of Chl *a* fluorescence can be done either in vivo, within living cells, or in vitro, after extraction of the Chl *a* pigment with a solvent such as acetone or ethanol. In this chapter we will only deal with measurements of in vivo fluorescence.

The fate of absorbed light energy has three potential pathways: (1) it drives the photochemical reaction, which provides the chemical energy and reducing power needed to fix carbon from CO_2 into biomass; (2) it may be converted to heat, or (3) re-emitted as fluorescence [1]. The energy content in light photons varies depending on the wavelength (color of the light), i.e., a blue light

Kristian Spilling (ed.), *Biofuels from Algae*, Methods in Molecular Biology, vol. 2996,
https://doi.org/10.1007/978-1-0716-5031-8_2,

photon carries more energy than red light photon. However, for the photochemical reaction and fluorescence emission it is only the energy difference between the ground state and singlet state 1 (S1) that is used, meaning that the energy difference between a blue and red photon is always lost as heat [1].

There is generally a positive correlation between Chl *a* and its fluorescence. However, Chl *a* per cell (or other biomass unit) will vary depending on the light environment. For example, the Chl *a* content typically increases in dim light as an acclimation mechanism to lower photon flux, and similarly decreases if the amount of light increases. When using constant light, the light acclimation will be stable and the fluorescent measurement is more closely correlated with biomass than for algae grown under natural light.

In addition to using single fluorescent measurements as proxy for biomass, more advanced measurements can be made when making some assumption about the loss of energy as heat (e.g., that it is constant over short time periods). One of the key measures is the variable fluorescence (F_v), which is the difference between the minimum (F_0) and maximum (F_m) fluorescence [1]. These variables can be used to calculate the photochemical efficiency: F_v/F_m, which can be used as a measure of stress to the algal cells. The minimum fluorescence can only be measured when all the reaction centers are open, and this is achieved by dark acclimation of the culture prior to measurement. Then the whole photosynthetic machinery will be ready for processing absorbed photons and all reaction centers are open. Maximum fluorescent is measured when all the reaction centers are closed. This can be induced either chemically, by adding an herbicide such as DCMU that blocks the photosynthetic electron transport chain, or by applying a strong light flash that saturates the electron transport chain. There are different approaches to the latter, the most common being Pulse Amplitude Modulated (PAM) Fluorometer, in which one strong pulse of light is provided, and Fast Repetition Rate Fluorometry (FRRF) where a sequence of rapid light flashes gradually saturates the electron transport chain [2]. Different variable fluorescence techniques, such as FRRF and PAM, can yield varying results for F_o and F_m due to the distinct methods used to saturate the photosystem. Some fluorometers also record the fluorescent transient from F_0 to F_m, often termed the OJIP curve [2, 3]. An OJIP curve is the build-up of fluorescence from the origin (O) through inflections (J and I) up to a peak (P) Fig. 1. The shape of the OJIP curve is influenced by the filling of the electron transport chain and can be divided into photochemical and thermal phases. It is affected by the reduction of intermediate electron acceptors within the electron transport chain, particularly Q_A [3]. Another example is presented in Fig. 2 where the build up of fluorescence is done with rapid flashes using FRRF.

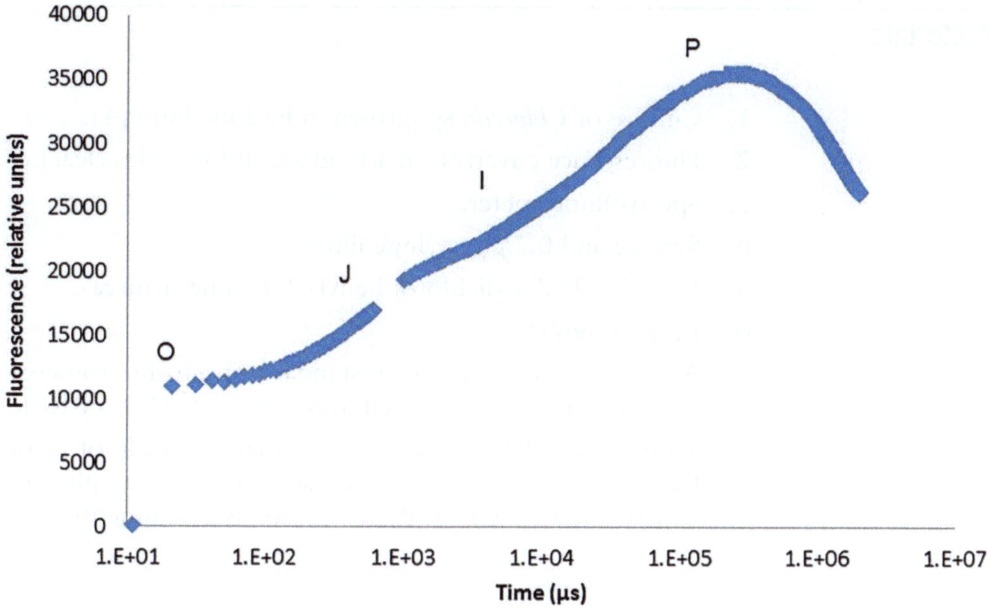

Fig. 1 An OJIP curve depicting the increase in fluorescence over 1 s (1E+06 µs). F_0 (O), J, I, and F_m (P) points. This is an example of a *Chlorella* sp. culture and the OJIP curve may have more prominent secondary peaks in other species [2, 3]

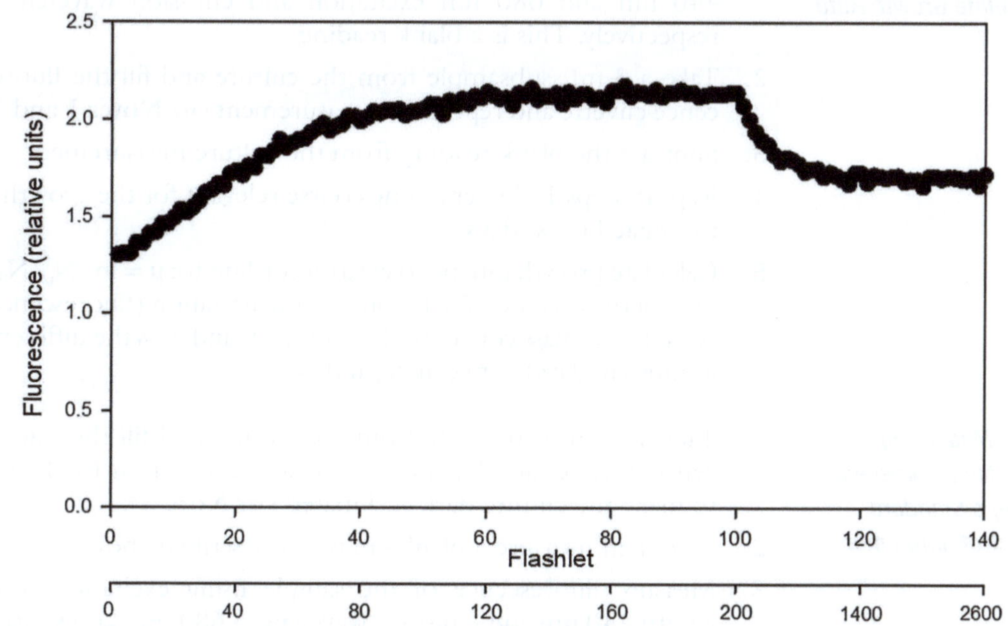

Fig. 2 An example of the fluorescence induction curve obtained by FRRF that is induced by a flashlet sequence comprising 100 saturation flashlets with a 2 µs interval and 40 relaxation flashlets with a 50 µs interval. Note the change in time scale between the saturation and relaxation phases

2 Materials

1. Culture of *Chlorella* sp. grown in F/2 medium [4].
2. Fluorescence cuvettes (quartz glass, all four sides clear).
3. Spectrofluorometer.
4. Syringe and 0.2 μm syringe filter.
5. DCMU (3-(3,4-dichlorophenyl)-1,1-dimethylurea).
6. Ethanol (96%).
7. An alternative to using a standard spectrofluorometer and DCMU is to use a variable fluorometer (PAM or FRRF).
8. To measure OJIP curves, a fluorometer capable of recording fluorescent transient must be used. In this example we have used an AquaPen from Photon Systems Instruments.

3 Methods

3.1 Measure Chl a Fluorescence and Calculate Growth Rate

1. Filter 3 mL of the sample, using the syringe and 0.2 μm filter, into the fluorescence cuvette and measure fluorescence using 440 nm and 680 nm excitation and emission wavelength, respectively. This is a blank reading.
2. Take a 3 mL subsample from the culture and fill the fluorescence cuvette and repeat the measurement (*see* **Notes 1** and **2**).
3. Subtract the blank reading from the culture measurement.
4. Repeat Steps 1–3 over a time course relevant for the growth of the algae hours- days.
5. Calculate growth rate per day (μ) according to: $\mu = \ln (N_t/N_0)/dt$; where N_0 is the initial biomass concentration (fluorescence), N_t is the biomass concentration at time t, and dt is the difference in time (in days) between N_0 and N_t.

3.2 Measuring Variable Fluorescence Using a Standard Spectrofluorometer

1. Take a 3 mL subsample from the culture and fill the fluorescence cuvette, and let the cuvette stay in darkness for 15 min to make the culture dark acclimated (*see* **Note 3**).
2. Filter and measure a blank sample as described above.
3. Measure fluorescence of the sample using excitation wavelength 440 nm and emission wavelength 680 nm, and subtract the blank reading. This is the F_0 value.
4. After the dark acclimated measurement take the cuvette and add 10 μL DCMU solution (*see* **Note 4**).

5. Let the cuvette stay in light for 2 min.
6. Measure fluorescence again using the same procedure as Step 3. This is the F_m value.
7. Calculate the variable fluorescence; $F_v = F_m - F_0$ (*see* **Notes 5** and **6**).
8. Calculate the photochemical efficiency index: $F_v / F_m = (F_m - F_0) / F_m$ (*see* **Note 7**).

3.3 Measuring Variable Fluorescence Using a Variable Fluorometer

1. Take a 3 mL subsample from the culture and fill the fluorescence cuvette.
2. Let the cuvette stay in darkness for 15 min to make the culture dark acclimated.
3. Measure fluorescence using excitation wavelength 450 nm and emission wavelength 680 nm, using the OJIP option.
4. The F_v will automatically be calculated from the measurements (*see* **Note 8**).

4 Notes

1. You might need to adjust your fluorometer to get fluorescence values that is not too low or over range. In most fluorometers, there are two ways this can be done; either by adjusting the emission and/or excitation slit widths, or by adjusting the voltage in the photomultiplier tube, i.e. increasing voltage increases sensitivity.
2. Before starting, it is advisable to test your instrument and the settings you are using at a wide range of biomass concentrations. The increase in fluorescence with increasing Chl *a* is initially linear, but at very high Chl *a* concentrations, it will start to level off, e.g., due to increasing reabsorption of fluorescent light. Finding the linear range can easily be done by running a dilution series where a dense culture is diluted with filtered (0.2 μm) water.
3. In case the cells are large and sink quickly, you might need to homogenize the solution before the measurement or have continuous stirring in the cuvette.
4. Dissolve DCMU in ethanol to a final concentration of 2 mM. The final ethanol concentration in the sample should be <1% to avoid problems with extraction of photosynthetic pigments.
5. The use of F_v / F_m as a measure of health/stress should be used with some caution as different species of algae will have

different range of F_v / F_m depending on, e.g., their photosynthetic pigments (e.g., cyanobacteria).

6. As a general rule of thumb F_v / F_m > 0.5 represents actively growing cells for eukaryotes, and F_v / F_m < 0.2 severely stressed cells. Cyanobacteria typically have lower F_v / F_m than eukaryote algae because of their different pigmentation not absorbing light well at 440 nm, which is the typical excitation light used in fluorometers. This could be taken into account by adjusting the excitation wavelength, or by using DCMU addition as described above [5].
7. All variables should be corrected by subtracting the blank first, so it is derived from: ((F_m − blank) − (F_0 − blank)) / (F_m − blank).
8. This will be done without blank corrections, which is no problem if the signal-to-blank reading is high. However, blank correction might have to be carried out if the blank is high (>10%) compared with F_m.

Acknowledgments

This updated chapter was supported by the AlgaeProBANOS project, co-funded by the European Union through the Horizon Europe program (project number 101112943), and used equipment part of the Finnish marine research infrastructure (FINMARI) consortium.

References

1. Falkowski PG, Raven JA (2007) Aquatic photosynthesis. Blackwell Science, Malden, p. 375
2. Suggett DJ, Prášil O, Borowitzka MA (eds) (2010) Chlorophyll a fluorescence in aquatic sciences: methods and applications. Springer, Dordrecht, p 323
3. Stribet A, Govindjee (2012) Chlorophyll a fluorescence induction: a personal perspective of the thermal phase, the J-I-P rise. Photosynth Res 113:15–61
4. Guillard RRL (1975) Culture of phytoplankton for feeding marine invertebrates. In: Smith WL, Chanley MH (eds) Culture of marine invertebrate animals. Plenum Press, New York
5. Raateoja M, Seppälä J, Ylöstalo P (2004) Fast repetition rate fluorometry is not applicable to studies of filamentous cyanobacteria from the Baltic Sea. Limnol Oceanogr 49:1006–1012

Check for updates

Chapter 3

Large Scale Cultivation of Microalgae in Open, Closed, and Hybrid Systems

Jacqueline Jerney and Kristian Spilling

Abstract

There are two main approaches for cultivating microalgae on a large scale: either in open or closed cultivation systems. The main difference between these systems is related to how they operate (e.g., cooling and gas exchange), vulnerability for outside influence (e.g., rainwater and introduction of unwanted species), and costs for building and running the system. Additionally, hybrid-systems can be used to combine the advantages of open and closed systems. In this chapter we introduce the main cultivation technologies and discuss their main advantages and disadvantages when cultivating microalgae, including multi-species approaches.

Keywords Gas exchange, Growth medium, Photobioreactor, Raceway pond

1 Introduction

When cultivating microalgae on a large scale, several problems arise, which are not relevant on a small scale: (1) It becomes more difficult to ensure even light distribution and mixing of large volumes of culture, but a good light distribution and/or effective mixing of the culture are crucial to ensure maximal productivity. (2) Another challenge is sufficient gas exchange with increasing size of the system: providing sufficient CO_2 for photosynthesis and removal of O_2 produced in this process. (3) Furthermore, soluble algal products (SAPs), which are released by microalgal cells into the culture medium during cultivation, can be a problem. SAPs may accumulate in the culture medium and can inhibit microalgal growth, raise the cost of microalgal recovery, and lower the performance of the downstream discharged water treatment process after microalgal cultivation [1]. In addition to problems associated with the cultivation itself, harvesting of the produced biomass can be very challenging, because separation of the biomass from the culture medium can be expensive and energy consuming. There are several different options to grow

Kristian Spilling (ed.), *Biofuels from Algae*, Methods in Molecular Biology, vol. 2996,
https://doi.org/10.1007/978-1-0716-5031-8_3,

Table 1
Overview of different cultivation systems. Further details in the text

	Open	Closed (photobioreactor)
Suspended	• Stagnant pond system • Raceway pond • Thin layer flow lanes	• Tubular • Sleeve bags • Flat panel
Biofilm	Algae turf scrubbers	

algae on a larger scale, depending on the purpose of cultivation and application of the algae biomass. Differences between cultivation technologies range from costs to productivity and controllability of the cultivation system.

Large scale cultivation systems can broadly be divided into open and closed systems (Table 1). Further differentiation can be made between suspended and biofilm systems, meaning that the algae grow either suspended (freely floating) in the water medium or attached to a surface forming a biofilm. See also the chapters in this book on aquaponics and biofilm cultivation.

2 Open Cultivation Systems

Microalgae cultivation in open ponds is the oldest and most simple approach to grow algae [2], and nowadays open systems are the most employed technology for outdoor cultivation [3]. Raceway ponds (Fig. 1), raceways, or "high-rate algal ponds" were developed in the 1950s for wastewater treatment and since the 1960s, outdoor open raceways have been used for commercial production of microalgae including also cyanobacteria [4, 5]. The main advantages of open cultivation systems are that they are cost efficient and easy to maintain [6]. Operational costs can be kept low by using only sunlight as the primary energy input and relatively little energy is needed to mix the culture. Stagnant pond systems are not mixed at all, which facilitates simple and cheap upscaling. Raceway ponds can also be constructed cost efficiently on a larger scale but require some form of mechanical movement of water, most often in the form of a paddle wheel, to keep the culture moving around the raceway [7].

Although open pond systems are most commonly used for commercial applications [8, 9], they have several drawbacks to consider (Table 2): (1) many growth relevant parameters like pH, temperature, mixing, and light availability cannot be regulated easily; (2) another disadvantage is that sparged CO_2 has a very short residence time, resulting in high losses and poor solubility of introduced CO_2;

Fig. 1 Open raceway pond where the algal cultivation is circulated through the track by paddle wheels. Picture: JanB46, Wikipedia, CC BY-SA 3.0

Table 2
Advantages and disadvantages of common cultivation systems, adapted from Pulz [7]

Cultivation system	Advantages	Disadvantages
Open	• Relatively cheap • Good gas exchange with the atmosphere (e.g., release of O_2) • Easy to operate • Easy to scale up • Cooling by evaporation	• High risk of contamination • High evaporation losses • Large area required • Light limitation if thick layers are used
Closed	• Good control of cultivation parameters • High productivity • Reduced contamination risk • Less CO_2 losses • Reproducible cultivation conditions	• Expensive to build and operate • Requires in most cases cooling • Scale up is difficult

(3) seasonal variations can contribute to high variation in biomass production; (4) during warm periods the water losses are high due to evaporation; (5) there is a major risk of contamination by other algae, or algal grazers, which can lead to poor productivity or even a total loss of the desired production strain [8, 10–13].

The bottlenecks that restrict commercial production of microalgal bioenergy have been summarized as "high cost" and "low efficiency" [14]. However, there are ways to optimize open systems. Microalgae cultivation in wastewater, recycling of water and nutrients, utilization of the whole biomass, selection of indigenous strains, and optimization of the harvesting technique are among the potential solutions to overcome these bottlenecks. Additionally, alternative open cultivation systems, like thin layer flow lanes or algae turf scrubbers (Table 1), can be applied to achieve high productivity or easier harvesting. Despite the drawbacks of open systems, most commercial production of microalgae takes place in open ponds due to the relatively low production costs.

3 Closed Cultivation Systems

There are several types of closed cultivation systems, often referred to as photobioreactors (PBRs; Figs. 2, 3, and 4). Most of the deviation between different systems is related to the design of the cultivation compartment, which can be tubular reactors, laminar (or flat panel) reactors, hanging plastic sleeves, or fermenter-like tank reactors [9, 11]. Several shapes and forms have been tested, each with advantages and disadvantages that need to be considered for the intended purpose [15, 16]. They may or may not be artificially illuminated depending on the location and intended product of the biomass. In most cases, the water is moved through the PBR either directly by a pump, by air/gas mixture in the case of an airlift or bubble column operated PBR, or by a combination of both pump and pressurized air.

The development of commercial scale photo-bioreactors has made good progress in the past decades and designs with relatively high efficiencies have been constructed [16]. However, due to much higher cost for building and operating closed systems compared with open systems, all the large-scale, closed systems operating at present produce algal biomass for high-value products such as cosmetics and neutraceuticals.

Closed systems offer a much better control of cultivation variables, compared to open systems (Table 2); for example, contamination with bacteria, algae, or grazers becomes less likely in a closed system [17]. Areal biomass productivity varies a lot between different cultivation systems but is generally higher in closed systems [17].

Photobioreactors do also have some challenges that are less problematic in open systems such as gas exchange and need for cooling.

Fig. 2 An example of a closed tubular photobioreactor. Water is pumped through the glass tube from a gas exchange unit and the algae collects light in the tube. Picture: IGV Biotech, Wikipedia, CC BY-SA 3.0

Gas exchange, getting enough CO_2 to the algae, is critically important for satisfying the carbon needed for running photosynthesis and for maintaining the pH. CO_2 dissolved in water is a weak acid and biological uptake of CO_2 by algae will increase the pH. In dense cultures, the pH may increase to critical levels, preventing further growth without sufficient gas exchange. Another critical aspect is removal of O_2, which rapidly builds up to supersaturated concentrations, slowing down or stopping algal growth. The temperature in a closed system rapidly increases when exposed to the sun and most algae will stop growing and start to decay in temperatures above 30–35°C. Proper cooling is therefore vital and there are several technical solutions ranging from spraying the cultivation unit with cooling water or by having water cooling units connected to the cultivation system.

From an operational standpoint, the main difference between open and closed systems is the costs for building and operating the systems. Closed systems consist of expensive parts and control units and typically require much more energy for operation than open systems. Closed systems typically reach higher areal productivity than open systems and this can offset some of the extra costs.

Fig. 3 Example of a lab scale photobioreactors with 100-liter cultivation volume (top) and 5 × 8-liter loops that can be used for growing inoculum (bottom). Photos by: K. Spilling

Fig. 4 Closed photobioreactor with a plate design. Similar to the tubular reactor in Fig. 2, the algal culture is pumped from a gas exchange unit to the panel unit where the algae collect light. Picture: IGV Biotech, Wikipedia, CC BY-SA 3.0

4 Hybrid Cultivation Systems

Hybrid cultivation systems combine the advantages of both open and closed systems to create a more efficient, cost-effective, and scalable solution for large-scale microalgae cultivation. These systems typically integrate the high productivity and controlled environment of PBRs with the lower operational costs and scalability of open pond systems [18, 19].

Microalgae are initially grown in closed photobioreactors under controlled conditions to achieve a high-density culture. This pre-growth phase minimizes contamination risks and accelerates biomass production. The dense algal culture is then transferred to open ponds for further growth under natural sunlight. Open ponds provide a cost-efficient environment for large-scale biomass production, utilizing the optimized inoculum from the PBR phase [20].

By starting the cultivation in PBRs, the risk of contamination during the critical early growth phase is significantly reduced, ensuring that only the desired microalgal strain dominates in the open pond. Hybrid systems take advantage of the higher biomass productivity achievable in PBRs while scaling up production in cost-efficient open ponds. By using PBRs only for the initial growth stages, the high capital and operational costs of closed systems are minimized, making the overall process more economically viable. Hybrid systems also allow for better utilization of resources such as water, CO_2, and nutrients by recycling culture media between the two stages, making it more environmentally sustainable [21].

Integrating PBRs with open ponds seamlessly requires careful planning to ensure the transition does not lead to contamination or loss of productivity. While hybrid systems offer improved scalability compared to standalone PBRs, challenges in scaling up open ponds for consistent yields remain. The open pond phase is still subject to external factors such as temperature fluctuations, evaporation, and contamination, which need to be managed. However, this could be used as an advantage if the environmental conditions in the open stage is beneficial for boosting components of interest such as lipids, which can be used to optimize production in a two-stage cultivation process [22].

Advances in automation have enabled smoother transitions between PBRs and open ponds, reducing contamination risks and optimizing culture conditions. Hybrid systems have been successfully integrated with wastewater treatment facilities, using PBRs for nutrient recovery and open ponds for biomass production [5]. Modular hybrid systems allow for flexibility in scaling operations based on demand, making them ideal for diverse applications such as biofuel production, aquaculture feeds, and high-value bioactive compounds [15, 18].

5 Multi-Species Cultures

Traditionally, research on microalgal cultivation has focused on single-species systems. However, studies suggest that combining multiple species can enhance productivity and improve biomass composition [23]. This positive effect is attributed to increased biodiversity and niche differentiation, where species with varying environmental optima complement each other to maximize overall productivity. Such complementarity can lead to overyielding, which means the collective yield of multiple microalgae species exceeds that of single species, in terms of whole biomass or a specific biomass component, such as lipids, when cultivated under the same condition [24]. Two types of overyielding can be distinguished: non-transgressive and transgressive. Non-transgressive overyielding occurs when the yield of a polyculture is greater than the weighted average of the yields of individual species in monoculture, yet still falls short of the best-performing monoculture. Transgressive overyielding, in contrast, describes scenarios where the productivity of a mixed community surpasses even the highest-performing monoculture [25].

Multi-species cultivation presents several potential advantages over monocultures in algal biofuel production, particularly in nutrient uptake, productivity, and system stability. Diverse algal communities optimize resource use efficiency by leveraging the complementary traits of different species, enabling more effective utilization of nutrients like phosphate and nitrate. This results in faster growth and greater biomass yields. Notably, multi-species systems often achieve transgressive overyielding in lipid production, a critical factor for biofuel feedstock quality. Additionally, these communities are more resilient to environmental stressors, pathogens, and grazers, ensuring stable and consistent yields [26]. These benefits underscore the potential of multi-species cultivation as a sustainable and cost-effective strategy for enhancing algal biomass and lipid production [27, 28].

While multi-species cultivation offers potential advantages, it also presents several challenges. Maintaining a balanced and functional community can be difficult, as species interactions may lead to competitive exclusion [29], where one species outcompetes others, reducing biodiversity and potentially negating the benefits of complementarity. Additionally, the dynamic nature of diverse communities can complicate process control, as environmental shifts or nutrient availability might favor one species over others, resulting in inconsistent biomass composition and yields, but this may be overcome by deliberate design of the community [30]. Especially in open systems, with little environmental control, the unpredictability of species dominance and their varying responses to environmental stressors may also impact lipid production and other desired traits, making it harder to optimize for specific biofuel feedstock qualities [31, 32]. The logistical challenges of identifying and cultivating compatible species with complementary traits require extensive research and trial-and-error, potentially increasing costs and slowing the development of large-scale applications.

6 Harvesting

Even in dense algal cultures, more than 90% of the weight is water, and harvesting the microalgae is a challenge for the economic feasibility of large-scale algal cultivation. There are ways to harvest the biomass quickly and efficiently, such as centrifugation, but these consume a lot of energy and are thus expensive. Current harvesting methods include chemical, mechanical, electrical, and biological harvesting methods [6, 33, 34] (Table 3).

Table 3
Overview of different harvesting methods [6] and dry solids output concentration [20] and references herein, n.a. = not available

Method	Process	Comments	Dry solids output concentration (%)
Chemical based	Precipitation/ flocculation	• Addition of electrolytes or synthetic polymers to neutralize negative surface charge • The use of metal salts for coagulation and flocculation is cautioned due to potential inhibition of the specific methanogenic activity of methanogenic and acetogenic microbes	3–8
Mechanical based	Centrifugation	• Centrifugal forces are utilized to separate based on density differences • Probably the most rapid and reliable method of recovering suspended algae • Easy to operate • High investment and operating costs	10–22
	Filtration	• Often used for filamentous strains • For small, suspended algae tangential flow filtration is considered to be more feasible • High costs and power requirements	2–27
	Sedimentation	• Low costs • Low reliability because of fluctuating density of algal cells • Slow	0.5–3
	Dissolved air flotation	• Air is released under high pressure and forms tiny bubbles in the water column, which adhere to the suspended matter causing the suspended matter to float • Has been proven in large scale • The additional use of flocculants might be problematic for further processing of the algae	n.a.
Electrical based	Separation based on electrophoresis	• No chemicals needed • High power requirements and electrode costs	n.a.
Biological based	Autoflocculation	• High pH and the consumption of dissolved CO_2 lead to co-precipitation of algal cells together with calcium phosphate	n.a.
	Bioflocculation	• Flocculation caused by secretion of polymers	n.a.
	Microbial flocculation	• Addition of flocculating microbes	n.a.

Most of these methods aim at flocculation and sedimentation of the microalgae. Flocculation is the process of small particles binding together forming larger particles, which then sink quicker. The sinking material can then be collected as a first step in the harvesting process. Usually, further steps are required afterwards to concentrate the biomass. Flocculation can be induced by addition of a chemical (flocculating agent) or it can be self-induced, e.g., by stopping the CO_2 supply and letting the pH increase to levels inducing flocculation [35].

Harvesting may also be less challenging depending on the cultivation system and the application of the biomass. Algal turf scrubbers are for example used to grow a biofilm which can be easily scraped of mechanically [36] or harvested by vacuuming [37]. If the biomass is used for biogas production, the optimal water content is typically much higher than for production of lipids for biodiesel.

There are also technologies for circumventing the traditional harvesting by getting the algae to release the desired product, for example lipids, into the water [38]. The lipids separate from the water naturally and can be collected easily. There would thus be no need for collecting the actual biomass, which could be left producing more of the desired product. For example, a lot of work has been done on "milking" the green algae *Botryococcus* spp. [39]. Although the milking method drastically reduces harvesting costs, its development still requires better understanding of the underlying mechanisms and knowledge of suitable microalgae strains [40].

7 Cost-Effective Technologies for Fuel Production

The development of algal biofuels focuses on utilizing microalgae as a renewable source of third-generation biofuels, including biodiesel, bioethanol, biogas, and biohydrogen. Microalgae seem like a particularly promising source for biofuel due to their rapid growth, high lipid content, and ability to grow in diverse water conditions, including wastewater. However, the commercial scalability of algal biofuels faces challenges such as high production costs, inefficient harvesting and processing, and environmental concerns. Despite intensive research in the past decades, the commercial cultivation of microalgae remains limited by numerous constraints due to its complexity. Thus, full-scale long-term productivities remain much lower than expectations from laboratory data and theoretical predictions [41].

There are, however, several interesting and promising developments, which might help reducing cost/increasing productivity. Innovations integrating cultivation sensors and camera systems with artificial intelligence running the production facility could be a way to optimize productivity [42]. Co-occurring heterotrophic bacteria can be important for the health and productivity of the microalgae, which could be optimized [43]. Also, recent advancements in genetic engineering and metabolic pathway optimization, such as CRISPR technology, have enhanced lipid and metabolite yields, addressing productivity issues [44]. Integrated cultivation systems combining PBRs and raceway ponds, alongside bio-refinery approaches, have been explored to reduce costs and improve efficiency. Despite these advancements, further research into cost-effective technologies and large-scale applications is critical to making algal biofuels a viable alternative to fossil fuels [44].

Acknowledgments

Updating this chapter for the second edition of this book was supported by the AlgaeProBANOS project, co-funded by the European Union through the Horizon Europe program (project number 101112943), and pictured equipment part of the Finnish marine research infrastructure (FINMARI) consortium.

References

1. Zhuang L-L, Wu Y-H, Espinosa VMD et al (2016) Soluble algal products (SAPs) in large scale cultivation of microalgae for biomass/bioenergy production: a review. Renew Sustain Energy Rev 59:141–148
2. Chaumont D (1993) Biotechnology of algal biomass production: a review of systems for outdoor mass culture. J Appl Phycol 5:593–604
3. Pruvost J, Cornet J-F, Pilon L (2016) Large-scale production of algal biomass: photobioreactors. In: Bux F, Chisti Y (eds) Algae biotechnology: products and processes. Springer, New York, pp 41–66
4. Chisti Y (2016) Large-scale production of algal biomass: raceway ponds. In: Bux F, Chisti Y (eds) Algae biotechnology: products and processes. Springer, New York, pp 21–40
5. Novoveská L, Nielsen SL, Eroldoğan OT et al (2023) Overview and challenges of large-scale cultivation of photosynthetic microalgae and cyanobacteria. Mar Drugs 21(8):445
6. Becker E (1994) Microalgae: biotechnology and microbiology. Cambridge University Press, Cambridge
7. Christenson L, Sims R (2011) Production and harvesting of microalgae for wastewater treatment, biofuels, and bioproducts. Biotechnol Adv 29(6):686–702
8. Pulz O (2001) Photobioreactors: production systems for phototrophic microorganisms. Appl Microbiol Biotechnol 57:287–293
9. Chen C-Y, Yeh K-L, Aisyah R et al (2011) Cultivation, photobioreactor design and harvesting of microalgae for biodiesel production: a critical review. Biores Technol 102(1):71–81
10. Brennan L, Owende P (2010) Biofuels from microalgae—a review of technologies for production, processing, and extractions of biofuels and co-products. Renew Sustain Energy Rev 14(2):557–577
11. Kumar A, Ergas S, Yuan X et al (2010) Enhanced CO2 fixation and biofuel production via microalgae: recent developments and future directions. Trends Biotechnol 28(7):371–380
12. Wang B, Li Y, Wu N, Lan CQ (2008) CO2 bio-mitigation using microalgae. Appl Microbiol Biotechnol 79:707–718
13. Ugwu C, Aoyagi H, Uchiyama H (2008) Photobioreactors for mass cultivation of algae. Biores Technol 99(10):4021–4028
14. Zhang T-Y, Hu H-Y, Wu Y-H et al (2016) Promising solutions to solve the bottlenecks in the large-scale cultivation of microalgae for biomass/bioenergy production. Renew Sustain Energy Rev 60:1602–1614
15. Sarker NK, Kaparaju P (2023) A critical review on the status and progress of microalgae cultivation in outdoor photobioreactors conducted over 35 years (1986–2021). Energies 16(7):3105
16. Penloglou G, Pavlou A, Kiparissides C (2024) Recent advancements in photo-bioreactors

for microalgae cultivation: a brief overview. *Processes* 12(6):1104
17. Carvalho AP, Meireles LA, Malcata FX (2006) Microalgal reactors: a review of enclosed system designs and performances. Biotechnol Prog 22(6):1490–1506
18. Narala RR, Garg S, Sharma KK et al (2016) Comparison of microalgae cultivation in photobioreactor, open raceway pond, and a two-stage hybrid system. Front Energy Res 4:29
19. Liyanaarachchi VC, Premaratne M, Ariyadasa TU et al (2021) Two-stage cultivation of microalgae for production of high-value compounds and biofuels: a review. Algal Res 57:102353
20. Bora A, Rajan AST, Kumar P et al (2024) Microalgae to bioenergy production: recent advances, influencing parameters, utilization of wastewater–a critical review. Sci Total Environ 946:174230
21. Adesanya VO, Cadena E, Scott SA, Smith AG (2014) Life cycle assessment on microalgal biodiesel production using a hybrid cultivation system. Biores Technol 163:343–355
22. Swain P, Tiwari A, Pandey A (2020) Enhanced lipid production in Tetraselmis sp. by two stage process optimization using simulated dairy wastewater as feedstock. Biomass Bioenergy 139:105643
23. Stockenreiter M, Graber A-K, Haupt F, Stibor H (2012) The effect of species diversity on lipid production by micro-algal communities. J Appl Phycol 24:45–54
24. Di S, Yang A (2018) Overyielding potential of microalgal polyculture with complementary light absorption spectra: a model-based analysis. Biomass Bioenergy 118:141–148
25. Vasseur DA, Messinger SM (2015) How does evolutionary history alter the relationship between biodiversity and ecosystem function? In: Belgrano A, Woodward G, Jacob U (eds). Aquatic functional biodiversity. Academic Press, San Diego, CA, pp 53–73
26. Thomas PK, Dunn GP, Coats ER et al (2019) Algal diversity and traits predict biomass yield and grazing resistance in wastewater cultivation. J Appl Phycol 31:2323–2334
27. Stockenreiter M, Haupt F, Graber AK et al (2013) Functional group richness: implications of biodiversity for light use and lipid yield in microalgae. J Phycol 49(5):838–847
28. Stockenreiter M, Haupt F, Seppälä J et al (2016) Nutrient uptake and lipid yield in diverse microalgal communities grown in wastewater. Algal Res 15:77–82
29. Louca S, Doebeli M (2016) Transient dynamics of competitive exclusion in microbial communities. Environ Microbiol 18(6):1863–1874
30. Mandal S, Corcoran AA (2022) A novel approach to build algal consortia for sustainable biomass production. Algal Res 65:102734
31. Kholssi R, Stefanova S, González-Ortegón E et al (2024) Population and functional changes in a multispecies co-culture of marine microalgae and cyanobacteria under a combination of different salinity and temperature levels. Mar Environ Res 193:106279
32. Sörenson E, Capo E, Farnelid H et al (2021) Temperature stress induces shift from coexistence to competition for organic carbon in microalgae-bacterial photobioreactor community–enabling continuous production of microalgal biomass. Front Microbiol 12: 607601
33. Deepa P, Sowndhararajan K, Kim S (2023) A review of the harvesting techniques of microalgae. *Water* 15(17):3074
34. Singh G, Patidar S (2018) Microalgae harvesting techniques: a review. J Environ Manag 217:499–508
35. Spilling K, Seppälä J, Tamminen T (2011) Inducing autoflocculation in the diatom Phaeodactylum tricornutum through CO2 regulation. J Appl Phycol 23:959–966
36. Higgins BT, Kendall A (2012) Life cycle environmental and cost impacts of using an algal turf scrubber to treat dairy wastewater. J Ind Ecol 16(3):436–447
37. Craggs RJ, Adey WH, Jenson KR et al (1996) Phosphorus removal from wastewater using an algal turf scrubber. Water Sci Technol 33(7):191–198
38. Vinayak V, Manoylov KM, Gateau H et al (2015) Diatom milking: a review and new approaches. Mar Drugs 13(5):2629–2665
39. Ennaceri H, Nwoba EG, Ogbonna CN et al (2023) Progress of non-destructive hydrocarbon extraction technology of Botryococcus braunii. Algal Res 73:103156
40. Grira S, Khalifeh HA, Alkhedher M, Ramadan M (2023) The conventional microalgal biofuel production process and the alternative milking pathway: a review. *Energy* 277:127547
41. Guieysse B, Plouviez M (2024) Microalgae cultivation: closing the yield gap from laboratory to field scale. Front Bioeng Biotechnol 12:1359755
42. Tummawai T, Rohitatisha Srinophakun T, Padungthon S, Sukpancharoen S (2024) Application of artificial intelligence and image processing for the cultivation of Chlorella sp. using tubular photobioreactors. ACS Omega 9(46):46017–46029
43. Rozenberg JM, Sorokin BA, Mukhambetova AN et al (2024) Recent advances and fundamentals of microalgae cultivation technology. Biotechnol J 19(3):2300725
44. Banerjee S, Banerjee S, Ghosh AK, Das D (2020) Maneuvering the genetic and metabolic pathway for improving biofuel production in algae: present status and future prospective. Renew Sustain Energy Rev 133:110155

Check for updates

Chapter 4

Growing Algae in Biofilms: Principles and Emerging Opportunities

Lara D. Mateos and Kristian Spilling

Abstract

Biofilm-based microalgal cultivation systems represent a different cultivation technology from suspended algal culturing. Unlike traditional suspended cultures, biofilm systems can offer higher biomass concentrations, simplified harvesting, and great potential for reduced energy consumption. Despite these advantages, challenges such as biofilm detachment, fouling, and scalability exist. These challenges must be addressed before large-scale biofilm cultivation can take place. Integrating biofilm-based systems into circular economy models presents an opportunity for resource recovery and climate change mitigation. This chapter explores the fundamentals of biofilm formation, including mechanisms of adhesion, biofilm structure, and important factors influencing growth dynamics. It provides a broad analysis of reactor designs, substrate materials, and operational parameters optimized for diverse applications.

Keywords Biofilm-based cultivation, Photosynthesis, Waste-water treatment, Diatoms, Biofuel, Algal turf-scrubber

1 Introduction to Biofilm-Based Microalgal Systems

Biofilm-based microalgal systems are cultivation platforms where microalgae grow as microbial communities attached to a surface, rather than as suspended cells in a liquid culture medium. These systems mimic natural environments, such as riverbeds and rocks in streams, where algae naturally form and grow in biofilms.

Most current technologies still cultivate microalgae in suspension [1]. However, a major bottleneck to commercial-scale implementation is the energy-intensive and costly harvesting process, which can account for up to one-third of total production expenses [2]. This is due to extensive dewatering steps, such as centrifugation or flocculation, as a consequence of the low solid concentrations (typically less than 1%), cell densities similar to water (1.08–1.13 kg L^{-1}), small cell size (5–25 μm), and strong negative surface charge [3–5]. Additionally, challenges faced by traditional suspension cultures include vulnerability to contamination, and significant water and energy requirements.

Kristian Spilling (ed.), *Biofuels from Algae*, Methods in Molecular Biology, vol. 2996,
https://doi.org/10.1007/978-1-0716-5031-8_4,

The adhesion of microalgae to surfaces allows for high cell density and efficient nutrient and light utilization. Therefore, biofilm-based systems can achieve significantly higher biomass densities compared to suspended cultures [6–8]. Additionally, harvesting the biomass in biofilm-based systems can be done by simple scraping or suctioning methods at a relatively high biomass-to-liquid concentration. The attachment-based growth reduces energy-intensive processes like centrifugation, making biofilm systems more sustainable and cost-effective than suspended cultures [6, 9, 10, 13]. Harvested biomass from biofilm cultivation systems often contains 80–85% moisture content, similar to the moisture content of centrifuged biomass [11, 12].

Microalgal biofilms are increasingly recognized for their potential in biotechnological processes such as wastewater treatment, biofuel production, and the synthesis of high-value products [14]. Moreover, the ability to design tailored surfaces and reactors for biofilm cultivation enhances their versatility, allowing for optimization across diverse industrial applications. However, while biofilm systems have demonstrated success at laboratory and pilot scales, scaling them up for industrial applications remains challenging. Innovations in reactor design, such as modular systems and advanced materials, are required to overcome these scalability limitations [15]. Furthermore, integrating biofilm systems with existing industrial processes requires interdisciplinary approaches to ensure compatibility and efficiency.

2 Mechanisms of Biofilm Formation in Microalgae

The initial stage of biofilm formation involves the attachment of microalgal cells to a substrate surface. The attachment can be through active or passive immobilization [16]:

Active immobilization requires adding chemicals for chemical attachment, such as flocculants, or for gel entrapment, where the biomass grows surrounded by the particles in the polymeric matrix [17]. This requires a dedicated matrix where the cells can be kept, which increases operational costs. Therefore, this technology is likely only relevant for processes where the end product is of very high value.

Passive immobilization, also called adsorption, relies on the natural ability of microorganisms to attach to substrates [18]. This process is mediated by physical and biochemical interactions, including van der Waals forces or electrostatic interactions, and is usually reversible. Adsorption then progresses into irreversible adhesion due to the production of extracellular polymeric substances (EPS). Passive immobilization is a natural process, relying on providing the surface for attachment, and is better suited for lower-value end products than active immobilization technologies (Fig. 1).

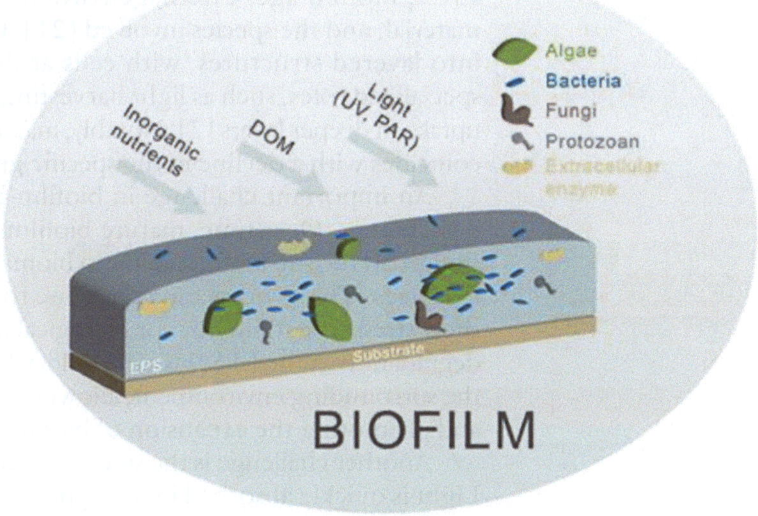

Fig. 1 Schematic of a natural biofilm. Reproduced from [81], CC-BY-SA-4.0

EPS are high-molecular-weight molecules, including proteins, phospholipids, polysaccharides, and nucleic acids that create a 3D polymeric network of structurally and compositionally diverse biopolymers, facilitating cellular interactions within the biofilm [19–21]. EPS play a key role in the structural and functional integrity of biofilms (see Fig. 1). In microalgal biofilms, the EPS matrix acts as a protective and structural scaffold which facilitates adhesion to surfaces and cohesion among cells [7]. This matrix not only anchors the cells to a substrate but also creates a microenvironment that supports nutrient retention, gas exchange, and protection from external stressors such as temperature fluctuations, changes in pH, and exposure to toxins [22]. EPS also retain water and nutrients, ensuring biofilm hydration and providing storage for essential resources [6, 23]. Additionally, biofilms are less susceptible to contamination by invasive species, as the dense EPS matrix provides a competitive advantage for the resident microalgal community. This resilience enhances the stability and longevity of biofilm systems, particularly in outdoor or industrial settings with variable conditions [24]. Specific surface characteristics, such as hydrophilicity, charge, and roughness, play critical roles in promoting or hindering adhesion [25, 26].

After initial attachment, microalgae undergo cell division and begin to secrete additional EPS, which serves as a supporting framework for additional biofilm growth. This phase involves the formation of microcolonies and the establishment of complex three-dimensional structures. The architecture of microalgal biofilms is species-specific [27]. EPS development is influenced by various factors which play a critical role in biofilm formation. These include nutrient availability, stress, biofilm age, external environmental conditions, substrate material, and the species involved [21]. Over time, biofilms mature into layered structures, with cells at different depths exhibiting specialized roles, such as light harvesting at the surface and nutrient uptake in deeper layers [7]. Notably, maximum EPS production often coincides with a decline in the specific growth rate [22].

An important challenge in biofilm-based systems is managing detachment. Over time, mature biofilms can become unstable and slough off the substrate, leading to biomass loss and operational inefficiencies. This process can occur due to mechanical forces, such as shear stress from fluid flow, or biological factors, including enzymatic degradation of the EPS. However, detachment also releases cells into the surrounding environment, allowing colonization of new surfaces and promoting the expansion of biofilm-based systems [28, 29].

Another challenge is the steep gradients that develop in biofilms. Light is quickly absorbed in the upper layer of the biofilm, and a diffusive boundary layer develops on the surface of the biofilm, which limits the uptake of elements needed for algal biomass growth, such as CO_2 and inorganic nutrients. This may also prevent waste products such as O_2 from photosynthesis from dissipating. All this puts some constraints both on the maximum growth rate of the biomass and the thickness of the biofilm. As the biofilm gets thicker, there can be a gradient from net autotrophic at the surface to net heterotrophic deeper in the biofilm.

2.1 Optimization of Factors Influencing Biofilm Formation

2.1.1 Environmental Conditions

Light intensity and temperature play significant roles in EPS production [30, 31]. Light availability and intensity are critical drivers for photosynthesis, which provides the energy required for cell growth, productivity, and EPS production. The design of biofilm-based systems must ensure that light penetrates as evenly as possible across the biofilm surface. Optimal light intensities vary depending on the species and biofilm thickness [7]. Excessive light at the surface can lead to photoinhibition, damaging the photosynthetic apparatus, while limited light penetration into deeper layers can restrict biomass productivity. Maintaining optimal light conditions has long been a recognized challenge in dense algal cultivation, prompting significant research efforts aimed at improving light distribution and utilization [32–34]. In addition to light intensity, the photoperiod (light–dark cycle) affects microalgal growth. During darkness, only respiration takes place, and it can be seen as a loss of biomass production. Some algal species can grow under continuous light, while others perform best when provided with a dark period. A balanced photoperiod, such as a 16:8 light-to-dark cycle, can enhance photosynthetic efficiency in these species, for example by allowing recovery from photodamage during dark phases [35].

Temperature affects enzymatic activities, metabolic rates, and overall microalgal physiology. Most microalgae thrive within a temperature range of 20–30°C, although specific species may exhibit unique temperature optima, e.g., some cold-adapted species thrive under the ice [36], while other species are adapted to warm water, i.e., thermophilic algae [37]. Deviations outside the temperature optimum slow growth, reduce photosynthetic efficiency, and may cause cellular damage or death in more extreme cases. Biofilm systems can integrate temperature control mechanisms, such as heaters or cooling systems to maintain stable temperature in the water flowing over the biofilm and surrounding air. However, this comes with added capital and operational costs.

Adequate and efficient nutrient supply is essential for biofilm productivity, ensuring sustained growth and metabolic activity. Nitrogen and phosphorus are essential macronutrients for microalgal growth. Nitrogen concentrations in the growth medium have been shown to affect both EPS and lipid production [21, 23, 38, 39]. Nitrogen supports protein and nucleic acid synthesis, while phosphorus is critical for energy transfer, membrane structure (phospholipids), and is also a critical element in RNA/DNA, important for growth. Maintaining an optimal nitrogen-to-phosphorus (N:P) ratio ensures balanced growth and prevents nutrient limitations or excesses. The commonly used Redfield ratio of N:P of 16:1 is often used as a guideline for balanced growth, but fast-growing species tend to have a lower N:P ratio due to higher requirement for ribosomal RNA [40]. Trace elements, such as iron, magnesium, and zinc, are vital cofactors for enzymatic activities and photosynthetic processes. Deficiencies in these micronutrients can impair cellular functions, reducing biofilm productivity. Supplementing trace elements in growth media ensures consistent availability for optimal biofilm development.

Hydraulic retention time (HRT) determines the contact time between the nutrient medium and the biofilm. Short HRTs may result in insufficient nutrient uptake, while excessively long HRTs can lead to nutrient depletion and reduced productivity [7]. Hydrodynamics, including flow velocity and turbulence, affect the delivery of nutrients and removal of waste products. While moderate flow rates support nutrient diffusion and biofilm stability, excessive shear forces can lead to biofilm detachment [41]. Flow dynamics should be optimized to prevent nutrient gradients or depletion zones. Similarly, it is critical to avoid the accumulation of compounds such as oxygen, which may have inhibitory effects at high concentrations within the biofilm [42]. Computational fluid dynamics (CFD) tools can be employed to optimize flow conditions, ensuring even nutrient distribution and minimizing shear stress [43]. Periodic flushing or pulsed flows can, in some cases, further enhance system performance by mimicking natural hydrodynamic conditions [44].

Carbon dioxide (CO_2) serves as the primary carbon source for photosynthesis, and its efficient delivery is essential for sustaining high productivity. Insufficient CO_2 limits photosynthetic rates, while excessive concentrations may lead to pH imbalances in the medium [45]. This is caused by CO_2 being a weak acid when dissolved in water, so when it is removed during photosynthesis, the pH increases. The pH of the surrounding environment impacts nutrient solubility and uptake, with a range of 7.5–8.5 being optimal for many species. CO_2 is commonly supplied through spargers, diffusers, or membrane-based systems to achieve uniform distribution [46]. Closed systems allow for recirculation of CO_2, improving utilization efficiency. However, this can incur in higher investment and construction costs. Coupling biofilm reactors with industrial CO_2 sources, such as flue gases, could provide a sustainable avenue for carbon capture and utilization.

2.1.2 The Biochemical Composition of the EPS

Chemical signals, including quorum sensing molecules, regulate the production of EPS and other biofilm components, enabling cells to respond collectively to environmental changes and substrate properties, influencing biofilm adhesion and cohesion, as well as its thickness and porosity [47]. Understanding the interplay between biofilm composition and structure, and also between the alga and accompanying heterotrophic bacteria, is likely important for optimizing biofilm-based cultivation systems. Despite the critical role of EPS in biofilm systems, the molecular-level interactions between EPS components, as well as between EPS and cells, remain poorly understood, and the contributions of individual polymeric components to the overall properties and functions of EPS need further investigation [48].

2.1.3 Substrate Materials

The materials show great variability in the biofilm attachment and productivity [49]. Natural substrates, such as wood and stone, have long been used for biofilm cultivation due to their availability and biocompatibility. Stone provides a durable and stable surface for biofilm attachment, while wood offers cost-effectiveness but is prone to degradation over time. However, cellulose-based materials, such as cotton, appear to support thicker biofilms than synthetic ones such as acrylic-based materials [50]. At the same time, some synthetic substrates, such as plastics, can offer greater control over surface properties and durability. There is no consensus on the best material, but important variables include cost, durability, reliability, or availability. Glass surfaces are particularly effective for laboratory-scale studies, providing the possibility of illumination on both sides of the biofilm. Innovations in synthetic materials include surface modifications, such as the incorporation of micro- and nanoscale textures or surface treatments to promote hydrophilicity. Although some of these innovations have been developed to prevent biofilms [51], this knowledge could be used also to promote biofilm formation.

Hydrophilic surfaces, such as glass or modified plastics, enhance cell attachment, while hydrophobic materials may require surface treatments to improve their compatibility [52]. Surface roughness also promotes adhesion by increasing the available contact area and providing microstructures that trap cells, preventing detachment under flow conditions [25]. However, this can also lead to darker spots where appropriate light and nutrient diffusion becomes more difficult. Additionally, microalgae often secrete specialized adhesion molecules, such as glycoproteins, which strengthen the bond between the cells and the substrate [48].

2.1.4 Microalgae Species and Bacterial Interactions

The development of microalgal biofilms is influenced more by the species involved than by surface properties [53]. Axenic (sterile) versus non-axenic (non-sterile) cultivation conditions significantly affect biofilm formation through the interactions between microalgae and co-existing bacteria. Symbiotic relationships can enhance nutrient availability, such as through bacterial nitrogen fixation, and contribute to EPS production; these interactions also contribute to improving biofilm stability and resilience under varying environmental conditions [54]. For instance, *Chlorella vulgaris* has been shown to produce minimal EPS on its own and grows significantly better in biofilms under non-axenic conditions in the presence of other organisms [53].

3 Biofilm-Based System Designs and Engineering

Attached cultivation bioreactors offer more flexibility in design in comparison to suspension-based microalgae cultivation, allowing for a greater range of reactor types. Also, biofilm-based photobioreactors are recognized for their potential for many applications, leading to the creation of new innovative designs of photobioreactors with unique shapes, i.e., conical, fractal tree-like, L-shaped, pyramid, ring, V-shaped, X-shaped, and others [6], open or closed. The adaptable nature of microalgal biofilms is key to their productivity in diverse applications, enhancing their functionality in industrial systems. Independently of the design of reactor, these can be open or closed. However, open systems have higher water evaporation rates, and therefore higher water footprints (0.5–7 L m^{-2} d^{-1}) than closed systems, influencing the environmental sustainability and cost of this technology [55, 56].

3.1 Fixed-Bed Reactors

3.1.1 Vertical or Inclined Reactors

These systems are designed with flat, vertical, or inclined surfaces. By maximizing the surface area-to-volume ratio, they support dense algal growth without significantly increasing land use, allowing for compact and efficient cultivation setups. They also often incorporate gravity-driven flow to distribute nutrients, reducing energy consumption (Fig. 2). Such designs are highly scalable and suitable for integration into urban and industrial settings [7, 57].

Fig. 2 Example of an experimental algal biofilm cultivation unit. This was built in plexiglass, and had a holding reservoir with a water pump, pumping the water onto the slope where the water flowed by gravity down into the reservoir again (upper left). After inoculation with different green algae, a biofilm quickly started to form (upper right). The biofilm got thicker (lower left) and was harvested by scraping it off (lower right). Photos by K. Spilling

These reactors use stationary surfaces for biofilm attachment and have been widely employed in wastewater treatment due to their robustness and ability to handle high nutrient loads. One well-known example is the Algal Turf Scrubber (ATS) [58]. The term

Algal Turf Scrubber (ATS™) refers to US patent No: 4 333 263, 5 851 398. ATS have a simple design and construction that make them more scalable and cost-effective. ATS have been commonly used as biofiltration systems where nutrient-rich water flows over surfaces where microalgae and bacteria form a mat attached to an inclined plastic mesh (typically on a 0.5–2% slope), and permeates down. These systems are also referred to as permeate systems because the media is supplied to the substrate either continuously or in pulses that imitate the movement of waves [59].

3.1.2 Perfused Systems or Porous Substrate Bioreactors (PSBR)

In these systems, a microporous substrate layer that holds the microalgae is placed over a semi-permeable macroporous membrane through which the liquid nutrient media flows. This system has been demonstrated to sustain biofilm production with a variety of microalgal species [60]. Nutrients and humidity diffuse to the microalgal cells that attach onto the surface of the substrate separated from direct contact with the culture medium, avoiding any biomass loss due to shear stress and helping with potential contamination issues [61]. One example are the twin-layer (TS) systems [62]. Microalgae are directly exposed to the air for more effective gas exchange and light (eliminating any attenuation found in submerged systems). However, the risk of dissection is higher. These systems have been designed horizontally and vertically. In vertical systems, the culture medium is commonly supplied at the top with the help of a pump. However, some vertical systems use capillarity and evaporation to drive the suction of new media from the bottom to the top, i.e., passive systems [62, 63]. In these cases, governing factors include capillary pressure, viscous pressure drop, and gravitational force [62].

3.2 Rotating Biofilm Reactors (RBRs)

These systems are one of the most recent innovations within intermittently submerged systems [57] (as opposed to systems constantly submerged in the medium). RBRs typically consist of a rotating cylindrical drum or disc partially submerged in a nutrient medium. As the drum rotates, the biofilm alternates between exposure to light and nutrient solution, mimicking natural wave currents and promoting gas exchange, i.e., CO_2 uptake and O_2 release [11, 65]. Geometries such as spirals, grids, or packed beds can enhance the available surface area for biofilm attachment, boosting productivity [11]. Considering the ease of harvesting, vertical or triangular conveyor belt configurations have been integrated into RBR designs. In these systems, flexible materials like cotton belts or rope move continuously around rotating shafts, allowing for the formation of biofilms on their surfaces. Again, choice of materials is a key variable and research comparing materials like cotton rope and flat cotton belts found that while cotton rope supported higher biomass yields, it also harbored more non-photosynthetic organisms, potentially complicating downstream processing. Conversely, flat cotton belts yielded biomass with higher photosynthetic purity, making them more suitable for applications requiring high-quality algal products. Studies have demonstrated that such conveyor belt RBRs can achieve biomass productivities of up to 3.51 g m^{-2} day^{-1}, highlighting their potential for scalable biofilm cultivation [64]. Harvesting is streamlined by scraping the biofilm from the belt as it moves, reducing labor and energy costs associated with traditional methods. At the same time, this system's high surface area and efficient mixing present a series of advantages, but the constant movement increases operating costs for energy and maintenance [66].

4 Applications of Biofilm-Based Microalgal Systems

4.1 Bioremediation

Biofilm-based microalgal systems can remove nutrients from waste streams and transform them into valuable resources, such as biofuels, bioplastics, and fertilizers, closing the loop in nutrient and energy cycles [67]. Microalgae in biofilm systems can efficiently uptake excess nitrogen (e.g., ammonia and nitrate) and phosphorus. There is commonly a need to remove these elements from municipal and industrial wastewater due to environmental concerns such as eutrophication of local waterways. The biofilm structure facilitates high nutrient removal rates due to its dense microbial community and localized nutrient retention [68, 69]. The integration of biofilm-based systems into circular bioeconomy frameworks presents significant opportunities for sustainable development.

Heavy metals such as cadmium, lead, and mercury pose significant environmental and health risks when discharged into water systems. Depending on the composition of the biofilm, some microalgal systems can sequester heavy metals from contaminated water through mechanisms like biosorption, bioaccumulation, and precipitation. The EPS in biofilms [70] contain functional groups (e.g., hydroxyl, carboxyl) that can bind to metal ions and facilitate metal complexation, enabling the bioremediation of water contaminated with heavy metals such as Ni, Co, and Mn [71].

The oxygen provided by microalgal photosynthesis can support bacterial degradation of complex organic pollutants such as pesticides, dyes, and pharmaceutical residues. This synergistic relationship accelerates pollutant breakdown and promotes system efficiency. For instance, biofilm systems have been shown to achieve over 90% removal of certain pharmaceutical contaminants under optimal conditions [72].

4.2 Bioenergy Production

Microalgae are a promising source of third-generation biofuels due to their high oil content, fast growth, and non-competition with food crops. However, commercial-scale production remains unviable due to high costs, technical challenges in cultivation and processing, and inconsistent feedstock supply. While interest has grown, microalgal biofuels are still not economically competitive [73]. However, biofilm-based systems already offer significant advantages for biofuel production due to their high biomass productivity and reduced harvesting costs. Microalgae cultivated in biofilms accumulate lipids, carbohydrates, and other energy-dense compounds in high density, which can be converted into biofuels such as biodiesel and bioethanol [64, 74]. Advanced engineering approaches, such as optimizing light exposure and nutrient availability, can further enhance lipid accumulation and productivity. Bioenergy production is inversely related to bioremediation. Maximizing lipid production in microalgae is commonly achieved by stressing the algae through nutrient-limiting conditions or manipulating light conditions, rather than exposing them to nutrient-rich wastewater. Therefore, a lower biodiesel production of 0.12 L/kg algal biomass is achieved when wastewater treatment is the primary function [75].

Biofilm reactors can also be integrated with anaerobic digestion systems. The dense structure of biofilms and their high organic content make them particularly suitable as feedstock for methane-rich biogas production. Moreover, the integration of anaerobic digestion of other organic waste materials, such as agricultural residues or food waste, with biofilm systems enhances methane yields and provides opportunities for nutrient recycling, where digestate from biogas plants can be used as a nutrient source for microalgal cultivation, closing the loop in bioenergy systems [76–78].

5 Prospects and Emerging Technologies in Biofilm Cultivation

Artificial intelligence (AI) and the Internet of Things (IoT) are poised to revolutionize biofilm-based microalgal systems. Advanced AI algorithms can optimize system performance by analyzing real-time data from IoT-enabled sensors, which monitor critical parameters such as light intensity, temperature, nutrient levels, and biofilm thickness. Predictive models driven by AI could also help preempt system failures and enhance operational efficiency. These smart systems are especially promising for large-scale industrial operations, where manual monitoring is impractical.

The development of advanced materials for biofilm support could become a key area of research. Innovations include nanostructured surfaces and biocompatible coatings that promote adhesion and enhance biofilm stability. Additionally, biodegradable substrates could be explored further to minimize environmental impact and align with sustainability goals.

Genetic engineering offers opportunities to tailor microalgal strains for improved biofilm formation, stress tolerance, and metabolite production. Strategies include modifying genes involved in EPS synthesis to enhance biofilm stability or engineering metabolic pathways to increase lipid or pigment accumulation in addition to other secondary metabolites [79, 80]. Combining engineered strains with biofilm systems could unlock new potential for high-value product synthesis and resource recovery.

Furthermore, coupling biofilm systems into circular economy frameworks such as carbon capture technologies, can enhance their role in mitigating climate change. Industrial CO_2 emissions can be fed into biofilm reactors, where microalgae utilize the carbon for growth, effectively sequestering CO_2 while producing valuable byproducts. Such integrated systems align with global goals for carbon neutrality and sustainable resource management, making biofilm-based cultivation a cornerstone of future bioeconomic strategies.

6 Conclusion

Biofilm-based microalgal cultivation systems represent a promising alternative to conventional suspended culture methods, offering new avenues for sustainable resource management and industrial biotechnology. These systems can achieve high biomass concentrations, enhanced nutrient utilization, and simplified harvesting processes, making them a more cost-efficient alternative. Additionally, advances in reactor design, substrate materials, and operational optimization have the potential to significantly improve their efficiency and scalability. Biofilm systems have proven their versatility and potential for applications ranging from wastewater treatment and nutrient recovery to biofuel production and the synthesis of high-value products. Additionally, their compatibility with various industrial processes underscores their potential as a robust platform for algal cultivation.

To realize the full potential of biofilm-based microalgal cultivation systems, further research and innovation are essential. Interdisciplinary collaboration among biologists, engineers, chemists, and environmental scientists can drive the development of more efficient reactors, advanced materials, and genetically optimized strains. Scaling up these systems for industrial applications will require overcoming challenges related to cost, fouling, and operational complexity. Additionally, integrating biofilm systems with digital technologies, such as AI and IoT, can enhance monitoring and control, paving the way for smart and sustainable operations.

The future of biofilm-based cultivation lies in its ability to integrate into broader bioeconomic strategies, addressing global priorities such as food and water security, energy sustainability, and climate change mitigation.

Acknowledgments

This study was supported by a personal grant to Lara D. Mateos from the Maj and Tor Nessling Foundation, and by the AlgaeP-roBANOS project, co-funded by the European Union through the Horizon Europe program (project number 101112943), and we have used equipment part of the Finnish Marine Research Infrastructure (FINMARI) consortium. Cultivation units in Fig. 2 designed by Gerbrand Grobler.

References

1. Ozkan A, Kinney K, Katz L, Berberoglu H (2012) Reduction of water and energy requirement of algae cultivation using an algae biofilm photobioreactor. Bioresour Technol 114:542–548
2. Milledge JJ, Heaven S (2013) A review of the harvesting of micro-algae for biofuel production. Rev Environ Sci Biotechnol 12:165–178
3. Kumar N, Banerjee C, Negi S, Shukla P (2023) Microalgae harvesting techniques: updates and recent technological interventions. Crit Rev Biotechnol 43(3):342–368
4. Shelef G, Sukenik A, Green M (August 1 1984). Microalgae Harvesting and Processing: A Literature Review. Technical Report No. NREL/STR-231-2396, Solar Energy Research Institute, US. 70 pp.
5. Singh G, Patidar S (2018) Microalgae harvesting techniques: a review. J Environ Manage 217:499–508
6. Berner F, Heimann K, Sheehan M (2015) Microalgal biofilms for biomass production. J Appl Phycol 27:1793–1804
7. Cheah YT, Chan DJC (2021) Physiology of microalgal biofilm: a review on prediction of adhesion on substrates. Bioengineered 12(1):7577–7599
8. Johnson MB, Wen Z (2010) Development of an attached microalgal growth system for biofuel production. Appl Microbiol Biotechnol 85:525–534
9. De Carvalho CC (2018) Marine biofilms: a successful microbial strategy with economic implications. Front Mar Sci 5:126
10. Gross M, Jarboe D, Wen Z (2015) Biofilm-based algal cultivation systems. Appl Microbiol Biotechnol 99:5781–5789
11. Gross M, Henry W, Michael C, Wen Z (2013) Development of a rotating algal biofilm growth system for attached microalgae growth with in situ biomass harvest. Bioresour Technol 150:195–201
12. Morales M, Bonnefond H, Bernard O (2020) Rotating algal biofilm versus planktonic cultivation: LCA perspective. J Clean Prod 257:120547
13. Zhang Q, Liu C, Li Y et al (2017) Cultivation of algal biofilm using different lignocellulosic materials as carriers. Biotechnol Biofuels 10:1–16
14. Kesaano M, Sims RC (2014) Algal biofilm based technology for wastewater treatment. Algal Res 5:231–240
15. Moreno Osorio JH, Pollio A, Frunzo L et al (2021) A review of microalgal biofilm technologies: definition, applications, settings and analysis. Front Chem Eng 3:737710
16. Bos R, Van der Mei HC, Busscher HJ (1999) Physico-chemistry of initial microbial adhesive interactions–its mechanisms and methods for study. FEMS Microbiol Rev 23(2):179–230
17. De-Bashan LE, Bashan Y (2010) Immobilized microalgae for removing pollutants: review of practical aspects. Bioresour Technol 101(6):1611–1627
18. Moreno-Garrido I (2008) Microalgae immobilization: current techniques and uses. Bioresour Technol 99(10):3949–3964
19. Christenson L, Sims R (2011) Production and harvesting of microalgae for wastewater treatment, biofuels, and bioproducts. Biotechnol Adv 29(6):686–702
20. Shen Y, Yuan W, Pei Z et al (2009) Microalgae mass production methods. Trans ASABE 52(4):1275–1287
21. Mantzorou A, Ververidis F (2019) Microalgal biofilms: a further step over current microalgal cultivation techniques. Sci Total Environ 651:3187–3201
22. Babiak W, Krzemińska I (2021) Extracellular polymeric substances (EPS) as microalgal bioproducts: a review of factors affecting EPS synthesis and application in flocculation processes. Energies 14(13):4007
23. Schnurr PJ, Allen DG (2015) Factors affecting algae biofilm growth and lipid production: a review. Renew Sustain Energy Rev 52:418–429
24. Kesaano M (2015) Characterization and performance of algal biofilms for wastewater treatment and industrial applications. Utah State University, Utah

25. Tong C, Derek C (2022) Membrane surface roughness promotes rapid initial cell adhesion and long term microalgal biofilm stability. Environ Res 206:112602
26. Oh JK, Yegin Y, Yang F et al (2018) The influence of surface chemistry on the kinetics and thermodynamics of bacterial adhesion. Sci Rep 8(1):17247
27. Fanesi A, Paule A, Bernard O et al (2019) The architecture of monospecific microalgae biofilms. Microorganisms 7(9):352
28. Paul E, Ochoa JC, Pechaud Y et al (2012) Effect of shear stress and growth conditions on detachment and physical properties of biofilms. Water Res 46(17):5499–5508
29. Petrova OE, Sauer K (2016) Escaping the biofilm in more than one way: desorption, detachment or dispersion. Cur Opin Microbiol 30:67–78
30. Wolfstein K, Stal LJ (2002) Production of extracellular polymeric substances (EPS) by benthic diatoms: effect of irradiance and temperature. Mar Ecol Prog Ser 236:13–22
31. Gao Y, Bernard O, Fanesi A et al (2024) The effect of light intensity on microalgae biofilm structures and physiology under continuous illumination. Sci Rep 14(1):1151
32. Béchet Q, Shilton A, Guieysse B (2013) Modeling the effects of light and temperature on algae growth: state of the art and critical assessment for productivity prediction during outdoor cultivation. Biotechnol Adv 31(8):1648–1663
33. Kumar K, Sirasale A, Das D (2013) Use of image analysis tool for the development of light distribution pattern inside the photobioreactor for the algal cultivation. Bioresour Technol 143:88–95
34. Bialevich V, Zachleder V, Bišová K (2022) The effect of variable light source and light intensity on the growth of three algal species. Cells 11(8):1293
35. Maltsev Y, Maltseva K, Kulikovskiy M, Maltseva S (2021) Influence of light conditions on microalgae growth and content of lipids, carotenoids, and fatty acid composition. Biol 10(10):1060
36. Spilling K (2007) Dense sub-ice bloom of dinoflagellates in the Baltic Sea, potentially limited by high pH. J Plankton Res 29(10):895–901
37. Saini N, Pal K, Deepak B, Mona S (2021) Thermophilic algae: a new prospect towards environmental sustainability. J Clean Production 324:129277
38. Shen Y, Wang S, Ho S-H et al (2018) Enhancing lipid production in attached culture of a thermotolerant microalga *Desmodesmus* sp. F51 using light-related strategies. Biochem Eng J 129:119–128
39. Boonchai R, Kaewsuk J, Seo G (2015) Effect of nutrient starvation on nutrient uptake and extracellular polymeric substance for microalgae cultivation and separation. Desalin Water Treat 55(2):360–367
40. Rees T, Raven JA (2021) The maximum growth rate hypothesis is correct for eukaryotic photosynthetic organisms, but not cyanobacteria. New Phytol 230(2):601–611
41. Soares A, Gomes LC, Monteiro GA, Mergulhão FJ (2022) Hydrodynamic effects on biofilm development and recombinant protein expression. Microorganisms 10(5):931
42. Polizzi B, Fanesi A, Lopes F et al (2022) Understanding photosynthetic biofilm productivity and structure through 2D simulation. PLoS Comput Biol 18(4):e1009904
43. Ranganathan P, Pandey AK, Sirohi R et al (2022) Recent advances in computational fluid dynamics (CFD) modelling of photobioreactors: Design and applications. Bioresour Technol 350:126920
44. Wei G, Yang JQ (2023) Microfluidic investigation of the impacts of flow fluctuations on the development of *Pseudomonas putida* biofilms. NPJ Biofilms Microbiomes 9(1):73
45. Qiu R, Gao S, Lopez PA, Ogden KL (2017) Effects of pH on cell growth, lipid production and CO2 addition of microalgae *Chlorella sorokiniana*. Algal Res 28:192–199
46. Zheng Q, Xu X, Martin GJ, Kentish SE (2018) Critical review of strategies for CO2 delivery to large-scale microalgae cultures. Chin J Chem Eng 26(11):2219–2228
47. Muhammad MH, Idris AL, Fan X et al (2020) Beyond risk: bacterial biofilms and their regulating approaches. Front Microbiol 11:928
48. Xiao R, Zheng Y (2016) Overview of microalgal extracellular polymeric substances (EPS) and their applications. Biotechnol Adv 34(7):1225–1244
49. Genin SN, Aitchison JS, Allen DG (2014) Design of algal film photobioreactors: material surface energy effects on algal film productivity, colonization and lipid content. Bioresource Technol 155:136–143
50. Christenson LB, Sims RC (2012) Rotating algal biofilm reactor and spool harvester for wastewater treatment with biofuels by-products. Biotechnol Bbioeng 109(7):1674–1684
51. Graham MV, Cady NC (2014) Nano and microscale topographies for the prevention of bacterial surface fouling. Coatings 4(1):37–59
52. Tang J, Liu B, Gao L et al (2021) Impacts of surface wettability and roughness of styrene-acrylic resin films on adhesion behavior of microalgae Chlorella sp. Colloids Surf B Biointerfaces 199:111522

53. Irving TE, Allen DG (2011) Species and material considerations in the formation and development of microalgal biofilms. Appl Microbiol Biotechnol 92:283–294
54. Wang L, Cui Y-W (2024) Mutualistic symbiosis of fungi and nitrogen-fixing bacteria in halophilic aerobic granular sludge treating nitrogen-deficient hypersaline organic wastewater. Bioresour Technol 394:130183
55. Murphy T, Berberoğlu H (2012) Temperature fluctuation and evaporative loss rate in an algae biofilm photobioreactor. J Sol Energy Eng 134:011002. https://doi.org/10.1115/1.4005088
56. osadas E, Muñoz A, García-González MC et al (2015) A case study of a pilot high rate algal pond for the treatment of fish farm and domestic wastewaters. J Chem Technol Biotechnol 90(6):1094–1101
57. Ennaceri H, Ishika T, Mkpuma VO, Moheimani NR (2023) Microalgal biofilms: towards a sustainable biomass production. Algal Res 72:103124
58. Murshid S, Antonysamy A, Dhakshinamoorthy G et al (2023) A review on biofilm-based reactors for wastewater treatment: recent advancements in biofilm carriers, kinetics, reactors, economics, and future perspectives. Sci Total Environ 892:164796
59. Adey WH, Kangas PC, Mulbry W (2011) Algal turf scrubbing: cleaning surface waters with solar energy while producing a biofuel. Bioscience 61(6):434–441
60. Nowack EC, Podola B, Melkonian M (2005) The 96-well twin-layer system: a novel approach in the cultivation of microalgae. Protist 156(2):239–251
61. Shi J, Podola B, Melkonian M (2007) Removal of nitrogen and phosphorus from wastewater using microalgae immobilized on twin layers: an experimental study. J Appl Phycol 19: 417–423
62. Naumann T, Çebi Z, Podola B, Melkonian M (2013) Growing microalgae as aquaculture feeds on twin-layers: a novel solid-state photobioreactor. J Appl Phycol 25:1413–1420
63. Shi J, Podola B, Melkonian M (2014) Application of a prototype-scale Twin-Layer photobioreactor for effective N and P removal from different process stages of municipal wastewater by immobilized microalgae. Bioresour Technol 154:260–266
64. Patwardhan SB, Pandit S, Ghosh D et al (2024) A concise review on the cultivation of microalgal biofilms for biofuel feedstock production. Biomass Convers Biorefinery 14(6):7219–7236
65. Nguyen V-T, Le V-A, Do Q-H et al (2024) Emerging revolving algae biofilm system for algal biomass production and nutrient recovery from wastewater. Sci Total Environ 912:168911
66. Choudhary P, Malik A, Pant KK (2017) Algal biofilm systems: an answer to algal biofuel dilemma. In: Algal biofuels: recent advances and future prospects. Springer Cham, Switzerland, pp. 77–96
67. López-Sánchez A, Silva-Gálvez AL, Aguilar-Juárez Ó et al (2022) Microalgae-based livestock wastewater treatment (MbWT) as a circular bioeconomy approach: enhancement of biomass productivity, pollutant removal and high-value compound production. J Environ Manage 308:114612
68. Popa MD, Simionov I-A, Petrea SM et al (2025) Efficiency of microalgae employment in nutrient removal (nitrogen and phosphorous) from municipal wastewater. Water 17(2):260
69. Stockenreiter M, Haupt F, Seppälä J et al (2016) Nutrient uptake and lipid yield in diverse microalgal communities grown in wastewater. Algal Res 15:77–82
70. Sattayawat P, Yunus IS, Noirungsee N et al (2021) Synthetic biology-based approaches for microalgal bio-removal of heavy metals from wastewater effluents. Front Environ Sci 9:778260
71. Palma H, Killoran E, Sheehan M et al (2017) Assessment of microalga biofilms for simultaneous remediation and biofuel generation in mine tailings water. Bioresour Technol 234:327–335
72. Ahmed S, Mujtaba G, Rizwan M et al (2024) Current progress on the microalgae-mediated treatment of pharmaceutical active compounds. Chem Biochem Eng Q 38(1):1–13
73. Hoang AT, Sirohi R, Pandey A et al (2023) Biofuel production from microalgae: challenges and chances. Phytochem Rev 22(4):1089–1126
74. Spilling, K, Seppälä J (2012) Photobiology and lipid metabolism in algae. In: The science of algal Fuels: phycology, geology, biophotonics, genomics and nanotechnoly. Springer Dordrecht, Netherlands, pp. 383–398
75. Craggs RJ, Heubeck S, Lundquist T, Benemann J (2011) Algal biofuels from wastewater treatment high rate algal ponds. Water Sci Technol 63(4):660–665
76. Sialve B, Bernet N, Bernard O (2009) Anaerobic digestion of microalgae as a necessary step to make microalgal biodiesel sustainable. Biotechnol Adv 27(4):409–416
77. Ward A, Lewis D, Green F (2014) Anaerobic digestion of algae biomass: a review. Algal Research 5:204–214
78. Molinuevo-Salces B, Mahdy A, Ballesteros M, González-Fernández C (2016) From piggery wastewater nutrients to biogas: microalgae biomass revalorization through anaerobic digestion. Renew Energy 96:1103–1110

79. Sreenikethanam A, Raj S, Gugulothu P, Bajhaiya AK (2022) Genetic engineering of microalgae for secondary metabolite production: recent developments, challenges, and future prospects. Front Bioeng Biotechn 10:836056
80. Dhokane D, Shaikh A, Yadav A et al (2023) CRISPR-based bioengineering in microalgae for production of industrially important biomolecules. Front Bioeng Biotechnol 11: 1267826
81. Sabater S, Timoner X, Borrego C, Acuña V (2016) Stream biofilm responses to fow Intermittency: from cells to ecosystems. Front Environ Sci Sec Freshw Sci 4:14

Chapter 5

Combining Aquaponics and Applied Phycology: From Bioremediation of Wastewater to Algal Products

Tobias Lipsewers

Abstract

This chapter contains an introduction to aquaponics and provides, for instance, technical solutions, species selection, and ways to reduce the operating costs to sustainably grow microalgae and fish in one artificial ecosystem, using the principles of "algaeponics" (combining aquaponics and algae cultivation). Possibilities for high-value products derived from microalgae biomass will be shown, with emphasis on the production of biofuels. To the best of my knowledge, this approach has not been used so far, neither in research nor commercially. Thus, initial experimentation is required to determine, for example, the feasibility for large-scale production of mainly microalgae biomass. Minding the fact that the cultivation of algae for multiple purposes is an expanding industry, algaeponics is a highly promising field to explore, since it has several advantages compared to more common cultivation methods. Promising/recommended technologies for producing biofuel-relevant microalgae using algaeponics are, e.g., recirculating decoupled aquaponic systems, cultivation of algae in open ponds (e.g., raceway type in a greenhouse), and harvesting the biomass by filtration using a semipermeable membrane. A suitable combination of biofuel-relevant microalgae and fish species (based on pH and temperature requirements) is the co-cultivation of the common carp (*Cyprinus carpio*) and the two green microalgae *Chlorella vulgaris* and *Chlamydomonas reinhardtii* to test the production of biofuels (e.g., biodiesel and biohydrogen, respectively). Depending on, for example, the location of the facility (heating vs. cooling of the system water), additional fish/algae combinations can be explored.

Keywords Algaeponics, Water-reuse, Bioremediation, Microalgae biomass, Biofuels

1 Introduction

1.1 The Concept of Aquaponics

Aquaponics (AP) is a production system of aquatic organisms and plants where >50 % of the nutrients sustaining optimal plant growth derive from waste originating from feeding the aquatic organisms [1]. It combines the principles of aquaculture and hydroponics. The latter is a widely used technique to grow plants in a nutrient solution without the use of soil. In AP, the nutrient-rich effluent ("waste") water from the animal rearing unit of the facility serves as the nutrient solution for the plants.

Kristian Spilling (ed.), *Biofuels from Algae*, Methods in Molecular Biology, vol. 2996,
https://doi.org/10.1007/978-1-0716-5031-8_5,

A variety of organisms can be cultivated—for example, fish, crustaceans, bivalves, edible plants, and algae. The well-being of farmed animals (e.g., stocking densities and quality of feed) and impacts on the environment (e.g., sustainability of the raw materials used in the feed and efficient resource saving) should be considered.

Microorganisms (the biofilter of the system) and phototrophic organisms (crops such as plants and algae) can efficiently bioremediate the water by growing and thereby reducing the nutrient concentrations before recirculating it to, for example, fish tanks.

Achieving the correct (species-specific) quality and quantity of the nutrient pool is a critical factor when aiming for biomass production. The basic flow of nutrients in an AP system follows the steps below (Fig. 1) [2]:

- Fish feed is the main nutrient source
- The fish eat the feed and release unused compounds
- This waste is partitioned between the different components of the system (microbes, plants, and water)

In fully recirculating aquaponic systems (FRAS), the water is constantly recirculated between the RAS (recirculating aquaculture systems) and hydroponic components. Water is only removed for filter cleaning or when the operator decides. Some systems waste only 1.5% of the water daily (on average) [2].

1.2 The Historic Use of AP

The concept of AP was already known by Ancient China and the Aztec Empire and was applied in so-called floating gardens (Fig. 2).

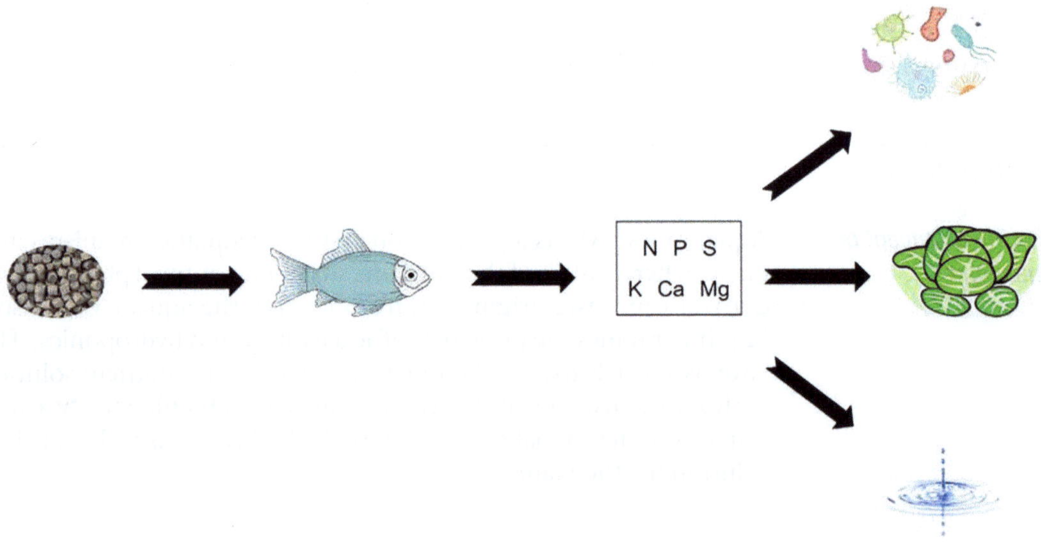

Fig. 1 A diagram showing the basic nutrient flows in an aquaponic system (from Lennard 2017 [2], used with permission)

Fig. 2 This image was created by artificial intelligence (DALL·E). It gives an impression of the Aztec chinampas of Central America (inspired by a diorama [6])

These societies combined the farming of fish with rice or vegetable production [1, 3–5].

Nowadays, various research and commercial AP facilities exist worldwide. One very large commercial farm is located in Northfield (Wisconsin) and uses de-coupled AP (a fish house and a greenhouse separated) to produce certified organic salmon and, for example, different kinds of lettuce [7].

AP can be realized in different ways. For instance, Fig. 3a shows a setup where the water is pumped up to potted plants, it flows through the pots, and trickles back into the fish tank, enriching the water with oxygen [8]. In contrast, Fig. 3b depicts a so-called deep

Fig. 3 An aquaponic system with fish (tilapia or perch) and edible plants (watercress and tomatoes, **a**) [8] and a deep flow hydroponics system using the floating raft technique (**b**) [9]

flow hydroponics system using the floating raft technique, where the roots of the plants grow directly into the nutrient-rich water, allowing a high planting density and thus an efficient use of the cultivation area [9]. AP can also be combined with vertical farming and rooftop gardening. Rooftop AP is suitable for buildings with flat roofs and could be especially interesting for science parks, universities, and research institutes, among others.

1.3 Opportunities and Challenges

1.3.1 Opportunities

The underlying principle of AP matches key aspects of a circular economy model, where, for example, waste production is decreased, the environmental footprint (greenhouse gas emissions) is improved, and resources (e.g., water) are saved.

Some of the aspects showing the potential and future relevance of land-based indoor AP farms are listed here:

- Extended product range and improved profitability for farmers
- Food safety: no use of, e.g., antibiotics and pesticides
- Increased local production leads to reduced emissions of greenhouse gases
- High-quality standards due to the highly controlled rearing environment
- Different types (e.g., from low to high tech) of systems exist
- Resource-saving: e.g., bioremediation and re-use of water and nutrients
- Improving the local economy: new possibilities for the labor market
- Interesting opportunities for research and industry (e.g., biofuel from "algaeponic" farms)

Efficient bioremediation of the wastewater allows a high percentage return of the water to the fish rearing unit. As algae are more efficient in taking up nutrients, producing these phototrophs (especially microalgae) instead of or in combination with plants in one system is a highly promising approach (algaeponics).

Depending on the national legislation, certified organic products can be produced using AP, which could be a motivation and benefit for future aqua farmers. For example, in the European Union (EU), plants that grow naturally in water (including algae) and fish can be certified organic when produced hydroponically and aquaponically [10–12].

1.3.2 Challenges

Depending on the scale of the farm, the technologies used, and the location (among others), the initial costs of building a tailor-made AP system to produce/study aquatic animals and microalgae for the production of high-value end products (such as biofuel) vary tremendously. There are ways to reduce the costs, such as the collaboration of multiple farms, joining forces of already existing producers (e.g., RAS and greenhouse farmers), and using waste heat. With regard to the profitability related to food production (fish, etc.), more suggestions on how to operate an economically viable AP farm are described in, for instance, Lipsewers 2023 [13]. Also, the fish produced in AP research facilities can be marketed as food, and the profit could be used to cover part of the operating costs and/or for further research, as done during a completed project ("fishglasshouse") at the University of Rostock in Germany [14]. Thus, certain challenges might be converted to opportunities.

For the nutrient content of the fish waste, which determines the growth rate of algae, the fish feed composition is more important than the fish species (Dr. Wilson Lennard, 2020, conference talk). Nutrient limitations affecting plant growth are applicable to algae as well. Insufficient nutrient concentrations for plant growth in AP systems arise from the fact that the fish waste alone does not provide sufficiently high concentrations of most macro-nutrients for the plants, especially in commercial terms [2]. The major macronutrient in fish waste is nitrogen (N) and the limited ones are [2]:

- Calcium (Ca)
- Potassium (K)
- Sulfur (S)
- Magnesium (Mg)
- Iron (Fe)

The remineralization of solid fish waste can reduce the limitations to some extent, for example for phosphorus (P). Many of the micronutrients are provided by the fish waste and because plants need only small amounts, it is assumed that limitations occur but are less common than for macronutrients. Thus, additional nutrient supplementation might be required to allow the efficient cultivation of plants [2] and algae. Precise fertilizing saves resources and requires regular measurements of the nutrient concentrations before the water enters the plant/algae cultivation unit.

1.4 New Directions

As an example of a very promising direction, the integration of AP with algal cultivation (algaeponics) is highlighted by this chapter. Fish and algae are a good match, also in artificial ecosystems, since they co-exist and interact in nature. The focus lies on combining, for example, fish and microalgae production, since the latter are widely used study objects in biofuel research and are already used in several industries (see "Biomass Use" below). The cultivation of algae for different purposes (e.g., in low-trophy aquaculture) is an expanding business and is suggested to contribute significantly to reach a circular economy. Cultivating them in recirculating systems using renewable energy and using waste heat from other industries will strengthen their role. There are several ways for using the produced fish. It can be marketed as food, used for research (e.g., new species for the aquaculture sector), and endangered species could be bred for restocking programs. Furthermore, ornamental fish (e.g., koi) can be cultivated for recreational fish keepers if it is not desired to produce fish for food.

2 Integration with Algal Cultivation

2.1 How Can It Be Done?

The production of microalgae using the system water of AP farms requires the combination/connection of three main components—RAS, cultivation of microalgae, and harvesting of the microalgae biomass. The separate technologies exist, their combination is less used and requires pilot studies (from small- to large-scale). However, research and development efforts and small-scale applications exist, such as applying the concept of multi-aquaponics, which involves, for instance, the cultivation of microalgae (e.g., *Chlorella vulgaris*), fish, and mushrooms [15].

Since the production of microalgae specifically for biofuel research/synthesis in AP systems is a new approach, the following part contains suggestions on which algae cultivation techniques could be promising for investigating their compatibility with AP systems.

Two common technologies to produce commercially relevant volumes of microalgae biomass are photobioreactors (PBRs) and open ponds. In principle, both approaches can be combined with AP, given the required technical solutions for, e.g., the recirculation of the water. Especially, the open pond (Fig. 4b) cultivation of pelagic microalgae (phytoplankton) is suitable/promising for AP, since it is, setup-wise, more similar to the so-called deep flow technique (Fig. 4a) used in hydroponic units of existing systems compared to, for example, tubular PBRs (Fig. 4c).

Fig. 4 Comparison of a deep-water floating raft culture (with lettuce in this case) used in aquaponics (**a**) [16], to an open pond for microalgae cultivation (**b**, from Shen et al. 2009 [17], used with permission), and a tubular photobioreactor (**c**, used with permission [18])

The raceway pond is one of the most frequently used open pond types for microalgae cultivation. It consists of a series of closed-loop channels (~30 cm deep) and a paddlewheel to ensure equal nutrient distribution and prevent sedimentation of the microalgae [19]. Also commercial fish production is often done in raceway systems. The combination of two separate raceways could be an interesting approach to test.

Co-cultivation of fish and algae has great potential for the whole salinity gradient of fish farming (fresh, brackish, and marine). As is also valid for culturing algae with common growth media in the laboratory, the type/design of the cultivation, harvesting techniques, and further processing to be combined with AP depend on the species, production volume, end product, among others.

2.2 Wastewater Sources that Could Be Used

Generally, microorganisms such as microalgae can be used to reduce the nutrient concentrations in different kinds of wastewater [e.g., 20].

The cultivation of several microalgae species using RAS wastewater was studied relatively recently [21, 22]. Stevčić et al. [21] found that the densities and specific growth rates of the green microalgae species tested did not differ between RAS wastewater and a reference medium. Rates and percentages of nitrate-N removal were significantly higher for *Acutodesmus* sp. and *Chlamydomonas reinhardtii* than for, e.g., *Haematococcus pluvialis*. The latter featured a significantly higher uptake rate of nitrate-N than some of the other tested species and a significantly higher cell uptake rate of phosphate-P than all other green microalgae. The non-green microalgae showed a poor growth potential and nutrient removal in both media [21]. A study by the same research group [22] has shown that all tested green microalgae (*H. pluvialis*, *Monoraphidium griffithii*, and *Selenastrum* sp.) can grow in unfiltered and filtered RAS wastewater. Independent of the medium, *Selenastrum* sp. and *M. griffithii* were able to remove ~99% (on average), and *H. pluvialis* ~75%, of the phosphate-P. The nitrate-N removal was below ~40% for all three species. The results obtained by Calderini et al. suggest that RAS wastewater is suitable to produce high-quality microalgal biomass without the removal of biological contaminants and suspended solids using certain algal species [22].

Since RAS is an integral part of AP, these species are, in principle, suitable for algaeponics as well. Certain green microalgae species seem to be especially promising, but also the potential of other species and algae groups, with emphasis on biofuel production, for example, could be studied further.

2.3 Opportunities and Challenges

2.3.1 Opportunities

The integration of new/additional cultivation/harvesting technologies (e.g., for microalgae production) with existing AP approaches creates a wide variety of possibilities for research and industry. For all aspects of, for example, RAS component design and engineering (e.g., fish tank sizing, shapes, and calculations) and hydroponics, I highly recommend a book by Wilson Lennard and the (partially highlighted) references therein [2]. However, some cost-relevant factors are highlighted under Sect. 3.2 below.

Using the principles of AP (potentially) allows the large-scale production of microalgae. After growing in the nutrient-rich effluent, thereby "cleaning" the system water (bioremediation), their biomass can be used for a large variety of purposes (see below), minding, for instance, the corresponding legislation and hygiene standards (see also Sect. 3.1).

Algae cultivated in indoorsystems can be certified organic in the EU (see above). Of course the national legislation might differ elsewhere.

2.3.2 Challenges

The suitability of the desired (maybe previously untested) species for the production of algal biomass with AP has to be investigated. All the organisms in an AP system experience the same abiotic conditions and their requirements in terms of, for example, pH and temperature must be similar enough (ideally identical) to successfully produce them in the same system.

For large-scale production, combining RAS and open-pond cultivation technologies to produce microalgae (especially for biofuel production) is a novel approach and requires experimentation to find a cost- and resource-efficient solution, upscaling, etc. (see also Sect. 2.1 above).

Potential nutrient limitations have to be managed (see Sect. 1.3 above).

2.4 Future Directions

Minding the benefits for saving resources (e.g., most prominently water) and thus, operating costs and environmental impacts of the production, I see AP as one of the most promising methods to produce large quantities of algal biomass. Resource saving (including money) during biomass production is especially important if it is processed further as a raw material for, e.g., the production of biofuel. Of course, also the sustainable production of other algae products (e.g., biomass for food and feed) is a significant advantage of using the principles of algaeponics.

3 Harvesting and Downstream Processing

3.1 Examples for Harvesting and Biomass Use

3.1.1 Harvesting Techniques: Examples

The cultivation of microalgae requires upscaling from about 20 ml (seed culture) to commercially relevant volumes [23]. Figure 5 shows the schematic process from a seed culture to the end product (harvesting included), according to Bauer et al. [23].

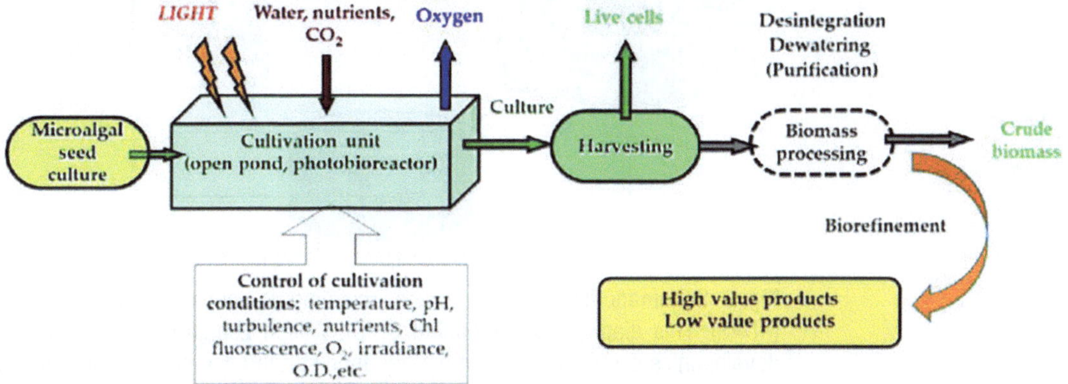

Fig. 5 Schematic overview: From microalgae seed culture to potential products [from Bauer et al. 2021 23]

The harvesting methods differ between different cultivation techniques and multiple options exist [e.g., 19]. For example, the biomass produced in open ponds can be extracted by filtration with a semipermeable membrane, which retains the microalgae while the liquid media pass through [24]. The mesh size of the membrane depends on the cell size of the cultured species. This approach could be a good choice for algaeponics as well. The membrane could be integrated into the pond to prevent cells from being washed out into an intermediate tank (for monitoring nutrient concentrations, etc.) before the water is recirculated to the fish tanks directly or diluted with the RAS water after biofiltration (ammonia conversion). Once ready for harvesting, the same membrane could be used (e.g., lifted by cranes) to extract the biomass.

A suitable AP setup is schematically shown in Fig. 6 [2]. The plant production unit (Fig. 6) could be replaced with an open pond for algae cultivation.

Furthermore, a system with two streams could be used, one for the hydroponic production of plants and the other for the production of algae. It is also possible to have different ponds with different species of algae in different growth phases. In both ways, the continuous nutrient reduction by phototrophic organisms and their production is ensured also after harvesting the biomass from, e.g., one pond, since it takes some time for newly inoculated algae to reach the exponential growth phase, where they take up the most nutrients.

3.1.2 Biomass Use: Examples

There are vast possibilities to use algae, and many species are already part of the food, feed, and fertilizer industries as well as medical research, among others.

For instance, the microalgae genera *Chlorella*, *Arthrospira*, *Haematococcus*, and *Dunaliella* have anti-tumor, anti-viral, and

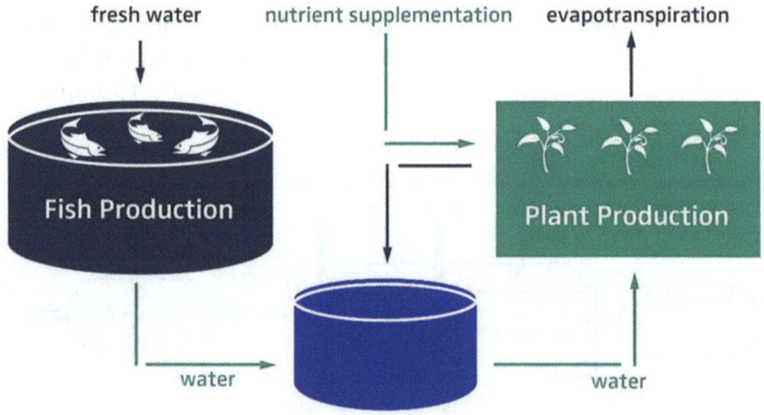

Fig. 6 Schematic flow diagram of a recirculating de-coupled aquaponic system (RDAS). This diagram was modified by Dr. Wilson Lennard from http://www.developonics.com/2017/04/namibia-aquaponics/ as adapted from http://www.developonics.com (from Lennard 2017 [2], used with permission)

anti-inflammatory properties for humans [25]. Furthermore, microalgae biomass can be used for direct feeding of the larvae of aquaculture-relevant species (e.g., fish, oysters, and prawns) and to produce feed for larger life-cycle stages [23]. The freshwater green alga *H. pluvialis* is especially interesting, since it is relevant for wastewater treatment in aquaculture (RAS), human nutrition, and feed (e.g., for salmonids). *H. pluvialis* is known to produce nature's strongest antioxidant, astaxanthin [26], and it can be cultivated using the nutrient-rich effluent water from RAS facilities, thereby reducing the nutrient concentration in this wastewater [21, 22]. In addition, diatoms produce silicate, which can be used to produce cosmetics and to increase the efficiency of solar panels [27].

Once the biomass is harvested (purified, etc.) it can be tested for undesired substances (if necessary for the desired end-product) before processing/analysing it further. The same downstream processing of the algal biomass as applied after cultivation with other methods can be used.

3.2 Algaeponics: A Viable Way to Produce Biofuel?

Learning from successful solutions proven by existing RAS and hydroponics operations benefits the design of viable AP systems [2]. Technologies widely used and popular in modern RAS comprise, for instance, round fish tanks (best fluid dynamics), self-cleaning rotating drum-screen filters (solids filtration), moving bed bioreactors (MBBs, biofiltration using, e.g., "bioballs" as a substrate for the bacteria), devices to ensure optimal dissolved oxygen (D.O.) concentrations, and optionally ozone generators or ultraviolet radiation for sterilization. When using screen filtration for solids removal (see above), the size of the filter (also valid for biofiltration) and its pore size depend on multiple factors and must be specifically calculated/chosen for individual setups. The stocking density of fish, amount of feed, the feed itself, feed conversion ratio, and production volume are important variables. Incorrect pore sizes lead to an increased waste of water and nutrients that are intended for cultivating the plants/algae [2]. MBBs and trickling biofilters are the recommended choices for AP. Commercially applied (large-scale) hydroponics technologies include, for example, several options for the plant cultivation unit (e.g., deep flow and nutrient film technique). The deep flow (or floating raft) technique is used in commercial AP farms in greenhouses or outdoors [2] and is comparable to the open pond cultivation of microalgae. Maintaining the water quality (temperature, pH, D.O., ammonia) is vital for successful aqua- and algaeponics.

These examples are applied in the industry because they are "easy" to construct, cost-efficient, and robust and thus result in reduced operating costs of a well-planned, well-designed, and well-managed facility [2].

There are multiple types of AP systems that are suitable for growing microalgae, such as recirculating de-coupled aquaponic systems (RDAS, Fig. 6). To cultivate fish and algae in one system requires the separation of the corresponding units for multiple reasons. This de-coupling allows, for instance, the cooling down of the fish rearing water before entering the algae unit if necessary. A complete list of systems and more details can be found in Lennard 2017 [2].

Sticking to proven and successful concepts and minding other recommendations mentioned in, e.g., this chapter (e.g., Sects. 1.3 and 2.3) can serve as a base to create a viable way to produce microalgae biomass as a raw material for, e.g., biofuel production.

Microalgae-based biodiesel production is considered a promising way to produce a renewable and sustainable source of energy. However, challenges such as high production costs and potential environmental impacts need to be addressed [28]. Both factors can be improved by using the principles of AP as described in this chapter.

3.2.1 Biofuel-Relevant Microalgae: Promising Species

The quality and quantity of fatty acids and lipids in the algal biomass are essential factors for the production of biodiesel. However, the lipid content of a microalgae cell depends on the species and the growth phase/conditions [e.g., 28], which makes the right choice of algae a key factor.

Under certain conditions (e.g., nutrient limitation), the lipid content of *Chlorella* strains (e.g., *C. vulgaris*) is increased to up to 50% or higher, which makes them interesting for biofuel production [29, 30]. The following list shows a selection of additional microalgae that may contain >50% lipids (percentage of dry weight) [30]:

- *Botryococcus braunii*
- *Crypthecodinium cohnii*
- *Skeletonema costatum*
- *Schizochytrium* sp.
- *Hantzschia* spp.
- *Nannochloropsis* sp.
- *Neochloris oleoabundans*
- *Prostanthera incisa*

A well-studied model organism in biofuel research (e.g., regarding biohydrogen production) is *Chlamydomonas reinhardtii* [e.g., 30].

3.2.2 Fish Species Commonly Used in AP Systems

A large number of freshwater fish species are suitable for AP. Basically any species that can be kept in tanks and standard RAS setups can be used [2]. Depending on, e.g., the local climatic conditions and consumer preferences, warm- or cold-water species can be cultivated. A selection of suitable species for AP/RAS at different temperatures is presented in Table 1.

Table 1
A selection of freshwater fish species that are suitable for aqua/algaeponics (AP = aquaponics; RAS = recirculating aquaculture systems)

Fish species (latin)	Common name	Suitable temperature (°C)	Comment	References
Cyprinus carpio	Common carp	20-25	AP-proof	[31]
Oreochromis niloticus	Nile tilapia	~27	AP-proof	[2]
Lates calcarifer	Barramundi	~27-30	AP-proof	[2]
Clarias gariepinus	African catfish	~27-30	AP-proof	[32]
Oncorhynchus mykiss	Rainbow trout	13-17	AP-proof	[2, 31]
Salvelinus alpinus	Arctic char	12-14	AP-proof	[2]
Coregonus lavaretus	European/common whitefish	17	RAS-proof	[e.g. 21]

Due to several reasons, one particular fish species, the common carp (*Cyprinus carpio*), will be highlighted here. This species features a broad temperature tolerance (15–30°C, optimum growth at 20–25°C, Table 1) and is robust [31]. Furthermore, *C. carpio* is the fourth most-produced fish species on a global scale and its production is especially popular in Europe and Asia [33].

A pH range of 6.7–7.5 (ideally 6.8–7.0) is suitable for AP, since it matches the needs of many fish and plant species as well as the biofilter bacteria. Staying below pH 7.0 also helps avoid high ammonia concentrations, potentially toxic for the fish [2]. Nevertheless, species-specific differences (also for algae) exist and need to be considered (also regarding temperature, dissolved oxygen, feed/nutrients, etc.).

3.2.3 Possible Combinations of Biofuel-Producing Microalgae and AP-Proof Fish Species

The main groups of phototrophic microorganisms targeted for, e.g., biodiesel production are diatoms, green algae, cyanobacteria, and the golden algae species producing oil and complex carbohydrates [34–36].

Depending on the growth conditions of the selected microalgae, warm- or cold-water fish species are available (Table 1). To warm the water, waste heat from other industries (e.g., biogas production) can be used to reduce operating costs, further increasing the sustainability of the farm, and thus contributing to a circular economy. When farming warm-water fish species above 25°C (e.g., *Tilapia* spp., *Lates calcarifer*, *Clarias gariepinus*, Table 1), the water might have to be cooled down before it enters the algae cultivation unit. Depending on the air temperature, this cooling might take place "naturally" between the different units of a de-coupled AP system (on the way from the fish house to the glasshouse). Some algae species that grow in "cold" RAS wastewater (17°C) are mentioned under Sect. 2.2, and previously untested ones could be tested for their biofuel potential. Vice versa, the potential of proven biofuel microalgae to grow at lower temperatures can be tested when a certain (cold-water) fish species is preferred (e.g., *C. reinhardtii* grows <20°C, see below).

C. carpio could be a good fish choice when growing algae between 20 and 25°C without having to cool down the water between fish tanks and the algae unit. Many microalgae species are known to grow well between 20 and 30°C and at a pH of 7–9 [28], and thus, various species should be suitable to be cultivated in the wastewater of a RAS unit inhabiting *C. carpio*, minding the mentioned pH range for AP (ideally 6.8–7.0).

As stated above, green microalgae species have been proven to grow in RAS wastewater [21, 22]. *C. vulgaris* can be cultivated at 25°C, and several species of the genus *Chlorella* are commonly produced using a pH range of 6.5–7.5 [37, 38]. By depleting the N-concentration, *C. vulgaris* would "clean" the wastewater and simultaneously create conditions of increased lipid production. This would be a win-win situation for the algaeponic production of biodiesel using this species. *C. reinhardtii* is well studied regarding its relevance for the production of biofuels and also grows well at 25°C [39]. The growth of *C. reinhardtii* in RAS wastewater has been studied previously. For these experiments, the temperature was kept at 17°C (temperature of the wastewater) and the pH range was 6.7–7.7 [21].

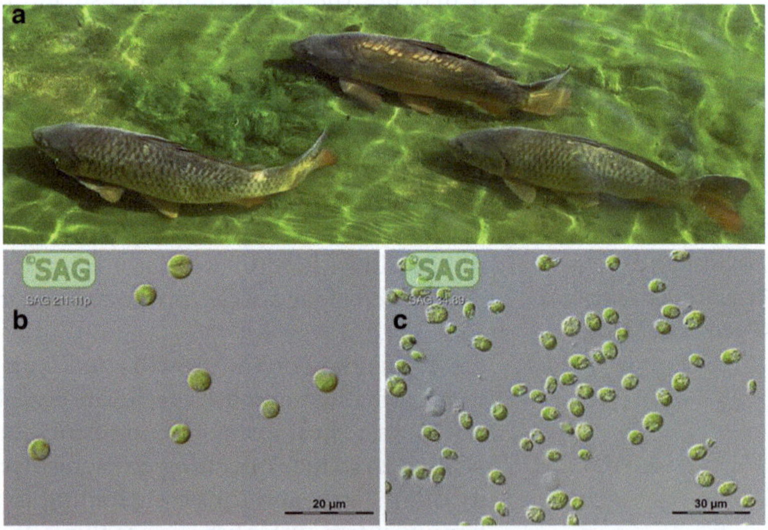

Fig. 7 The common carp (*Cyprinus carpio*, **a**) [40]. Two breeding varieties of this species are shown, a so-called mirror carp (in the middle) and two more naturally scaled ones; a strain (SAG 211-11p) of *Chlorella vulgaris* (**b**, used with permission [41]); and a strain (SAG 34.89) of *Chlamydomonas reinhardtii* (**c**, used with permission [42]). Note the different scale bars for the two micrographs

Thus, suitable pH ranges for both microalgae closely match the optimal range for AP and they grow well at a temperature within the optimal range for *C. carpio*. Therefore, cultivating *C. vulgaris* for the production of biodiesel and *C. reinhardtii* for the production of biohydrogen together with *C. carpio* in a (micro)algaeponics system is a highly promising combination for experimentation (Fig. 7).

Some ideas for potential/novel products derived from *C. carpio* are stated in, for instance, Lipsewers [13]. For example, the production of sustainable and affordable "caviar" [43] seems a very interesting direction for marketing carp-products to a broader consumer community.

Acknowledgments

Special thanks to Dr. Wilson Lennard, the author of books like "*Commercial Aquaponic Systems – Integrating Recirculating Fish Culture with Hydroponic Plant Production*" for his inspiration, and to Dr. Kristian Spilling (Marine and Freshwater Solutions, Finnish Environment Institute, SYKE, Helsinki) for inviting me to contribute to this book, suggesting the topic, providing the outline, commenting on the first draft, and for his support in choosing the final images.

References

1. Palm HW, Knaus U, Appelbaum S et al (2018) Towards commercial aquaponics: a review of systems, designs, scales and nomenclature. Aquac Int 26(3):813–842. https://doi.org/10.1007/s10499-018-0249-z
2. Lennard W (2017) Commercial aquaponic systems – integrating recirculating fish culture with hydroponic plant production. Produced and published in Australia, ISBN 978-1-64204-837-7
3. Ceder MC (2020) Aquaponics as a circular economy solution: opportunities and obstacles for aquaponics systems analyzed in a case study on Finland. Master's thesis. Centre for Food Policy, Department of Health Sciences, University of London
4. Kopsa P (2015) Aquaponics: practical thesis in Australia. Bachelor's thesis. HAMK Häme University of Applied Sciences. Available at: https://www.theseus.fi/handle/10024/103843. Accessed 11 Nov 2024
5. Goddek S, Delaide B, Mankasingh U et al (2015) Challenges of sustainable and commercial aquaponics. Sustainability 7(4):4199–4224. https://doi.org/10.3390/su7044199
6. EZ GRO Garden. https://ezgrogarden.com/history-of-hydroponics-2/aztec-chinampas-of-central-america/. Accessed 11 Nov 2024
7. SUPERIOR Fresh. https://www.superiorfresh.com/. Accessed 12 Nov 2024
8. Wikipedia. https://en.wikipedia.org/wiki/Aquaponics. This file is licensed under the Creative Commons Attribution-Share Alike 2.0 Generic license. Accessed 19 Dec 2024
9. Wikipedia. https://en.wikipedia.org/wiki/Aquaponics. By Bryghtknyght (own work), CC BY 3.0 (https://commons.wikimedia.org/w/index.php?curid=10754457). Accessed 19 Dec 2024
10. European Commission. https://agriculture.ec.europa.eu/farming/organic-farming/organic-production-and-products_en#rulesonwineaquacultureandhydroponics. Accessed 17 Nov 2024
11. European Commission. https://agriculture.ec.europa.eu/system/files/2022-09/organic-rules-faqs_en.pdf. Accessed 27 Feb 2023
12. European Union. https://eur-lex.europa.eu/legal-content/EN/TXT/PDF/uri=

CELEX:02018R0848-20220101&from=EN. Accessed 27 Feb 2023
13. Lipsewers T (2023) Aquaponics – The current state in Finland and trends for the future. Aquaculture Europe – the members' magazine of the European Aquaculture Society (EAS), Volume 48, September issue, pp 5–12. https://www.researchgate.net/publication/373578938_Aquaponics_The_current_state_in_Finland_and_trends_for_the_future
14. University of Rostock (Germany) – Faculty of Agricultural and Environmental Sciences. https://www.auf.uni-rostock.de/en/professuren/bau-und-umweltbereich/landschaftsoekologie-und-standortkunde/forschung/abgeschlossene-projekte/fischglashaus/. Accessed 28 Nov 2024
15. aquaponik manufaktur GmbH. https://www.aquaponik-manufaktur.de/en/von-aquaponik-zur-multiaquaponik/. Accessed 14 Nov 2024
16. Wikipedia. https://de.wikipedia.org/wiki/Datei:Deep_water_culture.jpg. This file is licensed under the Creative Commons Attribution-Share Alike 4.0 International License. Accessed 19 Dec 2024
17. Shen Y, Yuan W, Pei ZJ et al (2009) Microalgae mass production methods. Trans ASABE 52(4):1275–1287
18. Varicon Aqua Solutions. http://www.variconaqua.com/products-services/phyco-flow/. Accessed18 Nov 2024
19. Tan JS, Lee SY, Chew KW et al (2020) A review on microalgae cultivation and harvesting, and their biomass extraction processing using ionic liquids. Bioengineered 11(1):116–129. https://doi.org/10.1080/21655979.2020.1711626
20. Mohsenpour SF, Hennige S, Willoughby N et al (2021) Integrating micro-algae into wastewater treatment: a review. Sci Total Environ 752:142168. https://doi.org/10.1016/j.scitotenv.2020.142168
21. Stevčić Č, Pulkkinen K, Pirhonen J (2019) Screening of microalgae and LED grow light spectra for effective removal of dissolved nutrients from cold-water recirculating aquaculture system (RAS) wastewater. Algal Res 44:101681. https://doi.org/10.1016/j.algal.2019.101681
22. Calderini ML, Stevčić Č, Taipale S et al (2021) Filtration of Nordic recirculating aquaculture system wastewater: effects on microalgal growth, nutrient removal, and nutritional value. Algal Res 60:102486. https://doi.org/10.1016/j.algal.2021.102486
23. Bauer L, Ranglová K, Masojídek J et al (2021) Digestate as sustainable nutrient source for microalgae—challenges and prospects. Appl Sci 11(3):1056. https://doi.org/10.3390/app11031056
24. Al Hattab M, Ghaly A, Hammouda A (2015) Microalgae harvesting methods for industrial production of biodiesel: critical review and comparative analysis. J Fundam Renew Energy Appl 5(2):1000154
25. Hayes M, Skomedal H, Skjånes K et al (2017) Microalgal proteins for feed, food and health. In: Gonzalez-Fernandez C, Muñoz R (eds) Microalgae-based biofuels and bioproducts: from feedstock cultivation to end-products. Woodhead Publishing, Duxford, pp 347–368. ISBN 978-0-08-101027-3. https://doi.org/10.1016/B978-0-08-101023-5.00015-7
26. Algatech LTD. https://www.algatech.com/why-whole-food-sourced-from-microalgae-is-big-deal-david-foreman-the-herbal-pharmacist-explains/. Accessed 18 Nov 2024
27. Swedish Algae Factory. https://www.swedishalgaefactory.com/. Accessed 18 Nov 2024
28. Gaurav K, Neeti K, Singh R (2024) Microalgae-based biodiesel production and its challenges and future opportunities: a review. Green Technol Sustain 2(1):100060
29. Liu J, Hu Q (2013) Chlorella: industrial production of cell mass and chemicals. In: Richmond A, Hu Q (eds) Handbook of microalgal culture: applied phycology and biotechnology. Wiley, Oxford, pp 329–338. ISBN 978-1-118-56716-6
30. Ghasemi Y, Rasoul-Amini S, Naseri AT et al (2012) Microalgae biofuel potentials. Appl Biochem Microbiol 48:126–144
31. FarmingAquaponics. https://farmingaquaponics.com/best-fish-for-cold-water-aquaponics. Accessed 14 Nov 2024
32. Knaus U, Hübner DHD, Küchenmeister C et al (2024) Aquaponic growth of basil (Ocimum basilicum) with African catfish (Clarias gariepinus) in standard substrate combined with a Humicacid Fiber-Substrate (HFS). Sci Rep 14(1):17725
33. FAO (2020) The State of World Fisheries and Aquaculture 2020. Sustainability in action. Rome. https://doi.org/10.4060/ca9229en. Accessed 14 Nov 2024
34. Drapcho, CM, Nhuan, NP, Walker, TH, (2008). Biofuels engineering process technology. McGraw-Hill, New York, pp. 160–180
35. Morowvat MH, Rasoul-Amini S, Ghasemi Y (2010) Chlamydomonas as a "new" organism for biodiesel production. Bioresour Technol 101(6):2059–2062
36. Rasoul-Amini S, Montazeri-Najafabady N, Mobasher MA et al (2011) Chlorella sp: anew strain with highly saturated fatty acids for biodiesel production in bubble-column photobioreactor. Appl Energy 88(10):3354–3356

37. Deniz İ (2020) Determination of growth conditions for Chlorella vulgaris. Mar Sci Tech Bull 1:1
38. Borowitzka MA (2005) Culturing microalgae in outdoor ponds. In: Andersen RA (ed) Algal culturing techniques. Elsevier Academic Press, Burlington, MA, pp 205–218. ISBN 978-0-12-088426-1
39. Kong QX, Li L, Martinez B et al (2010) Culture of microalgae Chlamydomonas reinhardtii in wastewater for biomass feedstock production. Appl Biochem Biotechnol 160:9–18
40. Wikipedia. https://en.wikipedia.org/wiki/Common_carp. This file is licensed under the Creative Commons Attribution 3.0 Unported license. Accessed 19 Dec 2024
41. Culture Collection of Algae at Göttingen University (SAG, Germany), Chlorella vulgaris strain number SAG 211-11p, T. Darienko, SAG Culture Collection, published under license CC BY-SA 4.0 (https://creativecommons.org/). https://sagdb.uni-goettingen.de/detailedList.php?str_number=211-11p. Accessed 15 Dec 2024
42. Culture Collection of Algae at Göttingen University (SAG, Germany), Chlamydomonas reinhardtii strain number SAG 34.89, T. Darienko, SAG Culture Collection, published under license CC BY-SA 4.0 (https://creativecommons.org/). https://sagdb.uni-goettingen.de/detailedList.php?str_number=34.89. Accessed 02 Dec 2024
43. Binsi PK, Nayak N, Sarkar PC et al (2019) Conversion of carp roe mass to caviar substitutes: Stabilization with oregano extract. LWT 108:446–455

Check for updates

Chapter 6

Monitoring Microalgal Neutral Lipid Accumulation with Nile Red

Katariina Natunen

Abstract

The widely applied Nile Red (NR) method allows near real-time monitoring of microalgal neutral lipid accumulation. When added to a culture sample, optimally, the fluorescent dye NR penetrates the microalgal cell wall staining the intracellular neutral lipids, and the measured fluorescence is linearly correlated to the neutral lipid concentration. Here I describe an optimization protocol for determining the optimal staining parameters for each new microalgal species, followed by a basic NR staining protocol to be applied for monitoring of microalgal neutral lipid accumulation.

Keywords Dye, Fluorescence, Lipids, Microalgae, Nile Red, Solvents

1 Introduction

Traditional lipid quantification methods, e.g., lipid extraction and methylation followed by gas chromatographic analysis, are expensive and time-consuming, making them unsuitable for online monitoring of the accumulation of the refinable, neutral storage lipids in fast-growing microalgae cultures. Nile Red (NR) (9-diethylamino-5-benzo[α] phenoxazinone) is a fluorescent dye that stains both the polar cell wall lipids and the neutral storage lipids in microalgae cells. With the NR method, near real-time monitoring is possible due to quick fluorescence measurements that can be carried out in situ from living microalgae cultures. Due to its *solvatochromic* properties (i.e., the position, shape, and intensity of its absorption and emission bands depend on the polarity of its environment [1, 2]), it is possible to selectively measure neutral lipids by choosing the right wavelengths for fluorescence measurement.

The functionality of the NR method is optimally based on a linear relationship between the NR fluorescence intensity and the neutral lipid concentration, and the method has been widely applied for detecting and quantifying microalgal neutral lipids already since the 1980s [3]. Other applications of NR staining include,

Kristian Spilling (ed.), *Biofuels from Algae*, Methods in Molecular Biology, vol. 2996,
https://doi.org/10.1007/978-1-0716-5031-8_6,

e.g., identification and quantification of microplastics in environmental samples, which has emerged and increased strongly in the recent years [4]. Different means to measure NR fluorescence from stained microalgal cultures have included spectrofluorometry and fluorometry [3, 5–7], fluorescence microscopy [8], and flow cytometry [9, 10]. Neutral lipids can also be quantified at a cell-specific level by measuring the NR fluorescence of stained cells with the image-based particle analyser FlowCAM™ [11]. This opens new possibilities to study the effect of species diversity on lipid accumulation, as with the FlowCAM™ species with distinct morphologies can be distinguished from each other, and species-specific properties can be determined.

This method description focuses on spectrofluorometric NR fluorescence measurement of microalgae neutral lipids. The basic staining principle is relatively simple, but obtaining the different staining parameters requires optimization, as most of them are both species- and system-specific [12]. Therefore, I start by describing an optimization protocol to help choose the optimal staining parameters for each new species (*see* **Sect. 3.1**), and based on that, a basic staining protocol to be applied for daily monitoring of lipid accumulation (*see* **Sect. 3.2**).

2 Materials

Prepare and store all reagents at room temperature. The NR stock solution needs to be stored in darkness (*see* **Note 1**). Follow your local waste disposal regulations when disposing waste materials.

1. Microalgae culture where neutral lipid production has been induced by, e.g., nitrogen limitation.
2. Solvents: acetone and dimethyl sulfoxide (DMSO) (*see* **Sect. 3.1**).
3. Ethanol (96%).
4. Nile Red (NR) stock solution: Prepare a 0.25 mg/mL NR stock solution by dissolving NR powder (*see* **Note 2**) in acetone or in DMSO (*see* **Note 3**).
5. Spectrofluorometer and quartz cuvette(s) (10 mm light path, volume 3 mL) suitable for fluorescence measurements (*see* **Note 4**).

3 Methods

Carry out all procedures at room temperature. After using NR, rinse cuvette and all other reusable glassware with ethanol to remove traces of the dye. When measuring fluorescence, the fluorescence values should always stay in the linear range of your spectrofluorometer (*see* **Note 5**).

3.1 Optimization Protocol for Nile Red Staining Method

The optimization protocol described below is one of many ways to find out the optimal NR staining parameters. The protocol is iterative; thus, the steps below might have to be repeated until optimal results are achieved. They also need to be carried out separately for every new species and measuring equipment.

1. Verify that your microalgae culture has accumulated neutral lipids by determining the lipids, e.g., with gas chromatography (GC) [13] or by other means (*see* **Note 6**).
2. Choose a starting NR concentration (e.g., 1 μL/mL final concentration) and incubation time (e.g., 10 min) for measurements (*see* **Note 7**). Add NR into 3 mL of culture sample in a cuvette, and after incubating (*see* **Note 1**), place the sample into a spectrofluorometer and determine the optimal wavelengths for NR measurements by measuring excitation-emission fluorescence matrices (EEMs). Use an excitation range of approx. 450–550 nm and emission range of approx. 550–600 nm (*see* **Note 8**).
3. Optimize NR concentration and staining time with a series of kinetic fluorescence measurements, i.e., different volumes of NR stock solutions are added to culture samples and the evolvement of relative fluorescence values is followed as a function of time.
4. Place a cuvette with 3 mL of fresh (diluted) culture sample into spectrofluorometer. Add *x* μL of NR to the sample and follow the fluorescence for 20–40 min at the wavelengths determined in Step 2 (see **Note 9**).
5. Repeat Step 4 with a range of different additions of NR, e.g., 0.08–1 μg/mL (i.e., $x = 1–12$ μL per 3 mL sample). Choose the concentration/staining time combination that produces the highest and/or most stable fluoresce values (*see* **Note 10**).
6. Repeat Step 2 and see whether optimal excitation/emission wavelengths change with these new parameters.
7. If the culture has been verified to contain lipids (*see* **Step 1**), but no NR fluorescence is produced in spite of trying out different NR concentrations, one explanation is that the dye does not penetrate the rigid cell wall of the microalgae species [6]. The staining might be enhanced by adding a solvent to the algal sample, which makes the cell wall more porous for the dye to enter the cell, without breaking the cell [6, 12] (*see* **Note 11**). A commonly applied solvent is DMSO.
8. Test the effect of the following DMSO concentrations: 5, 10, 20, and 30% (v/v). Use the same volume of culture with every DMSO concentration. The highest concentration, 30 % (v/v) DMSO, equals 1 mL DMSO in 3 mL sample. Thus, also with lower DMSO concentrations, use 2 mL of culture

and compensate the difference between different DMSO concentrations by adding growth medium (*see* **Note 5**). For example, for the sample with 10% (v/v) DMSO, add 2 mL culture, 0.3 mL DMSO, 0.7 mL of growth medium, and the chosen concentration of NR.

9. Place a cuvette with 2 mL of fresh (diluted) culture sample into spectrofluorometer. Add *x* μL of DMSO (+ additional growth medium) and the chosen volume of NR to the sample and follow the fluorescence for 20–40 min (*see* **Notes 7 and 9**).

10. Repeat Step 9 with the different DMSO concentrations (*see* **Note 12**). Figure 1 shows the results of an optimisation experiment with a fixed NR concentration and three different solvents on the NR staining of *Phaeodactylum tricornutum*. The results show fluorescence values per total lipids determined by GC [12].

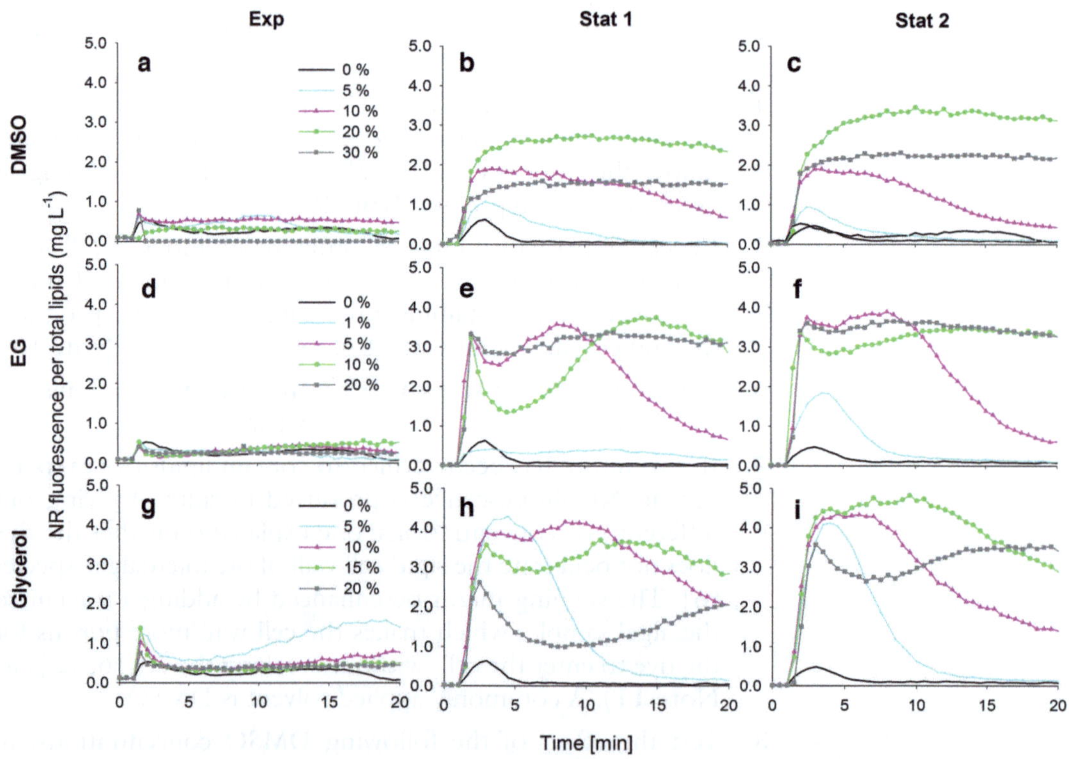

Fig. 1 The effect of dimethyl sulfoxide (DMSO), ethylene glycol (EG), and glycerol on the NR staining of *Phaeodactylum tricornutum* in exponential (Exp), early stationary (Stat 1), and late stationary (Stat 2) growth phases. The NR fluorescence values are divided by the total lipid concentration in the culture (mg/L). 20% (v/v) DMSO produced the highest and most stable signal [12]

3.2 Basic Nile Red Staining Protocol for Monitoring Lipid Accumulation

1. Take a fresh sample from your culture and dilute the sample if necessary (*see* **Note 5**).
2. Prepare a series of samples, where you have a 3 mL blank sample (i.e., growth medium, *see* **Note 5**), and parallel 3 mL (diluted) culture samples. If you are using DMSO (e.g., 10 % (v/v)), you have a 2.7 mL blank sample and 2.7 mL culture samples.
3. Add chosen volume (*see* **Sect. 3.1**) of NR stock solution one by one into all the samples (i.e., blank + culture samples) (*see* **Note 13**). If you are using DMSO (e.g., 10 % (v/v)), add 0.3 mL DMSO and chosen volume of NR one by one into all the samples (i.e., blank + culture samples).
4. Let the stained samples incubate in darkness (see **Note 1**) for the chosen time.
5. Measure NR fluorescence of each sample at the chosen wavelengths (*see* **Note 14**).
6. To obtain a proxy for cell-specific lipid concentration in the culture, divide the obtained fluorescence values by the culture's current cell concentration (cells/mL).
7. Carry out the measurements daily to follow lipid accumulation and choose the best days for harvesting the lipids. Figure 2 shows the results of monitoring cell-specific lipid accumulation in duplicate *Phaeodactylum tricornutum* cultures with NR fluorescence measurements and GC.

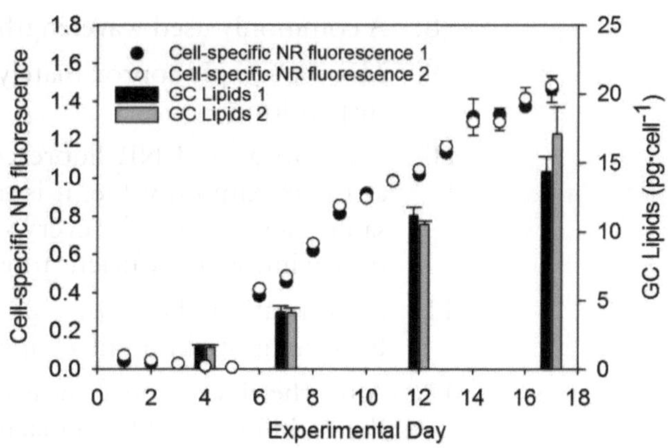

Fig. 2 Cell-specific lipid accumulation in duplicate *Phaeodactylum tricornutum* cultures measured with NR fluorescence (relative fluorescence units) and gas chromatographic analysis (pg/cell)

4 Notes

1. NR is photosensitive, i.e., the fluorescence intensity is quenched when exposed to bright light. Therefore, also samples stained with NR should be incubated in darkness.
2. Do not use technical grade NR.
3. When using acetone as a solvent, always close the stock solution bottle well and minimize the usage and storage time to prevent solvent evaporation. When working with NR dissolved in DMSO, a more stable solvent, wear special safety gloves, as DMSO easily penetrates through the skin.
4. It is important that the cuvette is suitable for fluorescence measurements. Do not use plastic cuvettes.
5. For checking the linear range of the instrument prepare a dilution series, measure and determine the linear range of the fluorescence output. If the sample is outside the linear range a dilution step must be added. If the growth medium is clear, purified water (Milli-Q or similar) can be used for dilutions and blank measurements. If the growth medium contains salt, remember to add same amount of salt to the purified water.
6. The measured NR fluorescence is always relative and system-specific. Thus, for absolute quantification of lipids with the NR method, a calibration step (i.e., connecting fluorescence values with actual lipid concentration) is needed.
7. Always check with microscopy or imaging flow cytometry, whether cells stay intact during staining. If cells break, decrease the dye and/or solvent concentration.
8. A commonly used wavelength combination is ~530/580 nm.
9. Mix the sample approximately every 2 min to keep the cells in suspension.
10. If the measured NR fluorescence stays stable for some time at the maximum value, it is not so crucial to apply the exact same staining time for every sample. Instead, hitting a certain time window is sufficient to get comparable results.
11. Also, heating the cells, e.g., with microwave treatment has been suggested for enhancing NR penetration [14].
12. After the fluorescence measurements, remember to consider the "additional" dilution factor caused by DMSO (+ growth medium) additions on top of the possible initial dilution of the culture.
13. When measuring a series of samples, add NR (and DMSO if needed) to the samples in, e.g., 1 min intervals to samples, and after incubation, also measure the samples using the

same time interval. Using a timer helps. For this, you need to prepare the samples in vials and transfer the samples to the cuvette before fluorescence measurement. This way, time is saved and all the samples are incubated the same time.

14. Due to spectral overlapping, NR also fluoresces in water at the optimal wavelengths. To get the final proxy for neutral lipids, extract the NR background fluorescence from the NR fluorescence value measured from the culture samples, and take into account the possible dilution factor.

References

1. Dutta AK, Kamada K, Ohta K (1996) Spectroscopic studies of nile red in organic solvents and polymers. J Photochem Photobiol A Chem 93(1):57–64
2. Murugan NA, Rinkevicius Z, Ågren H (2011) Modeling solvatochromism of Nile red in water. Int J Quantum Chem 111:1521–1530
3. Cooksey KE, Guckert JB, Williams SA, Callis PR (1987) Fluorometric determination of the neutral lipid content of microalgal cells using Nile Red. J Microbiol Meth 6:333–345
4. Shruti VC, Pérez-Guevara F, Roy PD, Kutralam-Muniasamy G (2022) Analyzing microplastics with Nile Red: emerging trends, challenges, and prospects. J Hazard Mater 423(Part B):127171
5. Elsey D, Jameson D, Raleigh B, Cooney MJ (2007) Fluorescent measurement of microalgal neutral lipids. J Microbiol Meth 68:639–642
6. Chen W, Zhang C, Song L, Sommerfeld M, Hu Q (2009) A high throughput Nile red method for quantitative measurement of neutral lipids in microalgae. J Microbiol Meth 77:41–47
7. Zheng X, Yin L, Qiang S, Li S, Chen Y (2022) Rapid method for lipid determination in Chlorella sp. based on Nile Red fluorescence. Bioresour Technol Rep 18:101077
8. Greenspan P, Fowler SD (1985) Spectrofluorometric studies of the lipid probe, Nile Red. J Lipid Res 26:781–789
9. Katayama T, Kishi M, Takahashi K et al (2019) Isolation of lipid-rich marine microalgae by flow cytometric screening with Nile Red staining. Aquacult Int 27:509–518
10. Satpati GG, Pal R (2015) Rapid detection of neutral lipid in green microalgae by flowcytometry in combination with Nile red staining – an improved technique. Ann Microbiol 65:937–949
11. Natunen K, Seppälä J, Koivula RJ, Pellinen J (2017) Monitoring cell-specific neutral lipid accumulation in Phaeodactylum tricornutum (Bacillariophyceae) with Nile Red staining – a new method for FlowCAM. J Phycol 53(2):396-404
12. Natunen K, Seppälä J, Schwenk D et al (2015) Nile Red staining of phytoplankton neutral lipids: species-specific fluorescence kinetics in various solvents. J Appl Phycol 27:1161–1168
13. Bligh EG, Dyer WJ (1959) A rapid method of total lipid extraction and purification. Can J Biochem Phys 37:911–917
14. Chen W, Sommerfeld M, Hu Q (2011) Microwave-assisted Nile Red method for in vivo quantification of neutral lipids in microalgae. Bioresource Technol 102:135–141

Chapter 7

Liquid Chromatography-Mass Spectrometry (LC-MS)-Based Analysis of Molecular Lipids in Algae Samples Using Data Dependent Acquisition (DDA)

Niels van Boxtel, Heli Nygren, Tuulikki Seppänen-Laakso and Heiko Rischer

Abstract

In this paper, an analytical method for the analysis of molecular lipids in algae samples is reported. The sample preparation is based on a modified Folch extraction, and the analysis is carried out with ultrahigh performance liquid chromatography combined with mass spectrometry (UPLC-MS). For the characterization of lipids, data dependent acquisition (DDA) analyses are carried out utilizing a high-resolution quadrupole-time-of-flight (Q-ToF) instrument. Throughput of the method is over 100 samples/day. The repeatability is good, and the relative standard deviation of spiked samples is <15%.

Keywords Lipid, Liquid chromatography, Mass spectrometry, Data dependent acquisition

1 Introduction

Lipids are a broad group of naturally occurring molecules, including fats, waxes, sterols, monoacylglycerols, diacylglycerols, triacylglycerols, phospholipids, sphingolipids, and glycolipids [1]. Lipids play important roles in biological systems as building blocks of cellular membranes including subcellular membrane compartments, in cell signaling and for energy production and storage.

Over the years, algae lipids have gained interest as potential sources of biofuel as well as high value products such as food supplements [1, 2]. Understanding the processing needs and properties of algae-derived bio-based products requires information of the acyl chains associated with different lipid classes. The algae lipid composition is highly diverse and depends on the species, its adaptation to the environment, and current growth conditions [1, 3]. Light, nutrient availability, and salinity stress have been reported to affect algal lipid metabolism and lipid composition, thus providing means to modify

Kristian Spilling (ed.), *Biofuels from Algae*, Methods in Molecular Biology, vol. 2996,
https://doi.org/10.1007/978-1-0716-5031-8_7,

the biomass composition for specific purposes [3]. Additionally, algal lipid composition has several special features, and certain algae species contain many unusual lipids and fatty acids [3]. Many microalgae species are enriched in certain polyunsaturated fatty acids (PUFAs) esterified to triacylglycolipids such as monogalactosyl diacylglycerols (MGDGs) and digalactosyl diacylglycerols (DGDGs) [4].

We have earlier utilized a lipidomics platform for clinical samples [6], and the UHPLC-MS-based lipidomics platform has shown to be a valuable tool for investigating lipids in many different sample matrices [7] including algae [8]. Here we describe a UPLC-MS-based global analysis platform for algae lipids using data dependent acquisition (DDA). Lipid composition of algal biomass consists mainly of neutral lipids such as triacylglycerols (TAGs), glycolipids MGDGs, DGDGs, and sulfoquinovosyl diacylglycerols (SQDGs) as well as phospholipids (mainly phosphatidylcholines, phosphatidylglycerols, phosphatidylinositols, and phosphatidylethanolamines [5]). The annotations of these lipid classes are based on accurate mass of the molecular ion and on MS/MS fragmentation. It is important to highlight that the method presented has a broad range of applications; for instance, it is well suited for the analysis of oleaginous yeasts, which are increasingly employed in the production of food lipids [9].

2 Materials

2.1 Samples

Samples: 5 mg freeze-dried algae material, e.g., *Euglena gracilis*, *Chlorella vulgaris*, or *Chlorella pyrenoidosa* biomass cultivated in bioreactors.

2.2 Standards and Chemicals

The internal standard mixture contains compounds from several different lipid classes: phosphatidylcholines (PC), phosphatidylethanolamines (PE), a ceramide (Cer), a phosphatidylglycerol (PG), a glucosylceramide (GlcCer), as well as di- and triacylglycerols (DAG and TAG, respectively). Lipids are denoted by their molecular composition as follows: <lipid class><number of carbon atoms in the first fatty acid moiety>:<number of double bonds in the first fatty acid moiety>/<number of carbon atoms in the second fatty acid moiety>:<number of double bonds in the second fatty acid moiety>. For example, the abbreviation PC (17:0/17:0) indicates a phosphatidylcholine comprising two C17 fatty acids with no double bonds.

1. Standard mixture 1 added to the samples before extraction: PC (17:0/0:0), PC(17:0/17:0), PE (17:0/17:0), PG (17:0/17:0) [rac], Cer (d18:1/17:0), GlcCer (d8:1/12:0), DAG (17:0/17:0/0:0) [rac], and TAG (17:0/17:0/17:0). The concentration level of each standard is approximately 0.2 μg/sample.

2. Standard mixture 2 added after extraction: PC (16:1/0:0-D3), PC (16:1/16:1-D6), and TAG (16:0/16:0/16:0-13C3). The concentration level of each standard is 0.2 μg/sample.
3. Extraction solvents: HPLC-grade chloroform/methanol (2:1, v/v).

2.3 UPLC-MS System

1. UPLC system (e.g., Acquity Ultra Performance LC™; Waters Corporation, Wexford, Ireland).
2. Sample organizer for automatic sampling.
3. LC column (e.g., Acquity UPLC® BEH C18 1.7 μm, 2.1 × 100 mm column, Waters Corporation).
4. Solvent A: 1% 1 M NH_4Ac and 0.1% HCOOH in milliQ.
5. Solvent B: AcCN/2-propanol (1:1, v/v), 1% 1 M NH_4Ac, and 0.1% HCOOH.
6. Quadrupole-time-of-flight (Q-ToF) mass spectrometer (e.g., ZenoToF 7600, Sciex).

2.4 Other Instrumentation

1. Ultrasonication bath.
2. Mixer mill homogenizer.
3. Vortexer.
4. Centrifuge.

2.5 Data Processing Software

MZmine 4 software for peak detection, alignment, and normalization is used to process raw MS data (http://mzmine.github.io/). Supported data formats are mzML, mzXML, mzData, NetCDF, Thermo RAW, Waters RAW, and Agilent CSV. Sciex Wiff and Wiff2 files need to be converted to a suitable file format such as mzML.

3 Methods

The UPLC-MS-based lipidomics platform consists of multiple steps (Fig. 1), including sample preparation, separation and detection, and finally data analysis. Total lipid extracts are obtained using a modified Folch extraction, and the extracts are analyzed in positive mode (ESI+). The ionization mode is chosen based on the studied lipids. If the sample set contains a large number of samples, it is important to randomize the samples prior to sample preparation and analysis in order to avoid potential bias. Peaks are detected, aligned, and calibrated with internal standards in specialized software packages such as MZmine 4 [10, 11] or MS dial (https://systemsomicslab.github.io).

The identification of lipids is performed automatically by comparing the obtained MS/MS spectra with an internal database containing

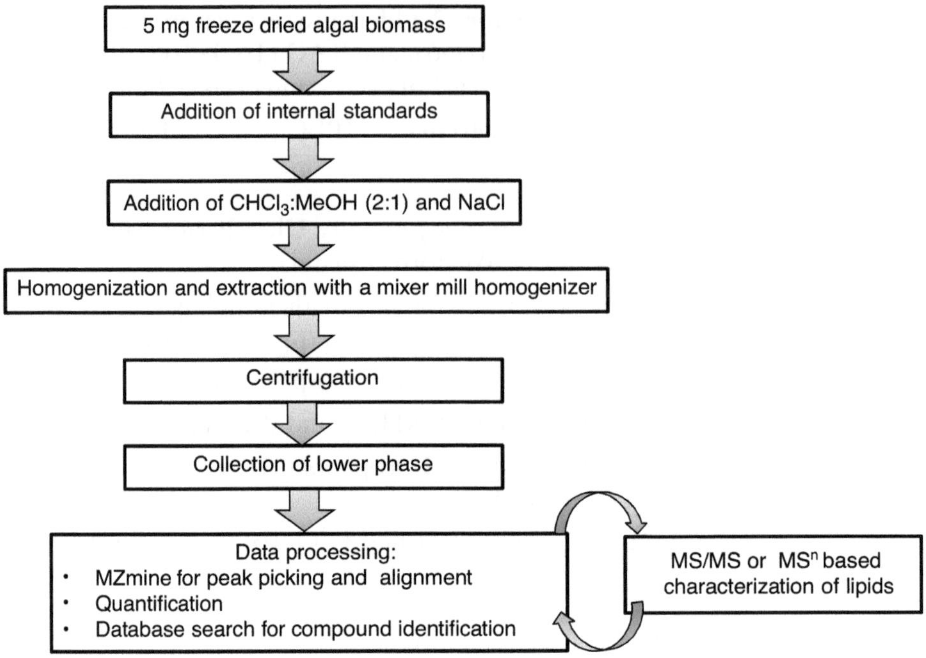

Fig. 1 Workflow of the lipid profiling analysis

MS/MS spectra of lipids from various classes. To minimize carry-over effects, relatively strong solvent is used as the UPLC eluent. Blank samples are analyzed between the real samples to detect possible carry-over. No significant carry-over was noticed. The repeatability of the analysis proved to be good, with RSD of the spiked standards varying from 4.9% (PC standard) to 8.1% (TAG standard) ($n = 12$).

3.1 Sample Preparation: Freeze-Dried Algal Biomass

1. Add 5 μL of standard mixture 1 and 200 μL of HPLC-grade chloroform/methanol (2:1, v/v) and 100 μL of 0.15M (0.9%) sodium chloride to 5 mg of freeze-dried algae sample and homogenize with three grinding balls (Ø 3 mm) for 2 min (20 Hz) with mixer mill homogenizer.
2. Allow to stand for 30 min.
3. Centrifuge the sample for 3 min at 7800 × g.
4. Collect the lower phase and add 10–20 μL of internal standard mixture 2.

3.2 Lipidomics Profiling Platform (UPLC-MS)

1. The UPLC gradient is as follows: set to start at 35% B, to reach 80% B in 2 min and 100% B in 7 min and remain there for 7 min. The total run time including a 4 min re-equilibration step at initial conditions is 18 min. An example of a DDA chromatogram is shown in Fig. 2.
2. Set the temperature of the column to 50°C.

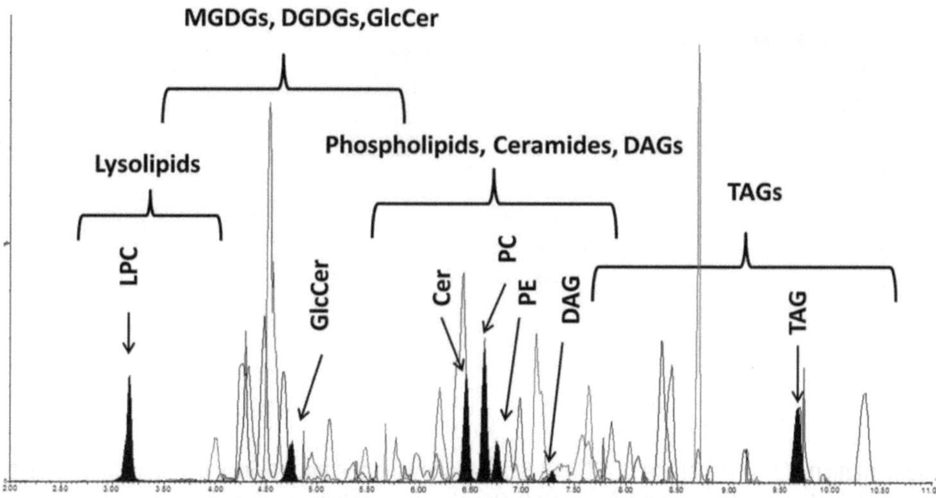

Fig. 2 UPLC-MS chromatogram (positive ion mode) showing traces of the internal standards (standard mixture 1) and some major lipids representing different lipid classes

3. Set the temperature of the sample organizer to 10°C.
4. Set the flow rate to 0.400 mL/min.
5. Inject 2 μL of lipid extract (chloroform phase).
6. Lipid compounds are detected by a Q-ToF mass spectrometer using electrospray ionization in positive or negative ion mode (*see* **Notes 1 and 2**).
7. Collect data in centroid mode over a mass-to-charge ratio range of m/z 300–1200 with a scan duration of 0.2 s.
8. Process data using the MZmine software (*see* **Note 3**) or equivalent. Data processing includes alignment of peaks, peak integration, normalization, and identification. Lipids are identified using an internal accurate mass spectral library based on their MS/MS spectra as shown in Fig. 3 and other suitable libraries such as NIST 17 or lipidome atlas in MS-DIAL 4.
9. The data is normalized using one or more internal standards representative of each class of lipid present in the samples: the intensity of each identified lipid is normalized by dividing it with the intensity of its corresponding standard and multiplying it by the concentration of the standard. All monoacyl lipids such as monoacylglycerols and monoacylglycerophospholipids are normalized with PC(17:0/0:0); all diacyl lipids except ethanolamine phospholipids are normalized with PC(17:0/17:0), all ceramides with Cer (d18:1/17:0), all diacyl ethanolamine phospholipids with PE(17:0/17:0), and TAG with TAG (17:0/17:0/17:0) (*see* **Note 4**). Other (also unidentified) molecular species are calibrated with

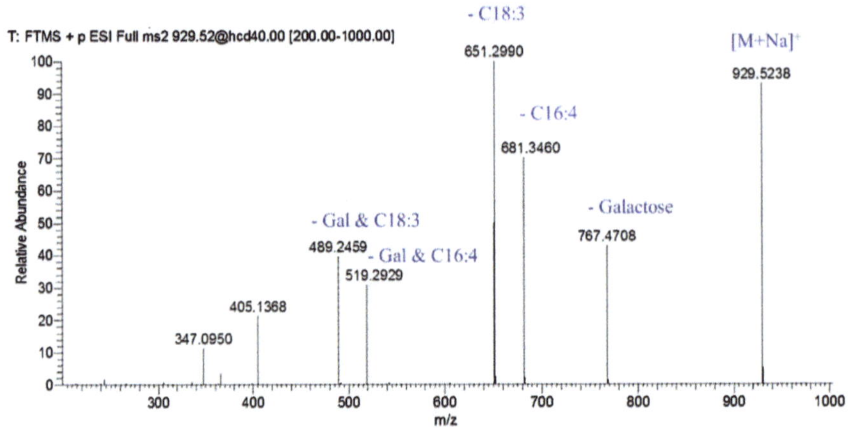

Fig. 3 Fragmentation of Na^+ ion of the DGDG (18:3/16:4) showing the neutral loss of a galactose moiety, 162 u and the losses of the fatty acyl groups 16:4 and 18:3. Spectra were obtained using nanoelectrospray and Orbitrap MS with HCD fragmentation (higher energy collisional dissociation) with a 40% normalized collision energy

PC (17:0/0:0) for retention time <300 s, PC (17:0/17:0) for retention time between 300 and 410 s, and TAG (17:0/17:0/17:0) for higher retention times. In cell cultivation samples, lipids are further normalized by dividing the normalized lipid concentration either with dry weight or the protein content of the sample.

10. If a more accurate quantitation of selected lipids is needed, standard curves should be used for quantification. If other reference compounds are not available, the standard curves are prepared using PC (16:1/0:0-D3), PC (16:1/16:1-D6), and TAG (16:0/16:0/16:0-13C3) and, as in the samples, standard mixture 1 as internal standard. Thereafter, the curves of PC (16:1/0:0-D3), PC (16:1/16:1-D6), and TAG (16:0/16:0/16:0-13C3) are used for quantitation of lysolipids, phospholipids, and triacylglycerols, respectively. The instrument vendor's software is used for quantitation. The linear range of the method is different for each lipid class and depends on the MS instrument used.

4 Notes

1. The described method is aimed at rapid screening of lipid molecular species across a broad range of lipid classes. While the use of positive ion mode is described, the same platform can also be applied using electrospray ionization in negative ion mode, which will lead to better sensitivity for specific phospholipid classes such as phosphatidylinositols and phosphatidylserines as

well as glycosylated sphingolipids and sulfoquinovosyldiacylglycerols (SQDGs). In negative ion mode, a better ionization efficiency is achieved when the UPLC eluents are prepared without formic acid. In the case of polar, glycosylated sphingolipids, another extraction method is needed in order to achieve sufficient extraction recoveries.

2. The method can be transferred to a different LC-MS system. Ideally, the system used should provide a high mass accuracy and have the capability to obtain MS/MS data in DDA mode. For example, a UPLC-Synapt TQ-S (Waters) instrument.
3. The updated version of the open source MZmine software, including tutorials, is available at http://mzmine.github.io/. The data processing parameters need to be optimized for a specific analytical system used in order to account for different peak shapes and lengths as well as different mass spectrometer resolution. Differential profiling of multiple samples requires steps such as peak detection, alignment (matching of peaks across multiple samples), and normalization using internal standards.
4. Some algae species/cells contain significant amounts of TAG (17:0/17:0). If new types of samples are analyzed, it is recommended first to run a sample extract without any internal standard addition and check the profile. If significant amounts of TAG (17:0/17:0) are detected, another internal standard should be chosen instead (preferably a stable isotope-labeled compound).

Acknowledgments

The authors thank Ulla Lahtinen and Airi Hyrkäs for excellent technical assistance in preparing extracts of algal samples.

References

1. Kumari P, Kumar M, Reddy CRK et al (2013) Algal lipids, fatty acids and sterols. In: Dominguez H (ed) Functional ingredients from algae for foods and nutraceuticals. Woodhead Publishing Ltd, Cambridge
2. Baldisserotto C, Popovich C, Giovanardi M et al (2016) Photosynthetic aspects and lipid profiles in the mixotrophic alga *Neochloris oleoabundans* as useful parameters for biodiesel production. Algal Res 16:255–265
3. Gray CG, Lasiter AD, Leblond JD (2009) Mono- and digalactosyldiacylglycerol composition of dinoflagellates. III. Four cold-adapted, peridinin-containing taxa and the presence of trigalactosyldiacylglycerol as an additional glycolipid. Eur J Phycol 44:439–445
4. Da Costa E, Silva J, Mendonca SH et al (2016) Lipidomic approaches towards deciphering glycolipids from microalgae as a reservoir of bioactive lipids. Mar Drugs. https://doi.org/10.3390/md14050101
5. Yao L, Gerde JA, Lee SL, Wang T, Harrata KA (2015) Microalgae lipid characterization. J Agric Food Chem 63:1773–1787
6. Nygren H, Seppänen-Laakso T, Castillo S et al (2011) Liquid chromatography-mass spectrometry (LC-MS)-based lipidomics for studies of body fluids and tissues. Methods Mol Biol 708:247–257

7. Zhao YY, Wu SP, Liu S et al (2014) Ultraperformance liquid chromatography–mass spectrometry as a sensitive and powerful technology in lipidomic applications. Chem Biol Interact 220:181–192
8. Cutignano A, Luongo E, Nuzzo G et al (2016) Profiling of complex lipids in marine microalgae by UHPLC/tandem mass spectrometry. Algal Res 17:348–358
9. Småros F, Vidgren V, Rondou K et al (2025) Microbial production of food lipids using the oleaginous yeast *Apiotrichum brassicae*. Food Res Int 200:115481
10. Pluskal T, Castillo S, Villar-Briones A, Orešič M (2010) MZmine 2: modular framework for processing, visualizing, and analyzing mass spectrometry-based molecular profile data. BMC Bioinformatics 11:395
11. Katajamaa M, Miettinen J, Orešič M (2006) MZmine: toolbox for processing and visualization of mass spectrometry based molecular profile data. Bioinformatics 22:634–636

Check for updates

Chapter 8

UPLC-ELSD Analysis of Algal Lipid Classes and Derivatization of Bound and Free Fatty Acids and Sterols for GC-MS Methods

Tuulikki Seppänen-Laakso, Heli Nygren and Heiko Rischer

Abstract

Constituents of microalgae and sample preparation for UPLC-ELSD and GC-MS analyses are described. Bound fatty acids from acylglycerols, alkylacylglycerols, galactosyldiacylglycerols, glycerophospholipids, and sterol esters are derivatized by using transesterification with sodium methoxide to form fatty acid methyl esters. Compounds containing free hydroxyl groups, either present originally or formed during previous step, like free fatty acids, sterols, α-tocopherol, phytol, and nonesterified alkoxyglycerols, are trimethylsilylated. The compounds in algal lipid extract are subsequently derivatized by these two steps.

Keywords Lipid classes, Evaporative light-scattering detection, Fatty acids, Gas chromatography mass spectrometry, Transesterification, Trimethylsilylation, Ultra performance liquid chromatography

1 Introduction

Both macro- and microalgae thrive in very diverse and extreme environments and produce specific bioactive substances and complex lipids and fatty acids [1]. The considerable variation is being utilized in selecting strains with adequate growth rate and accumulation of desired lipid quantities and compositions. Attention has also been paid to the possibility to obtain high-added value chemicals for nutritional, cosmetic, and pharmaceutical uses [2]. The use of versatile and novel extraction and analysis methods is a prerequisite for characterization and identification of the highly variable and heterogeneous lipid constituents in algae.

Considerable differences in the lipid content (1–75%) and composition appear among various microalgae species and within the same genus [2]. Besides synthesizing storage lipids, microalgae can be induced to accumulate them under stress conditions, e.g., nitrogen limitation. Screening of brackish and marine microalgae (19 species) grown under different conditions revealed a significant dependency between decrease in cell nitrogen and increase in lipids [3].

Kristian Spilling (ed.), *Biofuels from Algae*, Methods in Molecular Biology, vol. 2996,
https://doi.org/10.1007/978-1-0716-5031-8_8,

The biomass of the microalgae *Chlorella vulgaris*, for example, is composed of lipids (18%, dry weight), protein (28%), ash (11%), and carbohydrates (43%) [4]. Lipid class composition reveals abundance of neutral lipids, i.e., triacylglycerols (TAG, 27%) and free fatty acids (FFA, 31%) followed by polar lipids (1%). The proportion of phytol is 13%, chlorophyllides (non-phytol moiety of chlorophylls) 15%, and other constituents 14% [4]. The low total lipid content (2.6%) in the macroalgae *Ulva armoricana* consists of neutral lipids (56%), glycolipids (29%), and phospholipids (15%) [5]. The sterol composition of the unsaponifiable fraction (21% of total lipids) is characterized by high contents of cholesterol and iso-fucosterol (35% and 25%, respectively). Among chloroplast lipids, galacto- and sulfolipids (mono- and digalactosyl- and sulfoquinovosyldiacylglycerols) and phosphatidyl glycerol have been considered as distinctive lipids between plant and animal or yeast cells [6]. *Euglena gracilis* is considered a potentially useful cultivated organism and source of biofuels [7]. It can grow by photosynthesis but can also metabolize sugars. Under anaerobic conditions, when the culture medium contains organic carbon source, wax esters are synthesized [8].

Various mechanical, chemical, and enzymatic methods are used to extract lipids from algae biomass [9]. Solvent extraction procedures have been recently optimized in detail with the microalgae *Trachydiscus minutus* biomass. The subsequent extraction approach starting from water followed by 80% methanol and then by hexane/isopropanol (50:50, v/v) proved to be the preferred method to obtain the water-soluble metabolites, polar lipids, and nonpolar TAG in the separate fractions [10]. After mechanical disruption of the cells by lyophilization and grinding, e.g., solvent extraction can be intensified by using Soxhlet extraction or accelerated solvent extraction (ASE), which has been applied when studying TAG and FFA in algae [11]. Compared to the chloroform/methanol 2:1 mixture, 2-ethoxyethanol provides superior lipid and hydrocarbon recovery [12]. It is especially efficient in extracting wet biomass rather than dried algal pellets.

High-performance liquid chromatography (HPLC) with evaporative light-scattering detection (ELSD) provides a rapid and relatively simple method for monitoring the broad range of lipid classes of algae strains [13]. In this technique, a comprehensive separation of both nonpolar and polar lipids can be achieved on conventional silica [14, 15] or monolithic silica columns [16]. With careful optimization, diacylglycerol ethers (DAGE) can be separated from TAGs which have very similar structure and polarity [17]. It is known that responses in mass detection by light scattering are nonlinear and are better fitted to polynomial curves. However, to obtain comprehensive intact lipid profiles at subspecies level from microalgae extracts, ultrahigh performance liquid chromatography combined with high-resolution mass spectrometry (UHPLC-MS) is the method of choice [18].

Characterization of algal fatty acid compositions involves hydrolysis of triacylglycerols and other intact lipids, followed by derivatization of the fatty acids to methyl esters (FAME) prior to analysis by gas chromatography (GC) or GC-MS [18, 19]. After transesterification with sodium methoxide, the algal lipid extracts contain variable amounts of hydroxyl group-containing compounds like FFAs, nonesterified alkoxyglycerols, sterols, α-tocopherol, and phytol. Subsequent transesterification and trimethylsilylation enable the GC analysis of these compounds. The mixture contains then FAMEs, trimethylsilyl (TMS) derivatives, and squalene and other hydrocarbons. Notably this method is not specific for microalgae but widely used to analyze bacterial [20], fungal [21], and plant [22] material.

2 Materials

2.1 Samples, Standards, and Chemicals

1. 10 mg freeze-dried *Euglena gracilis* (single-cell eukaryote algae) or *Chlorella pyrenoidosa* (freshwater green algae) biomass from bioreactors.
2. External standards including squalene (SQ), wax ester (WE), sterol ester (SE), triacylglycerol (TAG), free sterol (ST), diacyl and monoacylglycerol (DAG, MAG), free fatty acid (FFA), alkyldiacylglycerol ether (DAGE), mono- and digalactosyldiacylglycerol (MGDG, DGDG), ceramide (Cer), cerebroside, cardiolipin (CL), phosphatidylethanolamine (PE), phosphatidylinositol (PI), phosphatidylserine (PS), phosphatidylcholine (PC), sphingomyelin (SPH), lysoPE, and lysoPC (*see* **Note 1**).
3. The internal standard mixture for GC-MS analyses of fatty acids and sterols consists of heptadecanoic acid (C17:0) in the form of free fatty acid (FFA) and triacylglycerol (TAG, triheptadecanoate). The samples are spiked with the standards before extraction at concentration levels of 10–30 μg FFA and 50–70 μg TAG. For identification, commercial fatty acid methyl ester (FAME) and FFA mixtures can be used.
4. Extraction solvents: chloroform/methanol (2:1, v/v), dichloromethane (DCM), petroleum ether bp. 40–60°C, hexane, and 15 mM saline solution (NaCl).
5. Derivatization reagents: sodium methoxide in methanol (NaOMe, 0.5 M), sodium hydrogen sulfate ($NaHSO_4$, 15%), *N*-methyl-*N*-(trimethylsilyl)trifluoroacetamide (MSTFA), and trimethylchlorosilane (TMCS).

2.2 UPLC-ELSD System

1. Ultra performance LC (UPLC) with a quaternary solvent manager (e.g., Acquity® H-class Waters Corporation, Wexford, Ireland).
2. Evaporative light-scattering detector (ELSD) (e.g., Waters Corporation).
3. LC column (e.g., Spherisorb Silica column, 2.1 × 100 mm, 3 μ m, narrow bore, Waters Corporation).

4. Solvent A: iso-octane (2,2,4-trimethylpentane) and tetrahydrofuran (99:1, v/v) (*see* **Note 2**).
5. Solvent B: 2-propanol and DCM (5:2, v/v).
6. Solvent C: 2-propanol and water (1:1, v/v) (*see* **Note 3**).
7. Nitrogen as nebulizing gas.

2.3 GC-MS System

1. Gas chromatograph (e.g., 7890A, Agilent Technologies Inc., Palo Alto, CA, USA).
2. Mass selective detector (e.g., 5975C Agilent Technologies).
3. Injection System (e.g., Gerstel MultiPurpose Sampler, GERSTEL GmbH & Co. KG, Mülheim an der Ruhr, Germany).
4. Column for esterified samples such as free fatty acid phase (FFAP) silica capillary column (e.g. 25 m × 0.2 mm × 0.3 μm from Agilent Technologies) or an equivalent.
5. Column for trimethylsilylated samples such as Rtx®-5MS column e.g., 15 m × 0.25 mm × 0.25 μm, Restek Corporation, Bellefonte, PA, USA) or DB-5 silica capillary column (e.g., 30 m × 0.25 mm × 0.25 μm, Agilent Technologies) or an equivalent.
6. Helium as carrier gas in GC-MS, 99.99%.

2.4 Other Instrumentation

1. Mixer mill, homogenizer (e.g., Retsch GmbH, Haan, Germany).
2. Vortexer.
3. Centrifuge (e.g., Eppendorf Zentrifugen GmbH, Leipzig, Germany).
4. Block heater (e.g., Stuart® SBH13OD/3, Bibby Scientific Ltd., Staffordshire, UK).
5. Sample concentrator (e.g., Stuart® SBHCONC/1 Bibby Scientific Ltd., Staffordshire, UK).

3 Methods

The study of the widely varying algal lipid constituents requires combinations of different analytical techniques (Fig. 1).

3.1 Sample Preparation

3.1.1 Homogenization and Extraction of Samples

1. Freeze-dried algal biomass samples (10 mg) are mixed with 500 μl of 15 mM saline solution and 1 ml of chloroform/methanol (2:1, v/v) and homogenized for 2 min (20 Hz, two steel grinding balls ø 3 mm). Samples for GC-MS analyses are spiked with C17:0 internal standards (FFA + TAG) before homogenization and extraction.
2. Vortex 5 min and allow to stand for 30 min.
3. Centrifuge for 3 min at 7800 × g.

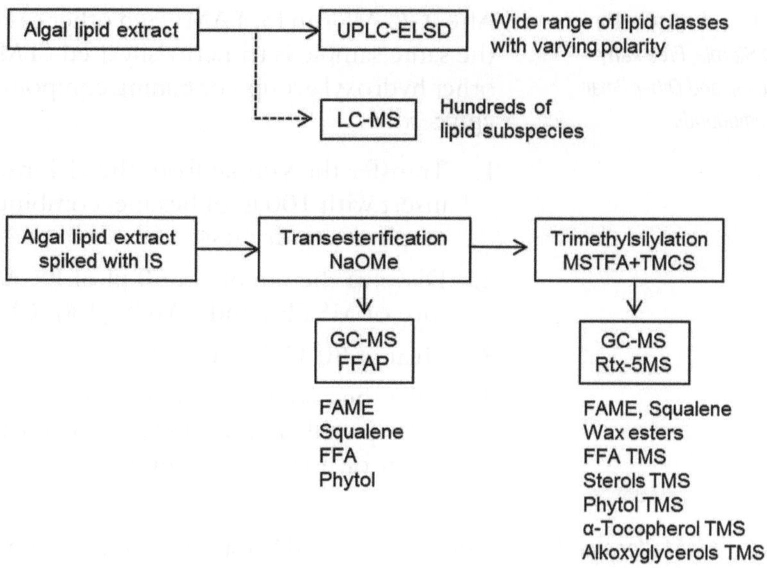

Fig. 1 Analytical techniques used in the present study to determine algal lipid constituents

4. Separate the lower organic layer into an Eppendorf tube (for UPLC-ELSD analysis) or into a glass tube (for transesterification). Repeat the extraction and combine the extracts.
5. Take 100 μl of extract from the Eppendorf tube into a LC vial insert (300 μl). The sample is ready for UPLC-ELSD analysis.
6. Evaporate the extract in the glass tube to dryness under nitrogen flow.
7. Dissolve the residue in 500 μl of petroleum ether (bp. 40–60°C). The sample is used for derivatizations and GC-MS runs.

3.1.2 Derivatization of Bound Fatty Acids for GC-MS

Transesterification with sodium methoxide is esterifying only fatty acids which are bound to glycerolipids, glycerophospholipids, or sterol esters, for example. Free fatty acids remain unesterified.

1. Add 250 μl of 0.5 M sodium methoxide solution (NaOMe in MeOH) and a couple of boiling stones to the sample in petroleum ether. Vortex the mixture.
2. Boil the mixture at 45°C for 5 min in a block heater (*see* **Note 4**).
3. Add 500 μl of 15% sodium hydrogen sulfate ($NaHSO_4$). Vortex the mixture.
4. Add 250 μl petroleum ether, vortex, and centrifuge for 1 min at 7800 × g. Separate the upper petroleum ether layer, evaporate to dryness under nitrogen flow, dissolve the residue in 100 μl of hexane, and transfer the sample to a GC vial insert (250 μl).

3.1.3 Derivatization of Sterols, Free Fatty Acids, and Other Polar Compounds

After GC-MS run for FAME and other easily evaporated compounds, the same sample is trimethylsilylated (TMS) for free fatty acids and other hydroxyl group-containing compounds, e.g., sterols and alkoxyglycerols.

1. Transfer the sample from the vial insert to a GC vial, rinse the insert with 100 μl of hexane, combine the hexane aliquots, and evaporate to dryness.
2. Dissolve the sample in 50 μl of DCM and add 25 μl of a mixture of MSTFA and TMCS (1%). Close the cap well.
3. Heat at 80°C for 20 min.
4. Take the cooled sample to a new vial insert (250 μl) and close the cap well. The sample is ready for GC-MS analysis on a semipolar or nonpolar column.

3.2 Lipid Class and Fatty Acid Analyses

3.2.1 Lipid Class Analysis by UPLC-ELSD

The UPLC-ELSD system is controlled by Empower 3 software.

1. Open the nebulizer gas (nitrogen, 50 psi), set the temperature of the nebulizer and drift tube to 40°C, respectively.
2. Set the temperature of the sample manager to 10°C.
3. Set the column temperature to 40°C.
4. The ternary gradient mobile phase composition is shown in Table 1 (*see* **Note 5**).
5. Inject 2 μl of lipid extract (in DCM).
6. Integrate the chromatogram for peak areas and calculate the concentrations by using equations of corresponding calibration curves (*see* **Note 6**) and by considering the weighed amount of freeze-dried algal powder.

Table 1
The ternary gradient mobile phase composition (see Sect. 3.2.1)

Time (min)	Flow (μl/min)	% A	% B	% C
Initial	800	100	0	0
0.5	800	100	0	0
3	800	75	20	0
3.5	800	46	52	2
7	800	32	52	16
10.5	800	32	52	16
12	800	30	70	0
16	800	100	0	0
20	800	100	0	0

3.2.2 Fatty Acid Methyl Esters by GC-MS

The GC-MS runs are controlled by MSD ChemStation 2.0 program (Agilent Technologies), and for identification the NIST Mass Spectral Search Program (Version 2.0f) is used.

1. The oven temperature program is from 70°C (2 min) to 235°C at a rate of 10°C/min; total run time is 30 min.
2. The temperatures of the injector and MS source are 220 and 230°C, respectively.
3. The samples (2 μl) are injected with the Gerstel MPS injection system.
4. The split ratio is 1:15–1:20.
5. The data are collected in electron impact (EI) mode (70 eV) at a mass range of m/z 40–600.
6. Heptadecanoic acid methyl ester (C17:0 FAME) derived from triheptadecanoate is used as internal standard for fatty acid methyl esters.

3.2.3 Free Fatty Acids, Sterols, and Other Polar Compounds as TMS Derivatives by GC-MS

Figure 2 shows a typical GC-MS total ion chromatogram (TIC) obtained from algal lipid extract. In this case, the level of sterols and squalene is very low. The hexadecatetraenoic acid (16:4n-3) is known to appear in MGDG and DGDG.

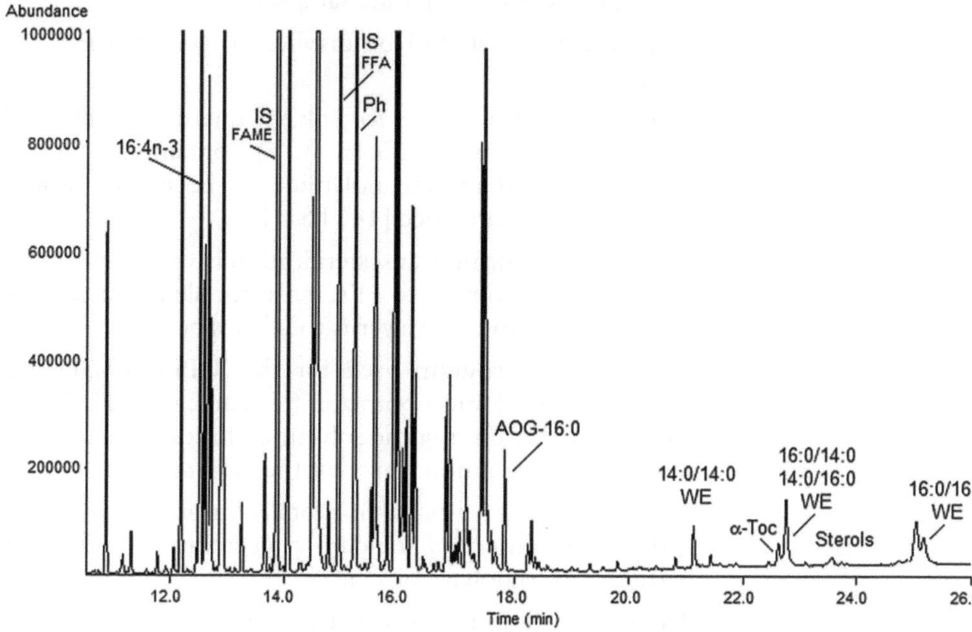

Fig. 2 Typical GC-MS total ion chromatogram obtained from transesterified and trimethylsilylated *Euglena gracilis* extract. Characteristic compounds for algae: 16:4n-3 (Methyl 4,7,10,13-hexadecatetraenoate), Ph (Phytol TMS); AOG-16:0 (1-*O*-hexadecylglycerol diTMS); 14:0/14:0 WE (tetradecanoic acid tetradecyl ester); 16:0/14:0 WE + 14:0/16:0 WE (hexadecanoic acid tetradecyl ester + tetradecanoic acid hexadecyl ester); 16:0/16:0 (hexadecanoic acid hexadecyl ester); α-Toc (α-tocopherol TMS). IS FAME (C17:0 methyl ester); IS FFA (C17:0 free fatty acid TMS). The majority of the peaks represent FAMEs, hydrocarbons, and FFA TMS-derivatives

1. The oven temperature program is from 70°C (1 min) to 270°C at a rate of 10°C/min; total run time is 30 min.
2. The temperatures of the injector and MS source are 220 and 230°C, respectively.
3. The samples (2 μl) are injected with the Gerstel MPS injection system.
4. The split ratio is 1:15–1:20.
5. The data are collected in electron impact (EI) mode (70 eV) at a mass range of m/z 40–600.
6. Trimethylsilyl ether of heptadecanoic acid (C17:0 FFA TMS) is used as internal standard for trimethylsilylderivatives (*see* **Note 7**).

4 Notes

1. Heptadecanoic acid (C17:0) as acyl group is used for quantification in UPLC-ELSD analyses of different lipid classes. For calibration curves, five to nine concentration points (approximately 0.02–20 μg/ml) with two to four replicates are used. Internal standards containing other than C17:0 as acyl groups could be selected for marine samples.
2. Tetrahydrofuran should be distilled to obtain a more stable baseline.
3. Different variations exist for solvent mixture C. The ratio of 2-propanol and water 85:15 (v/v) containing, e.g., 0.05% (v/v) acetic acid and ethanolamine has been used to improve the column performance [14, 16, 17].
4. Heating time during transesterification is very important since FAME can decompose to FFA if the sample is heated too long or the petroleum ether layer is totally evaporated.
5. The gradient program used for this UPLC-ELSD method has been applied and optimized from an earlier HPLC-ELSD method [15]. Running the solvent cycle continuously helps to obtain stable retention times for the analytes.
6. The calibration curves are not totally linear, and when dealing with a wide variety of standard concentrations, responses are better fitted to polynomial curves.
7. Trimethylsilylation improves the response and shape of FFA peaks which is reflected as lower relative standard deviation values for C16:0 and C18:0 FFA (average RSD 3.5 for TMS derivatives and 5.9 for non-silylated FFA, respectively).

Acknowledgments

This project was supported by the EU-funded SEABIOTECH project (FP7-SBT-KBBE-311932, http://www.seabiotech.org/). The authors thank Dr. Marilyn Wiebe who was kindly supplying the *Euglena gracilis* biomass. The authors owe thanks to Airi Hyrkäs for excellent technical assistance in sample preparation and optimization of the running conditions in UPLC-ELSD analyses, especially.

References

1. Kumari P, Kumar M, Reddy et al (2013) Algal lipids, fatty acids and sterols. In: Dominguez H (ed) Functional ingredients from algae for foods and nutraceuticals. Woodhead Publishing Ltd, Cambridge, pp 87–134
2. Amaro HM, Macedo AC, Malcata FX (2012) Microalgae: an alternative as sustainable source of biofuels? Energy 44:158–166
3. Schwenk D, Seppälä J, Spilling K et al (2013) Lipid content of 19 brackish and marine microalgae: influence of growth phase, salinity and temperature. Aquat Ecol 47:425–424
4. Yao L, Gerde JA, Lee S-L et al (2015) Microalgae lipid characterization. Agric Food Chem 63:1773–1787
5. Kendel M, Wielgosz-Collin G, Bertrand S et al (2015) Lipid composition, fatty acids and sterols in the seaweeds *Ulva armoricana*, and *Solieria chordalis* from Brittany (France): an analysis from nutritional, chemotaxonomic, and antiproliferative activity perspectives. Mar Drugs 13:5606–5628
6. Welti R, Wang X, Williams TD (2003) Electrospray ionization tandem mass spectrometry scan modes for plant chloroplast lipids. Anal Biochem 314:149–152
7. Matsuda F, Hayashi M, Kondo A (2011) Comparative profiling analysis of central metabolites in *Euglena gracilis* under various cultivation conditions. Biosci Biotechnol Biochem 75(11):2253–2256
8. Teerawanichpan P, Qiu X (2010) Fatty acyl-CoA reductase and wax synthase from *Euglena gracilis* in the biosynthesis of medium-chain wax esters. Lipids 45:263–273
9. Parmar A, Singh NK, Pandey A et al (2011) Cyanobacteria and microalgae: a positive prospect for biofuels. Bioresour Technol 102:10163–10172
10. Stranska-Zachariasova M, Kastanek P, Dzuman Z et al (2016) Bioprospecting of microalgae: proper extraction followed by high performance liquid chromatographic-high resolution mass spectrometric fingerprinting as key tools for successful metabolom characterization. J Chromatogr B 1015–1016:22–33
11. Samburova V, Lemos MS, Hiibel S et al (2013) Analysis of triacylglycerols and free fatty acids in algae using ultra-performance liquid chromatography mass spectrometry. J Am Oil Chem Soc 90:53–64
12. Jones J, Manning S, Montoya M et al (2012) Extraction of algal lipids and their analysis by HPLC and mass spectrometry. J Am Oil Chem Soc 89:1371–1381
13. Kobayashi N, Noel EA, Barnes A et al (2013) Rapid detection and quantification of triacylglycerol by HPLC-ELSD in *Chlamydomonas reinhardtii* and *Chlorella* strains. Lipids 48:1035–1049
14. Homan R, Anderson MK (1998) Rapid separation and quantification of combined neutral and polar lipid classes by high-performance liquid chromatography and evaporative light scattering mass detection. J Chromatogr B 708:21–26
15. Seppänen-Laakso T, Laakso I, Vanhanen H et al (2001) Major human plasma lipid classes determined by quantitative high-performance liquid chromatography, their variation and associations with phospholipid fatty acids. J Chromatogr B 754:437–445
16. Graeve M, Janssen D (2009) Improved separation and quantification of neutral and polar lipid classes by HPLC-ELSD using a monolithic silica phase: application to exceptional marine lipids. J Chromatogr B 877:1815–1819
17. Torres CF, Vázquez L, Señoráns FJ, Reglero G (2005) Study of the analysis of alkoxyglycerols and other non-polar lipids by liquid chromatography coupled with evaporative light scattering detector. J Chromatogr A 1078:28–34
18. MacDougall KM, McNichol J, McGinn PJ et al (2011) Triacylglycerol profiling of microalgae strains for biofuel feedstock by liquid chromatography-high-resolution mass spectrometry. Anal Bioanal Chem 401:2609–2616
19. Seppänen-Laakso T, Laakso I, Hiltunen R (2002) Analysis of fatty acids by gas chromatography, and its relevance to research on health and nutrition. Anal Chim Acta 465:39–62

20. Nyyssölä A, Ojala LS, Wuokko M et al (2021) Production of endotoxin-free microbial biomass for food applications by gas fermentation of gram-positive H_2-oxidizing bacteria. ACS Food Sci Technol 1:470–479
21. Marathe SJ, Hamzi W, Bashein AM et al (2020) Anti-angiogenic and anti-inflammatory activity of the summer truffle (*Tuber aestivum* Vittad.) extracts and a correlation with the chemical constituents identified therein. Food Res Int 137:109699
22. Nordlund E, Lille M, Silventoinen P et al (2018) Plant cells as food – a concept taking shape. Food Res Int 107:297–305

Chapter 9

Total Fatty Acid Content Determination of Whole Microalgal Biomass Using In Situ Transesterification

Stefanie Van Wychen and Lieve M.L. Laurens

Abstract

The aliphatic chains of fatty acids are the most prominent and potentially the highest value fuel precursor constituents of algal biomass, and thus accurately quantifying the algal biomass total fatty acid content is a prerequisite for comparing algal strains, growth conditions, and processes. Direct, acid-catalyzed transesterification of whole microalgal biomass is a simple, effective, strain-agnostic, and widely used method to determine the fatty acid content in whole algal biomass, with the precision needed to detect small but significant differences between strains. Such a direct transesterification procedure typically covers the following steps: first, solubilizing the lipids in the biomass matrix and then liberating the fatty acids to make these available for catalytic transesterification to fatty acid methyl esters (FAMEs), subsequent extraction into hexane, and then quantification by gas chromatography. The method we describe here requires less than 10 mg of biomass per sample and is considered high-throughput and highly accurate.

Keywords Fatty acid methyl ester, Fatty acids, In situ transesterification

1 Introduction

Some of the major challenges associated with algae as a biofuel/bioproduct feedstock are characterization of the biomass and the development of novel, cost-competitive conversion technologies that emphasize the valorization of the entire biomass and thus increase the value proposition of algal biofuels [1, 2]. The aliphatic chains of the fatty acids are perhaps the most auspicious constituents of algal biomass for both fuel and high-value products (e.g., polyunsaturated fatty acids as nutraceutical supplements or feedstocks for bio-derived polyurethane polymers), and thus profiling and quantifying the respective biomass fatty acid content and chemical composition (e.g., chain length and level of unsaturation) are imperative for comparing algal strains, growth conditions, extraction, and processing yields [3, 4]. There is a growing emphasis on whole biomass conversion for direct access to the lipid fraction, and thus a rapid, high-throughput analytical methodology for fatty

Kristian Spilling (ed.), *Biofuels from Algae*, Methods in Molecular Biology, vol. 2996,
https://doi.org/10.1007/978-1-0716-5031-8_9,

acid quantification is desirable. The application of a whole biomass, direct, in situ transesterification method is discussed in detail here and builds on a validated method, and exploratory work that compared different catalysts and theoretical and practical yields [5].

In the current and historical literature, wide ranges of methodologies are reported for determining lipid content and extraction yields [6]. The definition of lipids as compounds soluble in organic solvents has been the basis for the quantification of the total lipid fraction of algae as the total quantity of compounds soluble in a chloroform/methanol solvent mixture but also has led to challenges in the interpretation of these data [7–9]. A multitude of reports are present in the literature assessing the application of solvent extraction processes and the respective impact on the lipid quality and chemical composition [10], where it has been shown that gravimetric extraction yield for lipids is highly dependent on the polarity of the solvents used and the composition of the algal lipids [11–14]. The utilization of an extraction-based quantification of the lipid fraction is thus fraught with challenges, and the respective yields obtained are highly dependent on the species and nutritional status of the cells [10, 15].

In situ or direct transesterification of whole microalgal biomass is an alternative to lipid quantification and is routinely used to determine total lipid content in whole biomass samples. This method eliminates the need for timely and noncompound-specific extraction gravimetric-based analyses. Many methods exist for direct transesterification of biomass and lipid samples [5, 16–19]. These methods typically rely on a base, acid, or base/acid two-step catalysis to transesterify lipids. Base catalysts exhibit lower yields compared to acid catalysts, due to their inability to esterify free fatty acids, which can make up a significant portion of algae lipids [5]. Because of the likely and frequent presence of free fatty acids in microalgal lipid extract samples, we chose to focus on the use of an acid catalyst, which is known to be more effective at conversion of all lipid types [5, 16]. Methanolic HCl is a commonly employed catalyst for esterifying lipids in situ, and its effectiveness on microalgal lipids has been studied on both large and small scales. We have shown that it is an effective catalyst for the small-scale esterification of typical microalgal lipids within a biomass matrix, even in the presence of moisture content, such as live water samples [5, 18]. Using chloroform/methanol as a cosolvent and HCl/MeOH as a catalyst, it is possible to (trans-) esterify microalgal lipids into their fatty acid methyl ester (FAME) equivalents on a scale requiring less than 2 mL of total reagent/solvent and 10 mg of biomass. The process of transesterification, with a detailed description of quantification of the respective FAME by gas chromatography (GC).

2 Materials

Prepare all reagents in a fume hood. Use solvent-compatible analytical ware when completing all steps. Store reagents in a suitable location for flammable and toxic chemicals. Carefully follow all relevant chemical handling procedures.

2.1 Apparatus

1. Analytical balance, accurate to 1 mg or 0.1 mg.
2. Vacuum oven set to 40 ± 3°C or a vacuum desiccator.
3. Digital dry block, capable of maintaining 85 ± 3°C.
4. Volumetric flask (class A), 10 mL.
5. Gas-tight syringes with PTFE plungers and 22 gauge, point style 2 needles. Covering a volume range of 5–1000 μL.
6. Adjustable pipet covering 300 μL.
7. Clear, 11 mm crimp-top glass vials for GC analysis with 1.5 mL fill volume.
8. Clear, 11 mm crimp-top glass vials with inserts for GC analysis with 300 μL fill volume.
9. Aluminum crimp caps, 11 mm PTFE/silicone/PTFE septa for the GC vials.
10. Gas chromatograph (GC) equipped with a variable split-flow injector or equivalent device and a flame ionization detector (FID).
11. Capillary column with polyethylene glycol stationary phase: for example, Agilent J&W GC Column DB-WAX length 30 m, internal diameter 0.25 mm, film thickness 0.25 μm, or equivalent.

2.2 Transesterification Reagents

1. Prepare 2:1 (v/v) chloroform/methanol by adding 100 mL of HPLC grade methanol to 200 mL of HPLC grade chloroform. Use a glass graduated cylinder to measure out chloroform and methanol. Store the final solution in a glass bottle with a lid appropriate for organic solvent storage. This reagent may be prepared ahead of time and stored for up to 2 months in the dark. Chloroform is a trihalomethane and possible human carcinogen. Methanol is a slight health hazard and moderately flammable.
2. Prepare 0.6 M (2.1%, v/v) hydrochloric acid:methanol by adding 5 mL of concentrated ACS grade HCl to HPLC grade methanol. Use a graduated cylinder to measure out methanol into a glass bottle. Use a pipet to add hydrochloric acid directly to the methanol. Never add methanol to the acid, as this may cause an explosion. This reagent may be prepared ahead of time and stored for up to 2 months in the dark. Hydrochloric

acid is toxic and corrosive. Methanol is a slight health hazard and moderately flammable.

3. For dilutions and extractions, use HPLC grade n-hexane. Hexane is a significant health hazard and extremely flammable.
4. Hexanes are used for gas-tight syringe rinses and do not need to be HPLC grade quality.

2.3 Standards

1. Prepare a ~10 mg/mL tridecanoic acid methyl ester (C13:0ME) internal standard in hexane. Use a certified tridecanoic acidmethyl ester (C13:0ME) standard (e.g., Sigma-Aldrich #91558-5ML). To make up a 10 mg/mL solution, weigh out approximately 100 mg of the C13:0 ME into a 10 mL, class A, volumetric flask. Record the weight of the C13:0 ME to the nearest 0.1 mg. Bring to volume with HPLC grade hexane, and mix well. Transfer the C13:0ME-hexane solution into 1.5 mL GC vials, and seal immediately with PTFE/silicone/PTFE crimp caps. Take extra care while transferring to limit evaporation, as it is very important to maintain the recorded concentration. Record the date of preparation, concentration, and any other pertinent information on the vials, and place upright in a laboratory freezer (−20°C) for storage. Store sealed for up to 6 months.
2. The GLC 461C FAME standard from NuChek was prepared specifically for use with microalgal FAME analysis and contains a typical mix of microalgal FAMEs. If using another mix, use an analytically certified standard for quantitation. Do not use qualitative mixes or uncertified mixes.
3. To prepare the standards and create a stock of prediluted calibration mixes, tare a clean, 10 mL, class A, volumetric flask. Score the neck of the 461C ampule using a GC column cutter or glass scorer. Do not snap the top off yet. Warm the 461C neat mix ampule with your hands to be sure that all the mix is in the bottom of the vial and no compounds are crystallized in the top of the vial before breaking off the top. Using a glass Pasteur pipet, transfer the neat mix as quantitatively as possible to the tared 10 mL, class A, volumetric flask. Keep the Pasteur pipet and neat mix warm with your fingers to prevent the crystallization of saturated longer-chain fatty acids while transferring. Record the weight to the nearest 0.1 mg. Bring to volume with HPLC grade hexane, and mix well. Transfer approximately 1 mL of the mixture to the labeled 1.5 mL GC vials using a glass Pasteur pipet, working quickly and carefully to avoid evaporation. Seal vials immediately with PTFE/silicone/PTFE crimp caps. Record the date of preparation, concentration, and any other pertinent information on the vials, and place upright in a laboratory freezer (−20°C) for storage. Store sealed for up to 6 months.

3 Methods

Perform all work in a fume hood. Use solvent-compatible analytical ware when completing all steps. Carefully follow all relevant chemical handling procedures.

3.1 Sample Preparation

1. Label 1.5 mL GC vials using a permanent marker for each of the samples and quality control samples to be analyzed (*see* **Note 1**).
2. Record the weights of all labeled vials to the nearest 0.1 mg. Written labels should be dry before the weight is recorded.
3. Microalgal biomass: weigh between 5 and 10 mg of sample into the labeled and pre-weighed GC vials. Record the weight to the nearest 0.1 mg (*see* **Note 2**).
4. Dry samples in vials overnight to obtain an oven dry weight before proceeding with the transesterification (*see* **Note 3**).

3.2 Transesterification Reaction

1. Preheat a digital dry block, or equivalent, to 85 ± 3°C (*see* **Note 4**).
2. While the block is preheating, add the following to each of the sample vials in this order: 25 μL of the prepared ~10 mg/mL C13:0ME internal standard using a gas-tight syringe (*see* **Notes 5 and 6**), 200 μL of 2:1 chloroform/methanol using a gas-tight syringe, and 300 μL of 0.6 M HCl/methanol using a pipet with a plastic or glass-pipet tip. Do not use metal syringes for the HCl/methanol mixture.
3. Seal all vials with PTFE/silicone/PTFE crimp caps, and vortex well to mix the contents (*see* **Note 7**).
4. Place the sealed sample vials into the preheated block at 85°C for 1 h (*see* **Note 8**).
5. After 1 h on the digital dry block, remove the vials, and cool for at least 15 min, but no longer than an hour, at room temperature (*see* **Note 9**).

3.3 Isolation and Preparation of Sample FAME for GC Analysis

1. After samples have reached room temperature, add 1 mL of HPLC grade n-hexane to each of the vials using a gas-tight syringe (*see* **Note 10**).
2. Vortex well to mix the vial contents, and let vials stand undisturbed at room temperature for at least 1 h, but no more than 4 h, to allow the phases to separate (*see* **Note 11**). See Fig. 1 for an example of typical phase separation.
3. Label a new set of 1.5 mL GC vials for the final sample dilutions.
4. Add HPLC grade n-hexane to each of the new sample vials that require a dilution (*see* **Note 12**).

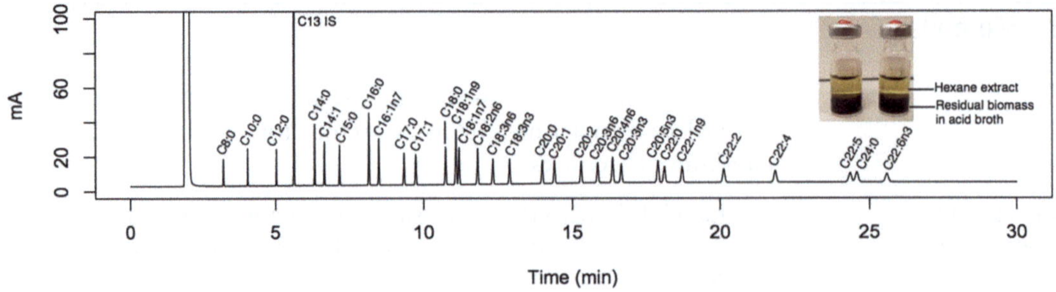

Fig. 1 Overview of a typical chromatogram for a calibration mix of fatty acid methyl esters, representing the complexity of fatty acids typically found in algae. The inset shows typical reaction vials after acid-catalyzed transesterification of whole biomass and hexane extraction of the resulting fatty acid methyl esters

5. Add the necessary amount of hexane upper phase from each of the samples to the new sample vials (Fig. 1) (*see* **Notes 12 and 13**).
6. Seal the 1.5 mL vials containing the dilutions with PTFE/silicone/PTFE crimp caps.

3.4 Preparation of FAME Standards for GC Analysis

1. Label a set of 300 μL insert vials for each level of FAME standard, 1 through 5 and a calibration verification standard (CVS) (*see* **Note 14**).
2. To each of the vials, add the following: 5 μL of the same C13:0ME internal standard (10 mg/mL) used in the transesterification of the samples (*see* **Note 15**) and 200 μL of the appropriate stock standard level or CVS (*see* **Note 15**).
3. Cap the vials immediately with PTFE/silicone/PTFE crimp caps.
4. Vortex the vials to thoroughly mix the contents.

3.5 FAME GC Setup and Data Analysis

1. Suggested GC analysis temperature and flow program for a DB-WAX GC column 30 m × 0.25 mm inner diameter × 0.25 μm film thickness is as follows: 1 μL injection at 10:1 split ratio; inlet temperature of 250°C; constant flow, 1 mL/min helium; oven temperature, 100°C for 1 min, 25°C/min up to 200°C, and hold for 1 min and 5°C/min up to 250°C and hold for 7 min (23 min total); and FID, 280°C, 450 mL/min zero air, 40 mL/min H2, and 30 mL/ min helium.
2. Create a calibration table with the amount of each compound in μg/mL based on the certificate of analysis (percent and purity) and the weight of the 461C working solution (or equivalent). See Fig. 1 for a typical fatty acid methyl ester chromatogram.
3. Update the C13:0 ME internal standard amount in μg/mL in the software to represent the exact concentration of the prepared standard.

4. Quantification by ChemStation (Agilent) or respective GC software (*see* **Note 16**).

3.6 FAME Calculations

1. Export the FAME concentrations in μg/mL for each individual FAME from the GC software, and normalize for the quantity of the internal standard C13:0ME (this may also be done by most software packages). The outcome is the total FAME content normalized for the recovery of the internal standard added at the start of the reaction:

$$\text{Total FAME}_{\text{C13 normalized}} = \sum_{\text{C4}-\text{C24}} \frac{\text{Amount}_{\text{Measured FAME C}_i}}{\text{Amount}_{\text{Measured FAME C13}}} \times \text{Amount}_{\text{Added FAME}_{\text{C13}}}$$

2. Calculate the total FAME as a percent of the dry weight of the sample. For the FAME analysis, the dry weight refers to the weight after drying the sample overnight at 40°C in a vacuum oven:

$$\%\text{Total FAME} = \frac{\text{Total FAME}_{\text{C13 normalized}}}{\text{ODW}_{\text{sample}}} \times 100$$

3. To report or calculate the root-mean-square deviation (RMSD) or the standard deviation (STDEV) of the samples, use the following calculation:

$$\text{RMS} = x_{\text{m}} = \text{mean} = \sqrt{\left(\frac{\sum_1^n x}{n}\right)^2}$$

$$\text{RMS deviation} = \sigma = \text{stdev} = \sqrt{\frac{\sum_1^n (x_i - x_{\text{m}})^2}{n}}$$

where x_{m} = the root mean square of all x values in the set, n = number of samples in set, x_i = measured value from the set.

4. To report or calculate the quality control metrics, use the relative standard deviation (RSD) of the triplicates, and use the following calculation (*see* **Note 17**):

$$\text{RSD} = \left(\frac{\text{stdev}}{X_{\text{mean}}}\right) \times 100$$

where stdev = RMS deviation and X_{mean} = the mean % total FAME of the triplicates.

4 Notes

1. Each sample should be analyzed in triplicate at a minimum, unless the amount of sample prohibits this. In addition, a quality control (QC) biomass or oil standard should be included in triplicate with each batch of samples analyzed. Ensure sufficient quantities of the standard are available for QC charting over multiple analyses.
2. All microalgal biomass samples should be freeze-dried/lyophilized and thoroughly homogenized (particle size <1 mm) before analysis. Care must be taken to ensure a representative sample is taken for analysis. Limit sample contact with oxygen, heat, and moisture to lessen the possibility of lipid degradation. Storage of samples should ideally be under an inert (nitrogen) atmosphere and at least at −20°C.
3. Dry samples in vials overnight in a 40°C oven under vacuum. After drying, take a final weight to calculate the moisture-free sample amount. Allow samples to cool in a desiccator under vacuum before recording a final weight. If no vacuum oven is available, samples can be stored overnight in a vacuum desiccator prior to analysis, given that a corresponding moisture measurement is available the same day for that material. Refer to Determination of Total Solids and Ash in Algal Biomass procedure for moisture measurement [20]. Because moisture content can be variable in samples, it is important to correct for moisture and report FAME content on a dry weight basis.
4. If using a digital dry block for heating, regularly check the temperature of each block using a certified thermometer as the heating coils used in these systems can fail over time.
5. Do not worry about capping the sample vials as you add the internal standard; the C13:0 ME will remain with the biomass, even if the hexane evaporates.
6. Recap the C13:0 ME vial after adding to all samples; it will be needed for standard preparation.
7. It is advisable to add all reagents in sets of ten samples, and then cap. Once an analyst is comfortable with the method, reagents can be added to 50 samples before capping. The key is to avoid excess solvent evaporation during preparation.
8. The block must be at temperature before adding samples. Once the reagents have been added to the samples, the samples must be heated. Do not allow samples to sit around as this may affect the reaction.

9. If the rest of the procedure cannot be completed on the same day, cooled samples may be stored overnight (no longer than 24 h) in a freezer (−20°C). Always store vials upright.
10. This can be done without removing the vial caps by using a small hollow-core needle to penetrate the cap while adding the hexane with the syringe through a second hole in the cap. Be sure the opening in the hollow-core needle faces away from the direction of the added hexane to avoid hexane being expelled back up through the needle.
11. If the rest of the procedure cannot be completed on the same day, vials must be recapped and then stored overnight (no longer than 24 h to avoid co-extraction of unwanted compounds) in a freezer (−20°C). Always store vials upright. If longer storage is required, transfer the hexane layer (after allowing for complete extraction of at least 1 h) to a new 1.5 mL GC vial, cap the vial, and store in a freezer (−20°C) for up to a week.
12. Use the following dilution guidelines when preparing to run the hexane (upper) phase of the transesterified samples. Estimated sample lipid content of <10%, no dilution. Estimated sample lipid content of 10–40%, 400 μL of HPLC grade hexane and 100 μL of the upper phase of the sample. Estimated sample lipid content of 40–80%, 450 μL of HPLC grade hexane and 50 μL of the upper phase of the sample.
13. The upper phase (hexane layer) of the sample can be withdrawn without removing the caps. Use caution not to disturb the lower phase when drawing the upper phase of the sample. Rinse the syringe thoroughly with hexanes between samples. Work in groups of ten samples or less to minimize evaporation of hexane, and cap immediately after adding the sample.
14. Prepare a series of calibration standard levels using the ~10 mg/mL 461C working solution. Use the recommendations in Table 1 to prepare levels 1–5 and a CVS (from an independent working solution). Label 1.5 mL GC vials with the standard level and date. Prepare each standard level before moving on to the next: add the necessary amount of HPLC grade hexane to the labeled 1.5 mL GC vial using a gas-tight syringe, and place a PTFE/silicone/PTFE cap on the vial (you do not need to crimp it yet). Add the necessary amount of FAME working solution (~10 mg/mL) using a gas-tight syringe, and crimp the cap. Then, move onto the next standard level. The prepared standard levels may be stored in PTFE/silicone/PTFE-capped (not punctured) 1.5 mL GC vials in a freezer (−20°C). Always store vials upright. Standards may be stored for up to 6 months.

Table 1
Standards prepared using the NuChek 461C ~10 mg/mL working solution

Standard level	~10 mg/mL 461C standard working solution (μL)	Hexane (μL)
5	500	500
4	250	750
3	100	900
2	30	970
1	10	990
CVS	90	910

15. Always recap standard levels and the C13:0 ME solution as soon as is feasible to prevent evaporation, which would affect the concentration. Keep the C13:0 ME solution used for a particular analysis until it has passed QA/QC.
16. Correct for internal standard recovery—in this case, C13:0 ME (the software will do this for you if you have entered an ISTD). Only work up data when calibration quality of correlation is 0.999 or better. Always check software integration and retention times to ensure they are correct. Response factors can be significantly different between different fatty acids; use only respective response factors for quantification and for novel fatty acids that are not present in the calibration mix. Determine respective response factors by transferring existing response factors from structurally similar compounds. Suggestions are given in Table 2.
17. Quality control: a report should always include FAME content on a dry biomass basis. The average of replicate measurements as well as the relative standard deviation (RSD) and the relative percent difference (RPD) between samples should be included in a report. Every analysis should include a reagent blank and triplicates of a QC (reference biomass material) sample in each run. Choosing QC samples should be done by matching the FAME concentration levels and matrices of QC samples to the samples in the run. An acceptable average standard deviation among replicated analyses for FAME in cell biomass ranges from about 0.10 for samples with 10% FAME to 0.20 for samples with 20% FAME, which results in warning limits (2× standard deviation) ranging from 0.20 to 0.40 and control limits (3× standard deviation) ranging from 0.30 to 0.60. Plot the results of the control sample(s) on a control chart, and examine the chart for trends. Results outside of

Table 2
Response factors for compounds not present in calibration standard

Compound	Compound in calibration curve response factor to be used
All C16:1	C16:1n7
C16:2	C18:2n6
C16:3/C16:4	C18:3n6
C18:4n3	C20:4n6
C20:4n3	C20:4n6

upper or lower warning limits, or 2× standard deviation (95% confidence limits), are evidence of possible problems with the analytical method (either sample preparation or GC instrumentation). Results outside of upper or lower control limits, or 3× standard deviation (99% confidence limits), indicate loss of control, and results of the run should be discarded. Two consecutive analyses falling on one side of the mean between the warning limits and the control limits also indicate loss of control and root cause investigation should be initiated.

Acknowledgments

This work was supported by the US Department of Energy under Contract No. DE-AC36-08GO28308 with the National Renewable Energy Laboratory. Funding provided by the US DOE Office of Energy Efficiency and Renewable Energy Bioenergy Technologies Office. The US Government retains, and the publisher, by accepting the article for publication, acknowledges that the US Government retains a nonexclusive, paid-up, irrevocable, worldwide license to publish or reproduce the published form of this work, or allow others to do so, for US Government purposes.

References

1. Laurens LML, Nagle NJ, Davis R et al (2015) Acid-catalyzed algal biomass pretreatment for integrated lipid and carbohydrate-based biofuels production. Green Chem 17: 1145–1158
2. Dong T, Knoshaug EP, Davis R et al (2016) Combined algal processing: a novel integrated biorefinery process to produce algal biofuels and bioproducts. Algal Res 19:316–323
3. Kruger JS, Schutter S, Knoshaug EP et al (2024) De-risking pretreatment of microalgae to produce fuels and chemical co-products. Energy Fuels 38:8804–8816
4. Maltsev Y, Maltseva K (2021) Fatty acids of microalgae: diversity and applications. Rev Environ Sci Biotechnol 20:515–547
5. Laurens L, Quinn M, Van Wychen S et al (2012) Accurate and reliable quantification

of total microalgal fuel potential as fatty acid methyl esters by in situ transesterification. Anal Bioanal Chem 403:167–178
6. Saini RK, Prasad R, Shang X, Keum Y-S (2021) Advances in lipid extraction methods—a review. Int J Mol Sci 22:13643
7. Christie WW (1993) Advances in lipid methodology-two. Oily Press, Dundee
8. Iverson SJ, Lang SLC, Cooper MH (2001) Comparison of the Bligh and Dyer and Folch methods for total lipid determination in a broad range of marine tissue. Lipids 36:1283–1287
9. Bligh EG, Dyer WJ (1959) A rapid method of total lipid extraction and purification. Can J Biochem Physiol 37:911–917
10. Rowland SM, van Wychen S, Dong T et al (2024) High-resolution lipidomics reveals influence of biomass and pretreatment process on the composition of extracted algae oils as feedstock for sustainable aviation fuels. Energy Fuels 38:6547–6552
11. Harwood JL (1998) Membrane lipids in algae. In: Siegenthaler P-A, Murata N (eds) Lipids in photosynthesis: structure, function and genetics, vol 6. Springer, Heidelberg, pp 53–64. https://doi.org/10.1007/0-306-48087-5_3
12. Rainuzzo JR, Reitan KI, Olsen Y (1994) Effect of short and long term lipid enrichment on total lipids, lipid class and fatty acid composition in rotifiers. Aquac Int 2:19–32
13. Guschina IA, Harwood JL (2006) Lipids and lipid metabolism in eukaryotic algae. Prog Lipid Res 45:160–186
14. Van Wychen S, Rowland SM, Lesco KC et al (2021) Advanced mass balance characterization and fractionation of algal biomass composition. J Appl Phycol 33:2695–2708. https://doi.org/10.1007/s10811-021-02508-x
15. Laurens LML, Van Wychen S, McAllister JP et al (2014) Strain, biochemistry, and cultivation-dependent measurement variability of algal biomass composition. Anal Biochem 452: 86–95
16. Lepage G, Roy CC (1986) Direct transesterification of all classes of lipids in a one-step reaction. J Lipid Res 27:114–120
17. Haas MJ, Wagner K (2011) Simplifying biodiesel production: the direct or in situ transesterification of algal biomass. Eur J Lipid Sci Technol 113:1219–1229
18. Griffiths MJ, van Hille RP, Harrison STL (2010) Selection of direct transesterification as the preferred method for assay of fatty acid content of microalgae. Lipids 45:1053–1060
19. Van Wychen S, Sowell A, Ramirez K, Laurens LML (2023) Determination of total lipids as fatty acid methyl esters (FAME) by in situ transesterification - Laboratory Analytical Procedure (LAP) NREL/TP-2700-87501. https://www.nrel.gov/docs/fy24osti/87501.pdf
20. Van Wychen S, Laurens LML (2023) Determination of total solids and ash in algal biomass: Laboratory Analytical Procedure (LAP). NREL/TP 2700-87520. https://www.nrel.gov/docs/fy24osti/87520.pdf

Chapter 10

Total Carbohydrate Content Determination of Microalgal Biomass by Acid Hydrolysis Followed by Spectrophotometry or Liquid Chromatography

Stefanie Van Wychen and Lieve M.L. Laurens

Abstract

The carbohydrate fraction of microalgal biomass is complex and consists of a combination of neutral, acidic, amino sugars, and sugar alcohols. These carbohydrate amalgamations can be difficult to accurately quantify with the most analytical carbohydrate determination techniques. The method described here provides a straightforward two-step sulfuric acid hydrolysis followed by soluble carbohydrate quantification by either a spectrophotometric method (based on aldehyde functional group derivatization), for an overall combined monomeric sugar concentration determination, or liquid chromatography, specifically high-performance anion exchange method, for a more specific monomeric sugar profile.

Keywords Carbohydrate, Liquid chromatography, Spectrophotometry

1 Introduction

A full understanding of microalgal biomass composition is crucial to develop options for the utilization and full valorization of algae [1–3]. Algal carbohydrates can form the basis for both fuel products such as ethanol and high-value products from a fermentation pathway such as succinic acid in a combined biorefinery approach to maximize the full potential of the biomass [4–7]. In this context, the term "carbohydrates" refers to both monomers and polymers of sugars and sugar derivatives such as uronic acids and amino sugars, where the respective monomeric molecular composition is important when determining which product pathways may yield viable options [6, 8]. Polymers can have widely varying molecular weights depending on the degree of polymerization but collectively make up the largest fraction in typical terrestrial and algae biomass sources [1, 3, 9, 10] and can have both a structural and storage biological function (such as cellulose and starch, respectively, in higher plants).

Kristian Spilling (ed.), *Biofuels from Algae*, Methods in Molecular Biology, vol. 2996,
https://doi.org/10.1007/978-1-0716-5031-8_10,

Microalgal carbohydrates are composed of various monomeric building blocks, including uronic acids (glucuronic and galacturonic acid), neutral sugars (e.g., glucose, galactose, rhamnose, fucose, arabinose, ribose, mannose, xylose), and amino sugars (e.g., glucosamine, galactosamine). Once liberated from the larger polymeric structures, accurately quantifying these sugars can be incredibly challenging due to suboptimal separation on most common chromatography systems. Innovations in separations and hydrolysis of the polysaccharides to monomers have improved and make direct detection of monomeric constituents more reliable. There are occasions where a rapid estimation of total carbohydrates are desired, and a method is described here for the rapid estimation of bulk carbohydrates based on spectrophotometric absorption of derivatives of the saccharide monomers.

Spectrophotometric determination of carbohydrates is a technique that is easy to adopt and does not require specialized equipment. One of the most commonly used spectrophotometric methods is the phenol–sulfuric acid (PSA) method [11, 12], though susceptible to varying sensitivity to microalgae-specific carbohydrates and does not produce consistent peak shapes under visible spectroscopy [13]. The PSA method is also prone to interference from other microalgal biochemical constituents, such as protein and lipids. Because of the above limitations, we developed a novel use for the spectrophotometric 3-methyl-2-benzothiazolinone hydrazone (MBTH), which has been shown to give an almost equal response of different aldehyde sugars in solution [13]. At the same time, we demonstrated that the MBTH quantification matched HPLC results for complex mixtures of microalgae-specific monosaccharides. Spectrophotometric methods are quick and can provide a researcher with an overall, bulk, quantitation of a component such as carbohydrates in an extracted biomass sample.

To design and optimize downstream conversion and fermentation of carbohydrates, a detailed understanding of the monomeric makeup of the carbohydrate fraction is necessary [5]. Identification and quantification of the respective monomers by chromatography often provide critical information on monomers and their fate in selective upgrading pathways, but separations on conventional chromatography systems can be challenging and are reliant on full hydrolysis of the polymeric forms of the carbohydrates to respective monomers [1]. The utilization of high-performance anion-exchange chromatography (HPAEC), where a high pH environment forces neutral carbohydrates into an ionic form, has proven instrumental in providing the chromatographic resolution of the unusual monosaccharides typically found in algae [1, 3, 10]. We discuss here a method for carbohydrate quantification that has been optimized for whole algae biomass analytical hydrolysis to monosaccharides and subsequent detection of total bulk carbohydrates by either MBTH derivatization and spectrophotometric detection, or monomer separation by HPAEC with pulsed amperometric detection (PAD) [14].

2 Materials

2.1 Hydrolysis Apparatus

1. Analytical balance, accurate to 1 mg or 0.1 mg.
2. Vortex mixer.
3. Water bath, set to 30 ± 3°C.
4. Autoclave, suitable for autoclaving liquids, set to 121 ± 3°C.
5. 10 mL glass tubes and caps, heavy walled and capable of withstanding high pressure in an autoclave (*see* **Note 1**).
6. Adjustable pipet(s), covering 250 μL and 7 mL.
7. Disposable plastic ~3 mL syringes and compatible 0.2 μm nylon filters, if planning to filter as opposed to centrifuging to remove solids.

2.2 Spectrophotometric Apparatus

1. Vortex mixer.
2. Adjustable pipet(s), covering 20 μL to 1 mL.
3. Cuvettes—glass, quartz, or plastic.
4. 13 × 100 mm glass test tubes or equivalent.
5. Glass marbles or caps to cover or plug the 13 × 100 mm glass test tubes.
6. Digital dry block, capable of maintaining 80 ± 3°C (compatible with glass tubes).
7. Spectrophotometer, set to measure at 620 nm.

2.3 Liquid Chromatography Apparatus

1. Vortex mixer.
2. Plastic graduated pipet, 5 and 10 mL with 0.1 mL graduations (*see* **Note 2**).
3. Plastic 2 L graduated cylinder or volumetric flask (*see* **Note 2**).
4. Screw-top vials and presplit screw caps compatible with the HPAEC-PAD injector.
5. High-performance anion-exchange chromatography (HPAEC) system equipped with pulsed amperometric detection (PAD) and appropriate saccharide-specific column (*see* **Note 3**).

2.4 Spectrophotometric Analysis Reagents and Standards

Use 18.2 MegaOhm (MΩ) water for all preparations requiring water. Carefully follow all relevant chemical handling and disposal procedures.

1. 02% (w/w, specific gravity 1.6338 at 20°C) sulfuric acid, used for sample two-step hydrolysis.
2. Prepare glucose standard and calibration verification standard (CVS) stock solutions (these must be made up separately as the CVS is used to check the accuracy of the calibration curve) using a high-purity glucose standard. Weigh approximately

25 mg of glucose into a clean, tared, class A, 100 mL volumetric flask. Record the weight to the nearest 0.1 mg. Bring to volume with water and cap, and mix thoroughly. The standard and CVS working solution concentrations should be approximately 0.25 mg/mL glucose. Due to balance limits, start with a higher glucose stock concentration, and dilute to 0.25 mg/mL. Always prepare the standards and CVS from separate stocks so that you can check the accuracy of your calibration with the CVS. Glucose stock solutions can be stored in a freezer (−20°C) for up to 2 months or 5 days in the refrigerator (4°C).

3. Prepare a 0.5 M NaOH solution using sodium hydroxide pellets. Weigh out 4 g of sodium hydroxide pellets. Add about 50 mL of water to a 200 mL volumetric flask. Slowly add pellets and swirl to dissolve. The mixture will become warm as the pellets dissolve. Add increments of water to help the pellets dissolve. Bring to volume with water. Mix well in stoppered flask. Store solution at room temperature in a plastic (HDPE) container, and remake once a precipitate forms (looks like silvery flakes)—typically this happens after 2–3 weeks. Sodium hydroxide is a moderate health hazard.
4. Prepare a 3 mg/mL MBTH (3-methyl-2-benzothiazolinone hydrazine) solution in water, and store in the dark at 4°C for up to 1 week. MBTH is a moderate health hazard.
5. Prepare a 1 mg/mL DTT (dithiothreitol) solution in water, and store in the dark at 4°C for up to 1 week. Dithiothreitol is a moderate health hazard with mild flammability and reactivity.
6. Prepare 0.25 M HCl (hydrochloric acid) in a 200 mL volumetric flask, and store at room temperature in the dark for up to 1 month. Hydrochloric acid is a moderate health hazard. Add a small amount of water to the volumetric flask. Pipet 4.16 mL of concentrated HCl into the flask. Bring to volume with water and cap, and mix thoroughly.
7. Ferric solution: 0.5% ferric ammonium sulfate dodecahydrate, 0.5% sulfamic acid (w/v), and 0.25 M HCl. Dissolve 200 mg of ferric ammonium sulfate dodecahydrate and 200 mg of sulfamic acid into 40 mL of 0.25 M HCl. This solution may be stored at room temperature in the dark for up to 1 month. Sulfamic acid and ammonium ferric sulfate dodecahydrate are moderate health hazards.

2.5 Liquid Chromatography Reagents and Standards

Use 18.2 MΩ water for all preparations requiring water. Carefully follow all relevant chemical handling and disposal procedures.

1. 72% (w/w, specific gravity 1.6338 at 20°C) sulfuric acid, used for sample two-step hydrolysis.

2. Prepare a master mix of all the sugars [L(-)fucose, L(+)rhamnose monohydrate, L(+)arabinose, D(+)galactose, D(+) glucose, D(+)mannose, D(+)xylose, D(-) ribose, D(+)galacturonic acid monohydrate, D(-)glucuronic acid] in a 10 mL, class A, volumetric flask by weighing out each sugar at ~20 mg (all sugars should be weighed into the same volumetric flask). Record the exact weight of each sugar to 0.1 mg. Add approximately 5 mL water to the volumetric flask, and swirl or vortex to dissolve all sugars. Bring to volume with water, and mix well in the stoppered flask. Transfer 1 mL aliquots of the ~2 mg/mL master sugar mix into vials that may be sealed and frozen. The master sugar mix vials may be stored for up to 2 months in a freezer (−20°C).
3. Follow Thermo Fisher Scientific/Dionex TN 71 instructions to prepare 100 mM NaOH and 1 M sodium acetate (NaOAc)/100 mM NaOH eluent solutions. Store in plastic (high-density polyethylene, HDPE) solvent bottles.

3 Methods

3.1 Sample Preparation for Hydrolysis

1. Label 10 mL heavy-wall glass tubes for each of the samples and quality control samples (*see* **Note 4**).
2. Record the weight of all labeled tubes to the nearest 0.1 mg. Written labels should be dry before the weight is recorded.
3. Weigh 25 ± 2.5 mg of freeze-dried algal biomass into a labeled and tared 10 mL glass tube (*see* **Note 5**). Record the weight of the sample to the nearest 0.1 mg.
4. Dry samples in vials overnight to obtain an oven-dry weight before proceeding with the hydrolysis (*see* **Note 6**).

3.2 Sample Hydrolysis

1. Add 250 μL of 72% (w/w) sulfuric acid to each tube (*see* **Note 7**).
2. Place tubes in a 30 ± 3°C water bath with a water level set to just above the sample/acid level in the tubes. Incubate samples for 1 h (*see* **Note 8**).
3. After 1 h, remove tubes from the water bath, and add 7 mL of 18.2 MΩ water to each tube. This will bring the sulfuric acid concentration to 4% (w/w) (*see* **Note 9**).
4. Autoclave the sealed samples for 1 h at 121°C using the liquids (slow) setting (*see* **Note 10**).
5. After completion of autoclave cycle, allow samples to cool (*see* **Note 11**).

3.3 Spectrophotometric Determination of Monomeric Sugars

1. Hydrolyzed samples for spectrophotometric analysis should be prepared by separating the solids from the hydrolysate (*see* **Note 12**). See Fig. 1b for a typical hydrolysate.
2. Preheat a digital dry block that fits 13 × 100 mm glass test tubes (or equivalent reaction vials) to 80 ± 3°C (*see* **Note 13**).
3. Prepare standard and sample dilutions directly in the 13 × 100 mm glass reaction vials. Keep batches to no more than 40 reaction vials (*see* **Note 14**).
4. Prepare the MBTH working solution in a beaker or plastic tube (*see* **Note 15**).
5. Add 500 μL of 0.5 M NaOH to each glass tube and vortex.
6. Add 500 μL of MBTH working solution (from Step 4 above) to each glass tube, vortex carefully to mix, and cover the tubes with a glass marble or cap. Immediately place the tubes with solutions in the preheated dry block at 80°C for 15 ± 1 min (*see* **Note 16**).
7. After the 15 min incubation time, turn off the block, and immediately add 1 mL of the ferric solution while the glass tubes are still on the block (*see* **Note 17**).
8. Once the samples have reacted with the ferric solution, add 2.5 mL 18.2 MΩ water, and mix well by pipetting or vortexing.
9. Place an aliquot of the samples and standards into the appropriate cuvettes and obtain an absorbance on the spectrophotometer at 620 nm (*see* **Note 18**).
10. Use the glucose calibration curve and linear regression to quantify the total carbohydrate concentration in mg/mL (*see* **Note 19**).

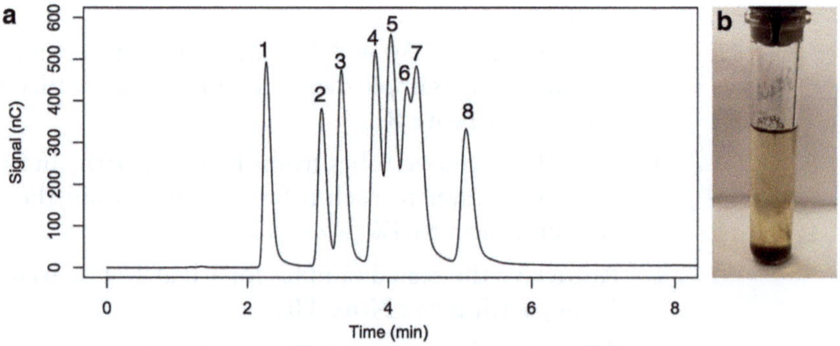

Fig. 1 (**a**) Representative HPAEC-PAD chromatogram of standards; 1. Fucose, 2. Rhamnose, 3. Arabinose, 4. Galactose, 5. Glucose, 6. Mannose, 7. Xylose, 8. Ribose. (**b**) Typical hydrolyzate of algal biomass in a 4% sulfuric acid solution in a total of 7.25 mL

3.4 Liquid Chromatography Determination of Monomeric Sugars

1. Hydrolyzed samples for spectrophotometric analysis should be prepared by separating the solids from the hydrolysate (*see* **Note 12**). See Fig. 1b for a typical hydrolysate.
2. Aliquot or dilute standards and samples into the labeled screw-top vials (*see* **Note 20**).
3. Prepare the HPAEC-PAD instrument for analysis (*see* **Note 21**).
4. Run standards and samples on HPAEC-PAD (*see* **Note 22**). See Fig. 1a for a typical chromatogram of the monomeric sugars.
5. Quantify the monomeric sugars by Chromeleon software (Thermo Fisher Scientific) (*see* **Note 23**).

4 Notes

1. Suggested tubes: VWR#89079-404—Do not use the caps that come with the tubes as they do not vent properly. Use a stopper and tear-away cap system. Stopper—VWR#15507-144. Tearaway aluminum cap—VWR#16171-851. The tubes may be washed and reused.
2. Use plastic (HDPE) pipets and LabWare when working with basic solutions.
3. HPAEC system equipped with pulsed amperometric detection (AgCl reference electrode, gold electrode), sodium hydroxide eluent generator (may also use manually prepared eluent if eluent generator is unavailable—follow Thermo Scientific TN 71), gradient pump and mixer, and the following column: PA-20 and PA-20 guard column (Thermo Scientific).
4. Run singlets or duplicates of each sample with a triplicate QC sample. Keep batches to a manageable size.
5. All microalgal biomass samples must be freeze-dried before analysis, and final moisture content should be <10%. Care should be taken to ensure a representative, and ground/homogenized (particle size <1 mm) sample is taken for analysis. Storage of samples should ideally be in an inert (nitrogen) atmosphere and at least at −20°C. Ensure sufficient quantities of the QC sample are available for QC charting over multiple analyses.
6. Dry samples in vials overnight in a 40°C oven under vacuum. After drying, take a final weight to calculate the moisture-free sample amount. Allow samples to cool in a desiccator under vacuum before recording a final weight. If no vacuum oven is

available, samples can be stored overnight in a vacuum desiccator prior to analysis, given that a corresponding moisture measurement is available for that material. Refer to 'Determination of Total Solids and Ash in Algal Biomass' for obtaining a moisture measurement [15]. Because moisture content can be variable in samples, it is important to correct for moisture and report carbohydrate content on a dry weight basis.

7. After all the samples have been weighed out, add 250 μL of 72% (w/w) sulfuric acid to each tube, and vortex to thoroughly mix the acid and biomass. Vortex samples carefully to ensure all solids remain at the bottom of the tube and immersed in the acid. If samples tend to clump, add sulfuric acid to one tube at a time, and vortex the tube immediately after the addition of acid. If the sulfuric acid is tinted yellow or brown, it needs to be replaced. Sulfuric acid is corrosive and is a moderate health hazard.
8. Vortex each tube vigorously every 5–10 min. Vortex samples carefully to ensure all solids remain at the bottom of the tube and immersed in acid.
9. Place a rubber stopper snugly in the top of the glass tube. Place a tear-away aluminum crimp cap on top of the rubber stopper. Crimp the caps tightly, and vortex to mix the contents thoroughly. Tubes may also be inverted to mix, ensure that the solids remain in the liquid, and do not end up stuck to the rubber stopper.
10. Place the tubes in a rack suitable for autoclaving. Be sure to place the rack in a tray for secondary containment. Secondary containment should contain any tubes and their contents if they were to break during the autoclave cycle.
11. After completion of the autoclave cycle, open the autoclave door, but allow tubes to cool for approximately 15 min before removing them from the autoclave. Once the samples are removed from the autoclave, allow the samples to cool for 30 min to 1 h or until tubes reach room temperature. Use caution when handling hot pressure tubes after removal from the autoclave. The pressurized tubes are a possible explosion hazard if not cooled properly.
12. Hydrolyzed samples for spectrophotometric and liquid chromatographic analysis should be prepared by separating the solids from the hydrolysate. Acidic solid-free hydrolysate may be stored in a refrigerator (4°C) for up to 24 h before being analyzed. If longer storage is necessary, store the hydrolysate (separated from the solids) in a freezer (−20°C) for up to 7 days. The sample should be filtered through a 0.2 μm nylon filter (use the disposable syringes and compatible filters) OR centrifuged to remove ALL solids.

13. If using a digital dry block for heating, regularly check the temperature of each block using a certified thermometer as the heating coils used in these systems can fail over time.

14. Create a six-point calibration with the 0.25 mg/mL glucose stock solution using the dilutions outlined in Table 1. Prepare dilutions directly into the glass reaction vials (13 × 100 mm or equivalent). Prepare the CVS at the center of the calibration range: 75 μL CVS glucose stock solution (0.25 mg/mL) and 425 μL water. Prepare the CVS dilution directly into the glass reaction vial. Prepare sample dilutions directly in the glass reaction vials: start with a 1:10 (50 μL hydrolysate) or greater dilution of the filtered/centrifuged acidic hydrolysate. Dilute the sample to a total volume of 0.5 mL. If the sample is known to be high in carbohydrates, start with a 1:20 dilution. The goal is to have the sample absorbance fall in the middle of the calibration curve. DO NOT use neutralized hydrolysate as the soluble salts leftover from the neutralization will affect the reaction. DO NOT use less than a 1:10 dilution on acidic hydrolysates as the excess acid can have an effect on the reaction. ALWAYS run a set of standards and a CVS with each batch of samples.

15. Prepare the MBTH working solution right before adding to samples: 1:1 (v/v) solution of 3 mg/mL MBTH and 1 mg/mL DTT. Prepare enough of the MBTH working solution to add 500 μL to each sample and standard in a batch. Keep batches to 40 total tubes or less.

16. Do not exceed the incubation time. Do not allow the samples and standards to sit around after addition of the NaOH or MBTH working solution—they must be placed on the hot block within a few minutes of adding these solutions.

Table 1
Suggested standard concentrations for the spectrophotometric method

Concentration of stock/ working (mg/mL)	Dilution stock + water (μL)	Final concentration (mg/mL)
0.25	0 + 500	0
0.25	20 + 480	0.010
0.25	30 + 470	0.015
0.25	50 + 450	0.025
0.25	75 + 425	0.0375
0.25	100 + 400	0.050

17. Once the ferric solution has been added, remove the glass tubes from the hot block, and carefully vortex to mix. Allow the samples and standards to react with the ferric solution for 10–15 min while the samples cool to room temperature. Tubes may remain open during this time. Allow them to cool in a hood to control fumes.
18. Once mixed, place an aliquot of the samples and standards into the appropriate cuvettes, and obtain an absorbance on the spectrophotometer at 620 nm. Take an absorbance within an hour after the final dilution step. Zero the spectrophotometer on 18.2 MΩ water. Do not zero the spectrophotometer on the 0 standard from the calibration standard set; this standard should be included as a point on the calibration curve.
19. Be sure to include the 0 standard as a point in the calibration. When building a linear calibration curve, do not force the curve through the origin. Remember to correct the sample carbohydrate concentration for the amount of hydrolysate that was used in the dilution (for a 1:10 in 0.5 mL, use 50 μL), and then get a total carbohydrate content by multiplying that concentration by the total volume from the hydrolysis—7.25 mL. Divide the carbohydrate content by the moisture-free weight of the sample to obtain the % monomeric sugar content on an ovendry weight basis. The target relative percent difference between duplicates or root mean square between triplicates should be <10%. CVS recovery should be between 95 and 105%. Track the results for the QC material over time.
20. Thaw a vial of the ~2 mg/mL master sugar mix. Dilute the master mix 1:10 to reach a final concentration of ~0.2 mg/mL. Create a six-point calibration with the ~0.2 mg/mL sugar mix using the dilutions outlined in Table 2. Prepare dilutions directly in the screw-top vials. Typical sample dilution for most microalgal samples (10–40% carbohydrates) is 1:20 or 50 μL filtered or centrifuged hydrolysate in 950 μL water.
21. Equilibrate the HPAEC-PAD at the following conditions: 4°C sample compartment temperature, 0.45 or 0.5 mL/min flow rate, 30–35 mM NaOH (either from the eluent generator or prepared from 100 mM NaOH eluent), isocratic, column temperature 35°C, and compartment/detector temperature 35°C. Equilibrate the HPAEC-PAD for at least 2 h before running standards and samples. Choose the isocratic condition that best separates the monomeric sugars eluting in the first 10 min; this may change based on column variability and column age.

Table 2
Suggested standard concentrations for the HPAEC-PAD method

Concentration of stock/ working (mg/mL)	Dilution stock + water (μL)	Final concentration (mg/mL)
0.2	10 + 990	0.002
0.2	20 + 980	0.004
0.2	50 + 950	0.01
0.2	100 + 900	0.02
0.2	200 + 800	0.04
0.2	300 + 700	0.06

22. Run standards and samples at the following conditions: sample compartment temperature of 4°C. Use a 0.45 or 0.5 mL/min flow rate, 30–35 mM NaOH isocratic eluent concentration for 10 min, followed by a ramping of the 1 M NaOAc/100 mM NaOH eluent from 2 to 19% (the remainder as water, 98–81%) with a linear (5) increase from 10 min until 30 min; turn off ramp; run 100 mM NaOH for 5 min; and then re-equilibrate for 10 min at the initial NaOH concentration (30–35 mM), with a total run time of 45 min. Column and compartment/detector temperature is 35°C. Use the Gold Standard PAD waveform (waveform A): E1, +0.1 V for 400 ms; E2, −2.0 V for 1 ms; E3, +0.6 V for 1 ms; and E4, −0.1 V for 6 ms. Run two to four water blanks before running standards and samples. Run a water and one of the standard levels (typically four) every ten samples to monitor system performance. Rerun the full calibration every 40–60 samples. Monitor sample areas during the run and adjust dilutions as necessary. This is very important as the calibration curves for the PAD are typically quadratic.
23. Do not force calibration curves through the origin. Monomeric sugar curves on the PAD are typically the best fit with a quadratic curve (second-order polynomial). Obtain the carbohydrate content as mg/mL for each sample. Remember to correct the sample carbohydrate concentration for the amount of hydrolysate that was used in the dilution (for a 1:20 in 1 mL, use 50 μL), and then get a total carbohydrate content by multiplying that concentration by the total volume from the hydrolysis—7.25 mL. Divide the carbohydrate content by the moisture-free weight of the sample to obtain the % monomeric sugar content on an oven dry weight basis. The target relative percent difference between duplicates or root mean square between triplicates should be <10%. CVS recovery should be between 95 and 105%. Track the results for the QC material over time.

Acknowledgments

This work was supported by the US Department of Energy under Contract No. DE-AC36-08GO28308 with the National Renewable Energy Laboratory. Funding provided by the US DOE Office of Energy Efficiency and Renewable Energy Bioenergy Technologies Office. The US Government retains, and the publisher, by accepting the article for publication, acknowledges that the US Government retains a nonexclusive, paid-up, irrevocable, worldwide license to publish or reproduce the published form of this work, or allow others to do so, for US Government purposes.

References

1. Templeton D, Quinn M, Van Wychen S et al (2012) Separation and quantification of microalgal carbohydrates. J Chromatogr A 1270:225–234
2. Williams PJLB, Laurens LML (2010) Microalgae as biodiesel & biomass feedstocks: review & analysis of the biochemistry, energetics & economics. Energy Environ Sci 3:554–590
3. Van Wychen S, Rowland SM, Lesco KC et al (2021) Advanced mass balance characterization and fractionation of algal biomass composition. J Appl Phycol 33:2695–2708. https://doi.org/10.1007/s10811-021-02508-x
4. Laurens LML, Nagle N, Davis R et al (2015) Acid-catalyzed algal biomass pretreatment for integrated lipid and carbohydrate-based biofuels production. Green Chem 17:1145–1158
5. Dong T, Knoshaug EP, Davis R et al (2016) Combined algal processing: a novel integrated biorefinery process to produce algal biofuels and bioproducts. Algal Res 19:316–323
6. Olguín EJ, Sánchez-Galván G, Arias-Olguín II et al (2022) Microalgae-based biorefineries: challenges and future trends to produce carbohydrate enriched biomass, high-added value products and bioactive compounds. Biology 11:1146
7. Margellou AG, Torofias SA, Iakovou G, Triantafyllidis KS (2024) Valorization of chlorella microalgae residual biomass via catalytic acid hydrolysis/dehydration and hydrogenolysis/hydrogenation. Catalysts 14:286
8. Moreira JB, Vaz BDS, Cardias BB et al (2022) Microalgae polysaccharides: an alternative source for food production and sustainable agriculture. Polysaccharides 3:441–457
9. Saeman JF, Bubl JL, Harris EE (1945) Quantitative saccharification of wood and cellulose. Ind Eng Chem Anal Ed 17:35–37
10. Hockin NL, Mock T, Mulholland F et al (2012) The response of diatom central carbon metabolism to nitrogen starvation is different from that of green algae and higher plants. Plant Physiol 158:299–312
11. Dubois M, Gilles KA, Hamilton JK (1956) Colorimetric method for determination of sugars and related substance. Anal Chem 28:350–356
12. Masuko T, Minami A, Iwasaki N et al (2005) Carbohydrate analysis by a phenol-sulfuric acid method in microplate format. Anal Biochem 339:69–72
13. Van Wychen S, Long W, Black SK, Laurens LML (2017) MBTH: a novel approach to rapid, spectrophotometric quantitation of total algal carbohydrates. Anal Biochem 518:90–93
14. Van Wychen S, Laurens LML (2023) Determination of total carbohydrates in algal biomass - Laboratory Analytical Procedure (LAP) NREL/TP-2700-87500. https://www.nrel.gov/docs/fy24osti/87500.pdf
15. Van Wychen S, Laurens LML (2023) Determination of total solids and ash in algal biomass: Laboratory Analytical Procedure (LAP) NREL/TP-2700-87520. https://www.nrel.gov/docs/fy24osti/87520.pdf

Chapter 11

Total Protein Content Determination of Microalgal Biomass by Elemental Nitrogen Analysis and a Dedicated Nitrogen-to-Protein Conversion Factor

Lieve M.L. Laurens, Jessica L. Olstad and David W. Templeton

Abstract

Accurately determining protein content is important in the valorization of algal biomass in food, feed, and fuel markets, where these values are used for component balance calculations. Using elemental nitrogen content to calculate protein concentration in the biomass is a well-accepted and widely practiced method and depends on developing a validated and appropriate nitrogen-to-protein conversion factor. The methodology reported here covers the quantitative assessment of the total nitrogen content of algal biomass and a description of the methodology that underpins the accurate *de novo* calculation of a dedicated nitrogen-to-protein conversion factor.

Keywords Amino acids, Nitrogen analysis, Nitrogen-to-protein conversion factor, Protein

1 Introduction

The protein content of algae has a strong influence in determining potential food, feed, co-product, and fuel uses for algal biomass. The protein content of microalgae can range from 7 to 40% [1–4] and can change dramatically over the course of the algal lifecycle. One source of biomass that is currently commercialized for food supplement sale is *Spirulina*, thanks to its high protein content (reported to be 50–60% of the biomass) [5].

Spectrophotometric protein measurement methods are ubiquitously used throughout the literature and can be useful for generating relative protein data or concentrations of protein in purified extract. However, colorimetric methods are significantly less useful for determining absolute protein values in complex materials such as whole algal biomass, where the values are important for component balance calculations. In the case of the Lowry spectrophotometric procedure, the color development is based on the reduction of the Folin reagent (Cu^{2+} to Cu^{+}) by aromatic residues and peptide bonds in protein,

Kristian Spilling (ed.), *Biofuels from Algae*, Methods in Molecular Biology, vol. 2996,
https://doi.org/10.1007/978-1-0716-5031-8_11,

after which the Cu^+ is chelated by bicinchoninic acid (BCA) to form the detected color [6–9]. Perhaps the biggest challenge to quantification of protein in whole biomass with a colorimetric method is the observation that the Folin reagent will react with other reducing substances in solution (including reducing sugars in solution), making this assay susceptible to algae species- and growth condition-specific interferences. These interferences also cause an incorrect high bias (up to twofold) in the calculated protein concentration data [9–11].

Perhaps the most direct method for protein determination is to measure the constituent amino acids (AAs) released after a harsh acid hydrolysis (often a 6 M HCl digestion for 24 h) followed by HPLC amino acid analysis [12]. This method has the advantage of breaking down the biological matrix and does not depend on selective removal of protein from the biomass. Free amino acids can also be detected in this method, in the same acid hydrolysis solution, along with AAs hydrolyzed from proteins. However, for complete AA analysis multiple (up to six) hydrolyses for each sample are needed to completely quantify the chemically diverse amino acids found in proteins [1, 3, 13]. Direct amino acid quantification after several hydrolyses per sample is useful for comprehensive total protein content and compositional analysis [14]; it is also expensive and time-consuming, and thus less applicable for screening or processing a large number of samples. There are literature reports that contain the amino acid profiles of many species of algae and represent different growth conditions [3, 15–18]. Several reports have come to the conclusion that a 4.78 custom nitrogen-to-protein conversion factor is appropriate and accurate for protein determination in microalgae [1, 15, 17].

Nitrogen analysis methods (%N) have been used to quickly determine crude protein values, thanks to the rapid and material-agnostic nature of a combustion method [19], though non-protein nitrogen can interfere with this calculation (*see* **Note 8**), and thus accuracy of this method relies heavily on an initial calibration of nitrogen content to measured amino acid content to determine a robust and material (or algae species) specific conversion factor. Reference methods are available to determine nitrogen in feedstocks such as animal feed, cellulosic biomass, and grains though the nitrogen-to-protein value used for this calculation tends to overestimate algal protein content [19]. Both a combustion (Dumas [20]) or a Kjeldahl method is acceptable [21], with the combustion method typically being preferred because of simplicity. In brief, for a combustion-based analysis of elemental nitrogen, weighed samples are combusted in oxygen at temperatures between 950 and 1100°C. The combustion products (including N and NO_x) are swept by a helium carrier gas through combustion catalysts, scrubbers, and a tube filled with reduced copper. The copper removes excess oxygen and reduces the NO_x to N_2. The N_2 is then separated from other gases on a chromatography column and measured with a thermal conductivity detector (TCD). By utilizing the completeness and specificity of direct AA analysis to determine algae-specific nitrogen-to-protein conversion factors and combining with a simpler %N method a useful high-throughput algae protein method emerges. Once a nitrogen-to-protein factor is determined, a rapid, total nitrogen determination can be carried out on a routine basis. The use of a dedicated nitrogen conversion factor then relates the nitrogen content back to a calculated total protein content.

2 Materials

2.1 Sample Preparation

1. Analytical balance, accurate to 1 mg or 0.1 mg.
2. Drying oven (40°C, vacuum) or lyophilizer (*see* **Note 1**).
3. Tin foil cups [e.g., Cat. No. 502-186-100 (small) or 502-397 (large), LECO].

2.2 Combustion Nitrogen Apparatus (See Note 2)

1. Furnace to maintain a minimum operating temperature of 950°C for combustion of the sample in pure (99.99%) oxygen.
2. An isolation system is typically in place to isolate nitrogen gas from other combustion products for subsequent measurement by a thermal conductivity detector (TCD). A device for converting NO_x products to N_2 or measuring N as NO_2 may be required and included in the system. In the case of the LECO TruSpec CHN Analyzer, the flow is passed through hot copper to remove residual oxygen and convert NO_x to N_2. Then, CO_2 and H_2O are removed via absorption using various sorbents (e.g., Lecosorb, Cat. No. 502-174-HAZ; Anhydrone, Cat. No. 501-171-HAZ, LECO), before the N_2 is analyzed via the TCD.

2.3 Reagents and Standards

Ethylenediaminetetraacetic acid (EDTA) is typically used as a reference standard to interpret the detector response as % nitrogen. The theoretical N concentration for EDTA is 9.586%.

3 Methods

3.1 Sample Preparation for Combustion Nitrogen Analysis

1. Record the weight of all labeled vials to the nearest 0.1 mg.
2. Weigh 50–100 mg (record exact weight to the nearest 0.1 mg) of freeze-dried and finely homogenized algal biomass into tared sample containers (*see* **Note 1**). Record the weight of the sample to the nearest 0.1 mg.
3. Dry samples in vials overnight to obtain an oven-dry weight before proceeding with the analysis to obtain a dry matter (DM) number to express the nitrogen concentration, alternatively after collecting a sample in the respective containers (*see* **Note 4**).

3.2 Instrument Operation and Quantification of Nitrogen Content

1. Turn on the furnace and bring all necessary delivered gases to their desired flow rates, then wait until furnaces have stabilized at their desired temperatures (*see* **Note 2**).
2. Enter sample identifiers in the software console and include in the sequence the reference standard, calibration check standard, and an appropriate number of blanks (*see* **Note 3**).
3. Transfer samples to the autosampler tray and set up the sequence run (*see* **Note 2**).

3.3 Data Analysis

1. The total nitrogen concentration of the biomass is determined via calibration of the instrument (*see* **Notes 5 and 6**), and the quality of the data is assessed via the rigorous quality control charting established by the respective laboratories (*see* **Note 7**).
2. The nitrogen content is used to determine the crude protein content using a nitrogen-to-protein conversion factor (*see* **Notes 8 and 9**). In the absence of a custom species-specific nitrogen-to-protein conversion factor, an average factor of 4.78 can be used for algal biomass:

$$^{1}\text{Algae Crude protein, } \%(w/w) = \%N \times 4.78$$

4 Notes

1. All microalgal biomass samples must be freeze-dried or oven-dried (40°C, vacuum oven) and fully homogenized before analysis and final moisture content should be <10%. Care should be taken to ensure a representative and ground/homogenized (ideally of a particle size <0.5 mm) sample is taken for analysis. Storage of samples should be under an inert (nitrogen) atmosphere and at least at −20°C.
2. The instrument used may differ between individual laboratories and available instrumentation. The methodology described here is instrument agnostic and can be translated to be used for any combustion nitrogen instrument available, assuming the respective analytical controls and QC guidelines are followed. Typical instrument settings for a LECO TruSpec CHN Analyzer are as follows; use of an external PC, which uses a Windows®-based software program; an autoloader; helium carrier gas (99.99% pure) at 35 psig; oxygen combustion gas (99.99% pure) at 35 psig; max oxygen flow of 6.3 L/min. In the case of our instrument, the furnace never gets turned off to

[1] This factor represents either an average factor from the literature [15, 17] or alternatively, a custom-calculated factor for the respective algae species can be used [1, 17].

avoid the risk of fracturing the combustion furnace tube after cooling and bringing back up to temperature; however, this may not be the case for other instrument manufacturers.

3. Run at least duplicates, ideally triplicates of each unknown sample with a triplicate QC sample. A quality control (QC) sample refers to a designated sample that is available in large enough quantities that replicate data can be collected over a long period of time and serves as a check for the analysis. A running average of the quantitative data should be kept as a control chart (*see* **Note 6**), which will make identifying and isolating outlier batches easier. Keep batches of new samples to a manageable size with respect to sample preparation time, and set up the instrument so that the reference standard (EDTA) and the QC biomass are run both at the beginning and at the end of a batch of samples, which will allow for the analysis and reporting of drift within a run. Good practice is to run five or more (up to ten) blanks until values are reproducible (standard deviation <0.01) and lower than 0.375% protein and then run four or more EDTA standards until three consecutive values agree within <0.15% protein of each other. Use the last three consecutive values to drift-correct for the EDTA.

4. Dry samples in vials overnight in a 40°C oven under vacuum. After drying, take a final weight to calculate the moisture-free sample amount. Allow samples to cool in a desiccator under vacuum before recording a final weight. If no vacuum oven is available, samples can be stored overnight in a vacuum desiccator prior to analysis, given that a corresponding moisture measurement is available for that material. Refer to other reference procedures for the determination of total solids and ash content in algal biomass for obtaining a moisture measurement. Because moisture content can be variable in samples, it is important to correct for moisture and report the nitrogen, and thus protein, content on a dry weight basis.

5. The nitrogen content is calculated based on the instrument calibration output and is expressed as a % of the dry weight of the sample that is used for combustion. A multi-point calibration curve is recommended, and when creating the calibration curve, at least three replicates of the standard (EDTA) should be used. Depending on the instrument and software in use, a drift factor may be used.

6. The combustion nitrogen system should be capable of measuring nitrogen in algal biomass materials containing between 0.2 and 20% nitrogen in the biomass, typically nitrogen concentration in algal biomass ranges from 6 to 12% for an early-harvest,

fully nutrient replete grown culture, to between 1 and 5% for biomass harvested from a nutrient-deplete grown culture.

7. Quality control: A report should always include both the nitrogen and the calculated protein content with the respective conversion factor. The average of replicate measurements as well as the relative standard deviation (RSD) and the relative percent difference (RPD) between samples should be included in a report. Every analysis should include a reagent blank, a standard such as EDTA, and ideally triplicates of a QC sample in each run. Choosing QC samples should be done by matching the nitrogen concentration levels and matrices of QC samples to the samples in the run. The overall accuracy of the system is demonstrated by making ten successive determinations of nitrogen in EDTA or another reference standard and ten successive determinations in a QC algal biomass material. If an EDTA sample within the run is more than ±0.2 from its calculated protein equivalent, a drift correction is performed. Means of determinations must be within 0.15 of the respective theoretical values, with standard deviations 0.15. An acceptable average standard deviation among replicated analyses for crude protein (CP) ranges from about 0.10 for samples with 10% CP to 0.20 for samples with 20% CP, which results in warning limits (2× standard deviation) ranging from 0.20 to 0.40 and control limits (3× standard deviation) ranging from 0.30 to 0.60. Plot the results of the control sample(s) on an X-control chart and examine the chart for trends. Results outside of upper or lower warning limits, or 2× standard deviation (95% confidence limits), are evidence of possible problems with the analytical system. Results outside of upper or lower control limits, or 3× standard deviation (99% confidence limits), indicate loss of control, and results of the run should be discarded. Two consecutive analyses falling on one side of the mean between the warning limits and the control limits also indicate loss of control.

8. Amino acid makeup and respective nitrogen concentration in each amino acid can be used to derive a nitrogen-to-protein conversion factor. Typically, for an algal biomass sample from a new species of algae, the total amino acid profile should be determined. In addition to a typical 24 h HCl hydrolysis for full protein hydrolysis to constituent amino acids, separate hydrolyses are needed for tryptophan (Trp) and for the sulfur-containing AAs (Met and Cys). Two additional hydrolysis time points (12 and 48 h) are run to account for AAs (Thr, Ser, and Tyr) that are partially degraded during hydrolysis plus a separate 2 h ammonia hydrolysis is run to determine the NH_3 released from Gln and Asn. Based on

the respective amino acid concentration of the new biomass sample the total nitrogen in the amino acids can be derived and thus form the basis for the conversion factor calculation. Furthermore, the nitrogen that is not captured in the amino acid portion of the protein in biomass is considered the nonprotein nitrogen (NPN) of the biomass and is taken into account in the utilization of a dedicated factor for algal biomass.

9. Basis for the calculation of a nitrogen-to-protein conversion factor: For use in food and feed nutritional analysis, it is suggested to use a method to determine a useful nitrogen-to-protein conversion factor [1, 13, 17]. It has previously been described in the literature how to determine upper (k_A) and lower limits (k_P) of such a factor and ultimately suggests combining these into a single, averaged factor (k). These factors can be easily defined mathematically as in Eqs. (1) and (2) below, though there are practical analytical and computational pitfalls to avoid when calculating these factors.

$$k_A = \left(\sum E_i\right) / \left(\sum D_i\right) \tag{1}$$

$$k_p = \sum E_i / \mathrm{N} \tag{2}$$

The term ΣE_i is the sum of the amino acid residues (the anhydrous form, accounting for the loss of H_2O during amino acid polymerization into proteins). Several tests are needed to completely analyze all AA found in biological samples, such as algae. The term ΣD_i is the sum of the nitrogen content of each of the AA residues plus the ammonia released from Gln and Asn during hydrolysis (see Table 1). A separate ammonia analysis using a 2 h HCl hydrolysis is needed to determine this value, which is normally included in the ΣD_i term. The term N refers to the %N found in the samples by combustion or Kjeldahl methods and includes any NPN found in the sample. In the absence of a dedicated and custom-calculated nitrogen-to-protein factor for a new species of algae, it is acceptable to use a previously published average factor of 4.78, which was determined as an average between at least 12 species, and at least 6 respective growth conditions for each species [15, 17]. We have determined that this factor is an appropriate average and takes the high NPN of algae into account and can be used throughout the microalgae composition field, though we do recommend as the matrix of the sample (biomass) origin changes, e.g. if wastewater is used to grow the algae, a periodic update to the nitrogen factor calculation is warranted.

Table 1

Values used to calculate nitrogen-to-protein conversion factors (k_A, k_P, k); AA = amino acid, AAA = anhydrous amino acid, residue = molecular weight of respective amino acid residue in a peptide or protein polymer

Amino acid	Abbr.	Molecular formula	Avg. Mass (Da)	Mass residue (Da)	ratio AAA/AA	%N in AA	%N in AAA
Aspartic acid	Asp	$C_4H_5NO_3$	133.1	115.1	86.50%	10.50%	12.20%
Threonine	Thr	$C_4H_7NO_2$	119.1	101.1	84.90%	11.80%	13.90%
Serine	Ser	$C_3H_5NO_2$	105.1	87.1	82.90%	13.30%	16.10%
Glutamic acid	Glu	$C_5H_7NO_3$	147.1	129.1	87.80%	9.50%	10.80%
Proline	Pro	C_5H_7NO	115.1	97.1	84.40%	12.20%	14.40%
Glycine	Gly	C_2H_3NO	75.1	57.1	76.00%	18.70%	24.60%
Alanine	Ala	C_3H_5NO	89.1	71.1	79.80%	15.70%	19.70%
Cysteine	Cys	C_3H_5NOS	121.2	103.1	85.10%	11.60%	13.60%
Valine	Val	C_5H_9NO	117.1	99.1	84.60%	12.00%	14.10%
Methionine	Met	C_5H_9NOS	149.2	131.2	87.90%	9.40%	10.70%
Isoleucine	Ile	$C_6H_{11}NO$	131.2	113.2	86.30%	10.70%	12.40%
Leucine	Leu	$C_6H_{11}NO$	131.2	113.2	86.30%	10.70%	12.40%
Tyrosine	Tyr	$C_9H_9NO_2$	181.2	163.2	90.10%	7.70%	8.60%
Phenylalanine	Phe	C_9H_9NO	165.2	147.2	89.10%	8.50%	9.50%
Tryptophan	Trp	$C_{11}H_{10}N_2O$	204.2	186.2	91.20%	13.70%	15.00%
Lysine	Lys	$C_6H_{12}N_2O$	146.2	128.2	87.70%	19.20%	21.90%
Histidine	His	$C_6H_7N_3O$	155.2	137.1	88.40%	27.10%	30.60%
Arginine	Arg	$C_6H_{12}N_4O$	174.2	156.2	89.70%	32.20%	35.90%
Asparagine	Asn	$C_4H_6N_2O_2$	132.1	114.1	86.40%	21.20%	24.60%
Glutamine	Gln	$C_5H_8N_2O_2$	146.1	128.1	87.70%	19.20%	21.90%
Ammonia	–	NH_3	17	16	94.10%	82.20%	87.40%
Water	–	H_2O	18	–	–	–	–
Mass hydrogen	–	H	1	–	–	–	–
Mass nitrogen	–	N	14	–	–	–	–

Acknowledgments

This work was supported by the U.S. Department of Energy under Contract No. DE-AC36-08GO28308 with the National Renewable Energy Laboratory. Funding was provided by the U.S. DOE Office of Energy Efficiency and Renewable Energy Bioenergy Technologies

Office. The U.S. Government retains, and the publisher, by accepting the article for publication, acknowledges that the U.S. Government retains a nonexclusive, paid-up, irrevocable, worldwide license to publish or reproduce the published form of this work, or allow others to do so, for U.S. Government purposes.

References

1. Templeton DW, Laurens LML (2015) Nitrogen-to-protein conversion factors revisited for applications of microalgal biomass conversion to food, feed and fuel. Algal Res 11:359–367
2. Pignolet O, Jubeau S, Vaca-Garcia C, Michaud P (2013) Highly valuable microalgae: biochemical and topological aspects. J Ind Microbiol Biotechnol 40:781–796
3. Lourenço SO, Barbarino E (1998) Distribution of intracellular nitrogen in marine microalgae: basis for the calculation of specific nitrogen-to-protein conversion factors. J Phycology 34:798–811
4. Lourenço SO, Barbarino E, Marquez UML, Aidar E (1998) Distribution of intracellular nitrogen in marine microalgae: basis for the calculation of specific nitrogen-to-protein conversion factors. J Phycology 34:798–811
5. Koru E (2012) Earth food Spirulina (Arthrospira): production and quality standards. In: El-Samragy Y (ed) Food addit intech https://doi.org/10.5772/31848
6. Garcia MDC, Lopez CVG, Fernandez FGA et al (2010) Protein measurements of microalgal and cyanobacterial biomass. Bioresour Technol 101:7587–7591
7. Servaites JC, Faeth JL, Sidhu SS (2012) A dye binding method for measurement of total protein in microalgae. Anal Biochem 421:75–80
8. Slocombe SP, Ross M, Thomas N et al (2013) A rapid and general method for measurement of protein in micro-algal biomass. Bioresour Technol 129:51–57
9. Cuchiaro H, Laurens LML (2019) Total protein analysis in algae via bulk amino acid detection: optimization of amino acid derivatization after hydrolysis with O-phthalaldehyde 3-mercaptopropionic acid (OPA-3MPA). J Agric Food Chem 67:5672–5679
10. Redinbaugh MG, Turley RB (1986) Adaptation of the bicinchoninic acid protein assay for use with microtiter plates and sucrose gradient fractions. Anal Biochem 153:267–271
11. Laurens LML, Van Wychen S, McAllister JP et al (2014) Strain, biochemistry, and cultivation-dependent measurement variability of algal biomass composition. Anal Biochem 452:86–95
12. AOAC (1997) Amino acids in feeds. In: Off method 994.12
13. Mosse J (1990) Nitrogen-to-protein conversion factor for ten cereals and six legumes or oilseeds: a reappraisal of its definition and determination. Var species seed protein content. J Agric Food Chem 38:18–24
14. Van Wychen S, Rowland SM, Lesco KC et al (2021) Advanced mass balance characterization and fractionation of algal biomass composition. J Appl Phycol. https://doi.org/10.1007/s10811-021-02508-x
15. Lourenço SO, Barbarino E, Lavín PL et al (2004) Distribution of intracellular nitrogen in marine microalgae: calculation of new nitrogen-to-protein conversion factors. Eur J Phycol 39:17–32
16. Lourenço SO, Barbarino E, De-Paula JC et al (2002) Amino acid composition, protein content and calculation of nitrogen-to-protein conversion factors for 19 tropical seaweeds. Phycol Res 50:233–241
17. Sägesser C, Kallfelz JM, Boulos S et al (2023) A novel approach for the protein determination in food-relevant microalgae. Bioresour Technol 390:129849
18. Wang Y, Tibbetts S, McGinn P (2021) Microalgae as sources of high-quality protein for human food and protein supplements. Foods 10:3002
19. AOAC (1990) Protein (crude) in animal feed: combustion method. Off Method 990:03
20. Ball MF, Meloni CR, Diaz MJ et al (1966) Nitrogen measurement in biological fluids and bone by an automatic Dumas technique. Anal Biochem 16:343–348
21. Kjeldahl J (1883) Neue Methode zur Bestimmung des Stickstoffs in organischen Körpern. Z Anal Chem 22:366–383

Chapter 12

Complete Acid-Based Hydrolysis Assay for Carbohydrate Quantification in Seaweed: A Species-Specific Optimized Approach

Emily T. Kostas, Stuart J. Wilkinson, Daniel A. White and David J. Cook

Abstract

Accurate quantification of the carbohydrate content of biomass is crucial for many biorefining processes. The most commonly followed protocol is typically a modification of the NREL-based assay (specifically designed for carbohydrate analysis from lignocellulosic biomass). However, this NREL protocol was revealed to be excessively thermochemically harsh for seaweed biomass. This can result in erroneously low total sugar quantification, as the reaction severity can degrade a proportion of the liberated sugars into decomposition products such as furans. Here we describe an optimization of the total acid hydrolysis protocol for accurate quantification of the carbohydrate content of seaweeds. Different species of seaweed can be accurately evaluated for their carbohydrate contents by following this optimized method.

Keywords Carbohydrate, Composition, Macroalgae, Mass balance, Optimization, Species

1 Introduction

In order to measure the carbohydrate content of seaweed biomass, modified versions of the NREL two-stage acid hydrolysis protocol [1] are typically applied [2–4]. The NREL carbohydrate assay is composed of two distinct stages and was initially developed for the quantification of total carbohydrate in lignocellulosic biomass (Fig. 1). The first stage of the assay is a low-temperature (37°C) treatment with concentrated acid (typically 12 M H_2SO_4), which induces the initial swelling of the biomass, followed by the fragmentation of the larger polysaccharide complexes found in the biomass. The second stage of the protocol involves using dilute acid (1 M H_2SO_4) but at a higher temperature (100°C), which then fully hydrolyzes the larger sugar fragments (oligosaccharides) into their constituent monomeric units (monosaccharides). These monosaccharide sugars can then be quantified either by ion-exchange chromatography, high-performance liquid chromatography, or gas chromatography–mass spectrometry

Kristian Spilling (ed.), *Biofuels from Algae*, Methods in Molecular Biology, vol. 2996,
https://doi.org/10.1007/978-1-0716-5031-8_12,

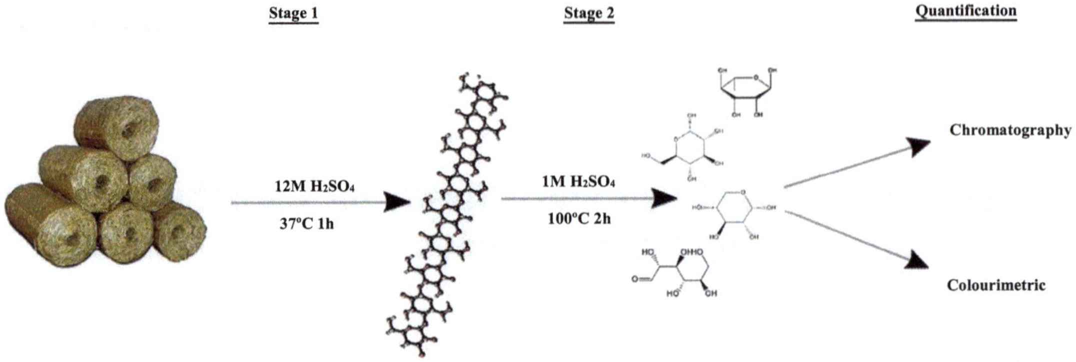

Fig. 1 Overview of the NREL assay [1] for determining carbohydrate analysis of lignocellulosic biomass. Stage 1: 1 mL of 12M H_2SO_4 is added to biomass (30 mg) and incubated at 37°C for 1h, liberating the larger polysaccharides from the biomass. Stage 2: acid strength diluted with distilled water to 1 M and incubated at 100°C for 2 h which hydrolyses the polysaccharides into their monomeric constituents. Quantification then achieved either using chromatographic or colorimetric methods

(ICS, HPLC, or GC-MS, respectively), or alternatively via colorimetric methods [5].

Seaweed polysaccharides are distinctly different from those of terrestrial plants in terms both of the sugar subunits (monomers) which are present and also the specific linkages between the monomers. In addition, seaweed-derived biomass is significantly less recalcitrant in nature when compared to lignocellulosic biomass. As such, the NREL-based protocol (which is specifically designed for lignocellulosic biomass) may be too thermochemically extreme for seaweed biomass. Therefore, use of the NREL assay in its original format may significantly underestimate the "true" carbohydrate content of the seaweed [6] through the potential degradation of liberated sugars into furan-based compounds [7]. We therefore evaluated the NREL protocol [1] to assess its suitability toward seaweed biomass and confirmed it to be thermochemically harsh [6]. Furthermore, we revealed that different species of seaweed require specific individual optimizations of the protocol for accurate total carbohydrate quantification. From our experimental work using *Laminaria digitata* (which was used as a benchmark species), we identified that optimization of stage 1 of the protocol (the low-temperature and concentrated acid phase) had a greater impact on the assay than was evident for stage 2 (the high-temperature dilute acid phase). The optimal conditions for obtaining the maximal sugar yields from *Laminaria digitata* required the use of 11 M H_2SO_4 originally rather than the 12 M H_2SO_4 used in the NREL protocol (Fig. 2), as this reduced the degree of furan generation. However, our experimental work concluded that stage 2 of the original NREL assay was already optimal and, as such, was not modified in any way (Fig. 3). The newly optimized stage 1 conditions

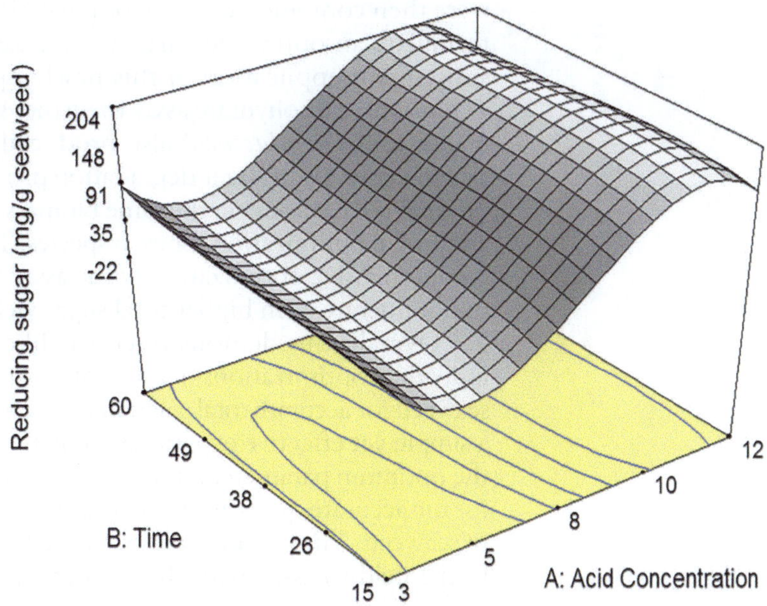

Fig. 2 3D response surface model showing the impact of different sulfuric acid concentrations (M) and reaction times (min) on the release of reducing sugars (mg/g) from *L. digitata*. Quantification using the phenol-sulfuric acid (Dubois) colorimetric assay. Model R^2: 0.56

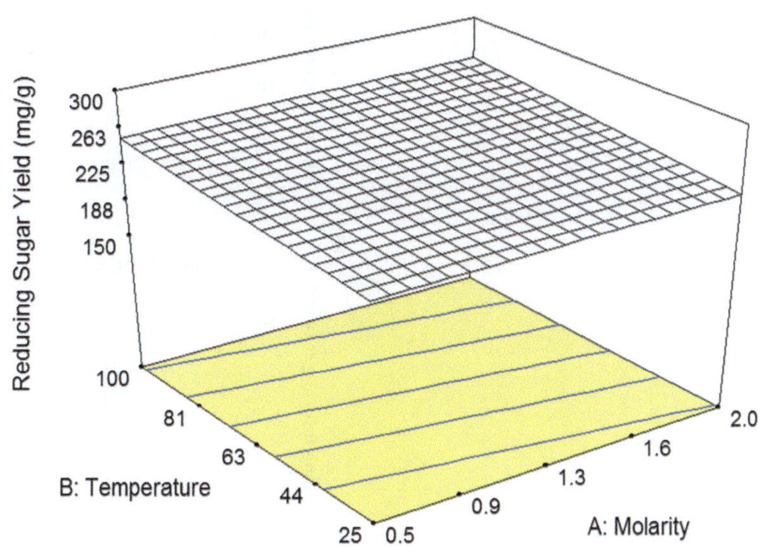

Fig. 3 3D response surface model showing the effect of simultaneous variation of acid concentration (M) and reaction temperature (°C) on the release of reducing sugars (%) from *L. digitata* during stage 2 of the total acid hydrolysis protocol. Stage 2 incubation time: 2 h. Reducing sugars quantified by phenol-sulfuric colorimetric assay. SCP involved using previously optimized stage 1 parameters: 11 M H_2SO_4, 37°C, 1 h. Model R^2: 0.17

were then combined with the original NREL stage 2 conditions to formulate an optimized carbohydrate assay (for *L. digitata*). The subsequent application of this newly optimized (specifically for *L. digitata*) carbohydrate assay to further seaweed species (*Chondrus crispus* and *Ulva lactuca*) also produced higher total sugar yields and lower levels of sugar degradation products than when using the original NREL assay on the same biomass (Fig. 4), even without any additional optimization for each species. This suggested that further comprehensive optimization of the assay for each individual species might liberate even higher total sugar yields.

Overall, this demonstrated the likely importance of specific individual optimizations of the protocol for each different species of seaweed for accurate total sugar quantification. Here, we demonstrate a simple yet effective experimental methodology to help determine the optimum parameters (for stage 1 of the acid hydrolysis protocol) for the accurate quantification of carbohydrate content in any species of seaweed. The protocol has been used by multiple research groups in the field to assist with the accurate quantification of sugars and

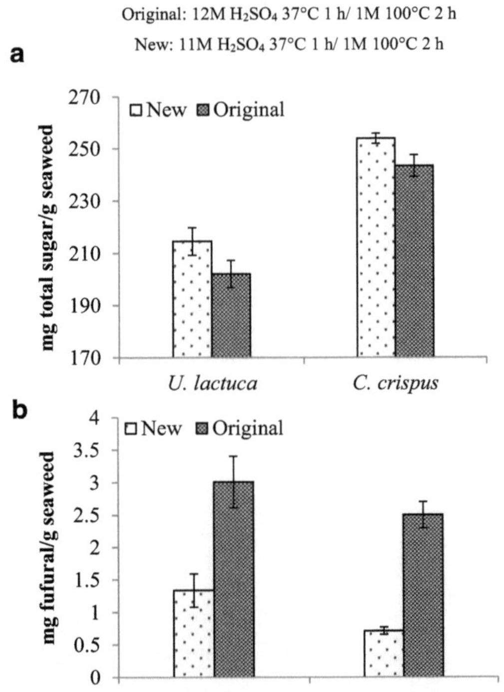

Fig. 4 Comparison between NREL assay and optimized carbohydrate quantification assay for total sugars measurement of the seaweed species *U. lactuca* and *C. crispus*. (**a**) Total sugar yields (sum of mannitol, fucose, arabinose, galactose, glucose and xylose; quantified by HPAEC-PAD) from both *U. lactuca* and *C. crispus*. (**b**) Furfural concentrations generated from both *U. lactuca* and *C. crispus* from both the original control protocol and the newly optimized protocol

carbohydrates from numerous species of seaweed, including *Sargassum* spp [8], *Alaria esculenta, Fucus distichus, Laminaria digitata, Saccharina latissimi,* and *Palmaria palmata* [9], and *Ascophyllum nodosum, Chondrus crispus, Fucus ceranoides, Fucus vesiculosus, Himanthalia elongata, Laminaria hyperborea, Pelvetia canaliculata, Rhizoclonium riparium, Sargassum muticum, Solieria chordalis, Ulva intestinalis,* and *Ulva lactuca* [10].

2 Materials

Prepare all reagents and solutions and perform all dilutions using ultrapure reverse osmosis (RO) water to achieve a sensitivity of > 18 MΩ cm (at 25°C) and using analytical-grade reagents unless otherwise stated. Caution must be used when handling hazardous reagents such as concentrated acid and phenol (if using the colorimetric sugar quantification). In addition, we advocate the use of a fume hood or cabinet when dispensing such hazardous reagents. All seaweed biomass must be dried in a fan-assisted oven at ca. 80°C for a minimum of 48 h prior to use. Dried biomass must be ground up or milled to ensure adequate homogeneity prior to analysis. Ball milling (or the use of a similar technique) is advised to produce a fine powder of the biomass that aids in the accurate weighing of small quantities of material. Once dried and milled, the biomass can be stored at room temperature in an airtight container. The suggested range of acid concentrations evaluated for optimization of stage 1 of the total acid hydrolysis is 3–12 M, with dilutions performed using RO water to prepare the reagents prior to use. For the optimization of stage 1 of the protocol for your specific biomass type, we recommend a minimum series of acid concentrations of 3, 6, 9, and 12 M (*see* **Note 1**). However, a more comprehensive optimization can be achieved using the experimental run conditions outlined in Table 1. Store reagents in a hazardous chemical cabinet, such as that designed for flammables.

1. RO water.
2. Algal biomass.
3. Sulfuric acid (H_2SO_4).
4. A 2.5% (w/v) phenol solution is prepared by adding 2.5 g phenol to 100 mL RO water (*see* **Note 2**). Prepare in a fume hood, and store in a hazardous chemical cabinet once prepared.
5. Acetic acid.
6. Methanol.
7. Analytical balance.
8. High-pressure liquid chromatography (HPLC).

Table 1

Experimental design used to optimize stage 1 (37°C) of the total acid hydrolysis methodology for quantifying carbohydrates in seaweed. Optimization conducted through screening different sulfuric acid concentrations (3–12 M) and reaction times (15–60 min) at 37°C, according to a *D-optimal* design space

	Factor 1	Factor 2		Factor 1	Factor 2
Run	A: H_2SO_4 Acid Conc	B: Time at 37°C	Run	A: H_2SO_4 Acid Conc	B: Time at 37°C
	(M)	(min)		(M)	(min)
1	12	15	22	10	60
2	3	15	23	10	15
3	3	15	24	12	60
4	7	15	25	12	60
5	7	35	26	12	15
6	3	60	27	12	15
7	5	45	28	8	15
8	12	35	29	8	15
9	7	35	30	8	60
10	5	25	31	12	35
11	12	60	32	8	35
12	10	25	33	10	15
13	3	35	34	10	25
14	3	60	35	10	25
15	12	15	36	10	45
16	7	60	37	10	45
17	12	60	38	10	60
18	7	35	39	9	20
19	7	35	40	9	50
20	10	45	41	11	20
21	5	60	42	11	50

9. C18 Techsphere ODS column (5 μm, 4.6 mm _ 250 mm; HPLC Technologies, UK).
10. Water bath(s) (0–100°C).
11. Pyrex reaction tubes.
12. Test-tube rack.
13. Luer-Lok-type syringes.
14. Syringe filters (0.45 μm).

15. 15 mL plastic centrifuge tubes.
16. Suitable HPLC vials corresponding to the HPLC used.

3 Methods

Carry out all procedures at room temperature unless otherwise specified.

3.1 Total Acid Hydrolysis Assay

1. Accurately weigh 30 mg of biomass into a 50 mL screw-capped reaction tube (e.g., Pyrex, UK) using an analytical balance accurate to four decimal places.
2. Carefully add 1 mL of the required concentration of H_2SO_4 (suggested acid concentration range is 3–12 M; see Sect. 2, item 2 for guidelines) into each reaction vessel (*see* **Note 3**) ensuring the biomass is completely covered or submerged within the acid. Carefully ensure the lids are tight on all reaction vessels while ensuring that the biomass stays submerged within the acid.
3. For stage 1 of the assay, place all reaction vessels in a test-tube rack and place the rack in a 37°C water bath (or incubator) for 1 h (*see* **Note 4**).
4. Upon completion of the 1 h incubation period at 37°C, remove the test-tube rack and add the required aliquot of RO water (*see* **Note 5**) to achieve dilution of the acid concentration to 1 M.
5. For stage 2 of the assay, carefully return the reaction vessel caps, return the vessels to the test-tube rack, and then place the rack in a 100°C water bath (or incubator) for 2 h.
6. After 2 h incubation at 100°C, remove the test-tube rack and place it into an ice-cold water bath (or a similarly suitable tray) to cool, and allow any particulates or suspended biomass to settle.
7. After cooling, unscrew the reaction vessel caps and carefully remove an aliquot (ca. 2 mL) of the liquid phase using a disposable plastic Luer-Lok-type syringe (2–5 mL capacity typically), while attempting to minimize uptake of solid particles.
8. Syringe filter each sample into a clean glass test tube using a Luer-Lok-type syringe filter of <0.45 μm retention size.
9. For the HPLC quantification of sugar degradation products (see Sect. 3.2 for details), remove an aliquot (0.5–1.0 mL) of each of the syringe-filtered samples, and place them into the appropriate HPLC vials for the system to be used.

10. For the HPAEC-PAD quantification of sugars (see Sect. 3.3), the syringe-filtered samples will need to be diluted due to the sensitivity of the detection system as care must be taken not to overload the detector. The precise dilution factor required is dependent upon the concentration of the sugars within the sample, which is of course unknown at this stage. However, typically, a 1000× dilution is first evaluated (*see* **Note 6**).
11. For achieving a 1000× dilution for HPAEC-PAD, mix 100 μL of each of the filtered samples with 9.9 mL of RO water in a 15 mL plastic centrifuge tube or a similar screw-capped test tube that can be inverted to ensure suitable mixing.
12. Transfer a 1 mL aliquot of this dilution into a suitable HPLC vial for HPAEC-PAD analysis.
13. If using the simpler colorimetric-based determination of total reducing sugars, a 100 μL aliquot of each filtered sample is required (see Sect. 3.4).

3.2 Quantification of Sugar Degradation Products by HPLC

1. This protocol utilizes the method described in [11]. The HPLC system requires UV detection at 270–290 nm.
2. The use of a PDA (photodiode array) variant of UV detection is highly recommended in order to provide additional spectral data to further aid the identification of any peaks detected rather than relying purely on comparison of retention times with those of authentic standards.
3. The HPLC column used is a C18 Techsphere ODS column (5 μm, 4.6 mm × 250 mm; HPLC Technologies, UK) at ambient temperature and using gradient elution.
4. The mobile phase is a binary mixture of 1% acetic acid (solvent A) and methanol (solvent B) running at a flow rate of 0.5 mL/min.
5. The gradient elution ramp is from 20 to 50% methanol over 30 min, with a 100% methanol column-cleaning phase (for 1 min) and a 9 min re-equilibration period (at 20% methanol) prior to the next injection.
6. The sample injection volume is 10 μL.
7. Quantification is performed by comparison of peak areas of authentic standards (0.1–1.0 g/L concentration range, dissolved directly in RO water), including 5-HMF, furoic acid, and furfural (*see* **Note 8**).

3.3 Quantification of Sugars by HPAEC-PAD

1. This protocol utilizes the method described in [12].
2. Dilute samples 1000× prior to analysis (see Sect. 3.1, Step 11).
3. The HPAEC system uses pulsed amperometric electrochemical detection (PAD).

4. Separation is achieved using a pa20 column (150 mm × 3.0 mm; Dionex, USA).
5. The system is operated using isocratic elution with 10 mM NaOH at a 0.5 mL/min flow rate with a column regeneration step using 200 mM NaOH at 0.5 mL/min after each injection.
6. Quantification is performed by comparison of peak areas of authentic standards of mannitol, fucose, galactose, arabinose, galactose, glucose, and xylose (0.0625–1 g/L concentration range, dissolved directly in RO water; *see* **Note 8**). Dilute standards 1000× (with RO water) prior to analysis.

3.4 Quantification of Total Reducing Sugars Using a Colorimetric Assay

1. This protocol utilizes the phenol-sulfuric acid method described in [5].
2. Transfer a 100 μL aliquot of each sample into a 5 mL test tube.
3. Carefully add 1 mL phenol solution (2.5% v/v) and 2.5 mL concentrated (72% v/v) H_2SO_4 (*see* **Note** 7).
4. Transfer 2 mL aliquots of this reaction mixture into quartz cuvettes.
5. Read absorbances at 490 nm using a spectrophotometer (after zeroing the instrument using water as a blank).
6. Quantification is then achieved by comparison to the absorbance (at 490 nm) of authentic glucose standards (0, 200, 400, 600, 800, and 1000 μg/mL).

4 Notes

1. Concentrated H_2SO_4 is highly hazardous, and we advocate that all handling to be conducted in a fume hood (cabinet). In addition, the dilution process for the acid to achieve the desired acid molarities is highly exothermic, and care should be taken to allow solutions to cool before use.
2. Phenol is also a highly hazardous compound and should always be handled in a fume hood (cabinet).
3. The use of automatic Dispensette pipettes (BrandTech Scientific, USA) is highly recommended for the rapid, reproducible, and safe dispensing of reagents.
4. Set all water baths or incubators to the correct temperature prior to commencing the assay, as attemperation for the 100°C water bath may take >2 h.
5. Example: dilution of acid from 12 to 1 M would involve careful addition of 11 mL RO water.

6. If using the 1000× dilution factor for samples (for HPAEC-PAD-based total sugar analysis) and the subsequent detection response is poor (poor signal-to-noise ratio through inadequately small peak sizes), then a more concentrated sample dilution can be evaluated, such as 100×.
7. The reaction of H_2SO_4 with phenol is highly exothermic; therefore, care should be taken. We advocate that all additions of reagents (reactions) are conducted in a fume hood.
8. A large stock of HPLC and HPAEC-PAD standards may be produced, filtered through <0.45 μm filters, labeled, and stored at −20°C for up to 3 months if frequent analysis is required. The stock can be removed from the freezer, thawed, and vortexed prior to use.

References

1. Sluiter A, Hames B, Ruiz R et al (2008) Determination of structural carbohydrates and lignin in biomass. NREL Laboratory Analytical Procedure 1617
2. van der Wal H, Sperber BL, Houweling-Tan B et al (2013) Production of acetone, butanol, and ethanol from biomass of the green seaweed *Ulva lactuca*. Bioresour Technol 128:431–437
3. Ge L, Wang P, Mou H (2011) Study on saccharification techniques of seaweed wastes for the transformation of ethanol. Renew Energy 36(1):84–89
4. Trivedi N, Gupta V, Reddy C, Jha B (2013) Enzymatic hydrolysis and production of bioethanol from common macrophytic green alga *Ulva fasciata* Delile. Bioresour Technol 150:106–112
5. Dubois M, Gilles KA, Hamilton JK et al (1956) Colorimetric method for determination of sugars and related substances. AnalChem 28(3):350–356. https://doi.org/10.1021/Ac60111a017
6. Kostas ET, Wilkinson SJ, White DA, Cook DJ (2016) Optimization of a total acid hydrolysis based protocol for the quantification of carbohydrate in macroalgae. J Algal Biomass Utln 7(1):21–36
7. Wilkinson S, Smart K, Cook D (2014) A comparison of dilute acid and alkali catalysed hydrothermal pre-treatments for bioethanol production from brewers spent grains. J Am Soc Brew Chem 72(2):143–153
8. Vázquez-Delfín E, Freile-Pelegrín Y, Salazar-Garibay A et al (2021) Species composition and chemical characterization of *Sargassum influx* at six different locations along the Mexican Caribbean coast. Sci Total Environ 795:148852
9. Singh A, Pal B, Singh KS (2024) Carbohydrate and pigment composition of macroalgae in a kelp-dominated Arctic fjord. Regional Stud Mar Sci 77:103644
10. Raikova S, Le CD, Beacham TA et al (2017) Towards a marine biorefinery through the hydrothermal liquefaction of macroalgae native to the United Kingdom. Biomass Bioenergy 107:244–253
11. Wilkinson S, Smart K, Cook D (2014) Optimisation of alkaline reagent based chemical pre-treatment of brewers spent grains for bioethanol production. Ind Crop Prod 62:219–227
12. Kostas ET, White DA, Du C, Cook DJ (2016) Selection of yeast strains for bioethanol production from UK seaweeds. J Appl Phycol 28(2):1427–1441

Check for updates

Chapter 13

Analysis of Brown Algal Sugars Using Ion-Exchange High-Performance Liquid Chromatography

Ingrid Maribu and Jan Arne Arnesen

Abstract

Several chromatographic methods have been developed for carbohydrate analysis. The difference in refractive index (RI) is often used for sugar detection due to the lack of chromophores or fluorophores in native sugars. This method uses anion-exchange chromatography and an RI detector to analyze sugars in brown algae. The identification and quantification of sugars in an unknown sample are possible due to the linear regression of sugar standards used to estimate the concentration of specific sugars. This method effectively separates and quantifies sugars in hydrolyzed brown algae samples. Regular calibration and maintenance are crucial for accurate results. The method gives a good estimation of the sugar composition and is reliable, provided that proper calibration and maintenance are performed.

Keywords Carbohydrates, Brown algae, High-performance liquid chromatography, Refractive index detector, Sugars, Ion-exchange

1 Introduction

Several chromatographic methods have been developed for carbohydrate analysis. Ion exchange chromatography is most widely used to separate the monosaccharides. As native sugars do not contain a chromophore or fluorophore, the difference in refractive index (RI) is often used in sugar detection [1]. RI detector is a universal bulk property detector that measures the difference in RI against a reference [1, 2]. RI detectors are often preferred as they detect compounds well despite being present in low amounts in the sample. However, RI detectors have a high sensitivity where a small change in temperature, pressure or mobile phase can cause a baseline shift. This can result in difficulties with reproducibility with gradients. Therefore, other detectors sometimes replace RI detectors [1].

Ion exchange chromatography (IEX) separates ions and molecules based on their difference in charge. It uses a stationary phase and a mobile phase. The stationary phase is a charged solid surface, often made up of resin or a gel, that will interact with molecules of

Kristian Spilling (ed.), *Biofuels from Algae*, Methods in Molecular Biology, vol. 2996,
https://doi.org/10.1007/978-1-0716-5031-8_13,

opposite charge from the mobile phase. IEX is divided into cationic or anionic, where the molecules are separated based on their charge [3].

The carbohydrate content is analyzed in hydrolyzed material, and the choice of optimal hydrolysis method will vary depending on the material. Hydrolysis with HCl is preferred for materials rich in uronic acids, while hydrolysis with H_2SO_4 is preferred for materials rich in lignocellulose. For quantitative determination of carbohydrates in seaweed, two-step hydrolysis with H_2SO_4 is suggested to be the best method [4]. The two-step sulfuric acid hydrolysis method was modified and adapted from Sterner et al. and Nguyen et al. [5, 6].

The method is adapted from Sharma et al. [7]. This analysis uses anion-exchanged high-performance liquid chromatography equipped with a LC-20AT apparatus, a CTO-20A column oven, and a RID-20A refractive index detector to determine the sugar content in seaweed samples (all from Shimadzu, Kyoto, Japan). The sugars are separated using a 300 × 7.8 mm Rezex ROA-Organic acid H+ (8% cross-linked hydrogen) analytical column equipped with a Carbo-H 4 × 3.0 mm guard column (all from Phenomenex, Torrence, CA, USA). The column's stationary phase consists of sulfonated styrene divinyl benzene, giving the column a negatively charged environment. This allows molecules with a neutral or positive charge to pass through the column and be separated based on their pKa [3, 8].

2 Materials

2.1 Hydrolysis

1. Glass tubes that can withstand 121°C.
2. Precision balance.
3. Concentrated H_2SO_4 (95–97%).
4. Heating block.
5. Distilled water.
6. Autoclave.
7. Centrifuge.

2.2 HPLC-RID

1. HPLC vials.
2. HPLC apparatus.
3. Column oven.
4. Refractive index detector.
5. 300 × 7.8 mm Rezex ROA-Organic acid H+ column analytical column.
6. Carbo-H 4 × 3.0 mm guard column.
7. Precision balance.
8. Glass bottles.
9. Concentrated H_2SO_4 (95–97%).

10. Distilled water.
11. Falcon tubes (50 or 15 mL).
12. Sugar standards: Glucose, xylose, mannitol, fucose.
13. Vortex.
14. Freezer (–20°C).

3 Methods

All solutions are stored at room temperature unless otherwise stated.

3.1 Hydrolysis (see Note 1)

1. Prepare a 72% H_2SO_4 (w/w) solution, by adding 72 g H_2SO_4 to 28 g distilled water.
2. Add 300 μL of 72% (w/w) H_2SO_4 to 30 mg lyophilized material.
3. Incubate the samples in a heating block at 30°C for 1 h.
4. Dilute the samples to 4% (w/w) H_2SO_4 by adding 5.4 mL distilled water.
5. Autoclave the samples at 121°C for 40 min (*see* **Notes 2 and 3**).
6. Centrifuge samples at 3000 *g* for 10 min.
7. Add 100 μL of the hydrolyzed samples to HPLC vials.

3.2 HPLC-RID

1. Make a 0.5 M H_2SO_4 stock solution by adding 26.6 mL H_2SO_4 to 973.4 mL distilled water and store at room temperature.
2. Prepare 5 mM H_2SO_4 mobile phase (*see* **Note 4**), by adding 10 mL from the 0.5 M H_2SO_4 stock solution to 990 mL distilled water and store it at room temperature.
3. Make 50 mg/mL stock solutions of sugar standards (*see* **Note 5**). Add first 2.5 g of the sugar and adjust the volume to 50 mL with distilled water and freeze them in several smaller tubes at −20°C until next time (to avoid freezing and thawing of the same solution several times).
4. Thaw sugar stock solution if needed (*see* **Note 6**).
5. Dilute the stock solution to make standard solutions of 2.5, 1.0, 0.5, 0.25, and 0.1 mg/mL of each sugar (*see* **Note 7**):

   2.5 mg/mL: Add 50 μL stock solution to 950 μL distilled water.

   1.0 mg/mL: Add 20 μL stock solution to 980 μL distilled water.

   0.5 mg/mL: Add 10 μL stock solution to 990 μL distilled water.

   0.25 mg/mL: Add 5 μL stock solution to 995 μL distilled water.

   0.1mg/mL: Add 2 μL stock solution to 998 μL distilled water.
6. Heat the column oven to 65°C (*see* **Note 8**).

7. Set the injection volume to 10 μL.
8. Choose isocratic mode with a flow rate of 0.6 mL/min.
9. Make a batch file (*see* **Note 9**).
10. Add 100 μL blank/standard/sample to separate HPLC vials.
11. Place the tray into the LC apparatus.
12. Check the lamp voltage (total.en between 7000 and 7200) (*see* **Note 10**).
13. Start the run.
14. Open data in the appropriate software.
15. Use linear regression to calculate the formula for a straight line based on the area of the peaks (Table 1) (*see* **Note 11**).

Table 1
An example of the retention times in minutes and area for the different standards at concentrations ranging from 0.1 mg/mL to 2.5 mg/mL, which is used to estimate the amount of sugars present in an unknown sample

Sugar	Concentration (mg/mL)	Retention time (min)	Area
Glucose	2.5	10.730	331519
	1.0	10.735	134346
	0.5	10.738	65834
	0.25	10.741	31867
	0.1	10.741	16305
Xylose	2.5	11.429	312578
	1.0	11.430	124028
	0.5	11.430	60382
	0.25	11.428	30036
	0.1	11.431	12870
Mannitol	2.5	11.794	330027
	1.0	11.794	129957
	0.5	11.792	63554
	0.25	11.796	33193
	0.1	11.800	12163
Fucose	2.5	13.027	1365890
	1.0	13.027	537743
	0.5	13.031	266346
	0.25	13.031	130211
	0.1	13.038	51482

16. Identify specific sugars based on their retention times (Fig. 1) (*see* **Notes 12–14**).
17. Calculate sugar concentration using the calculated formula for the linear regression (Fig. 2). An example is given below:

$$\left(\frac{Area + 6784.6}{548413}\right) = mg/mL\ Fucose \tag{1}$$

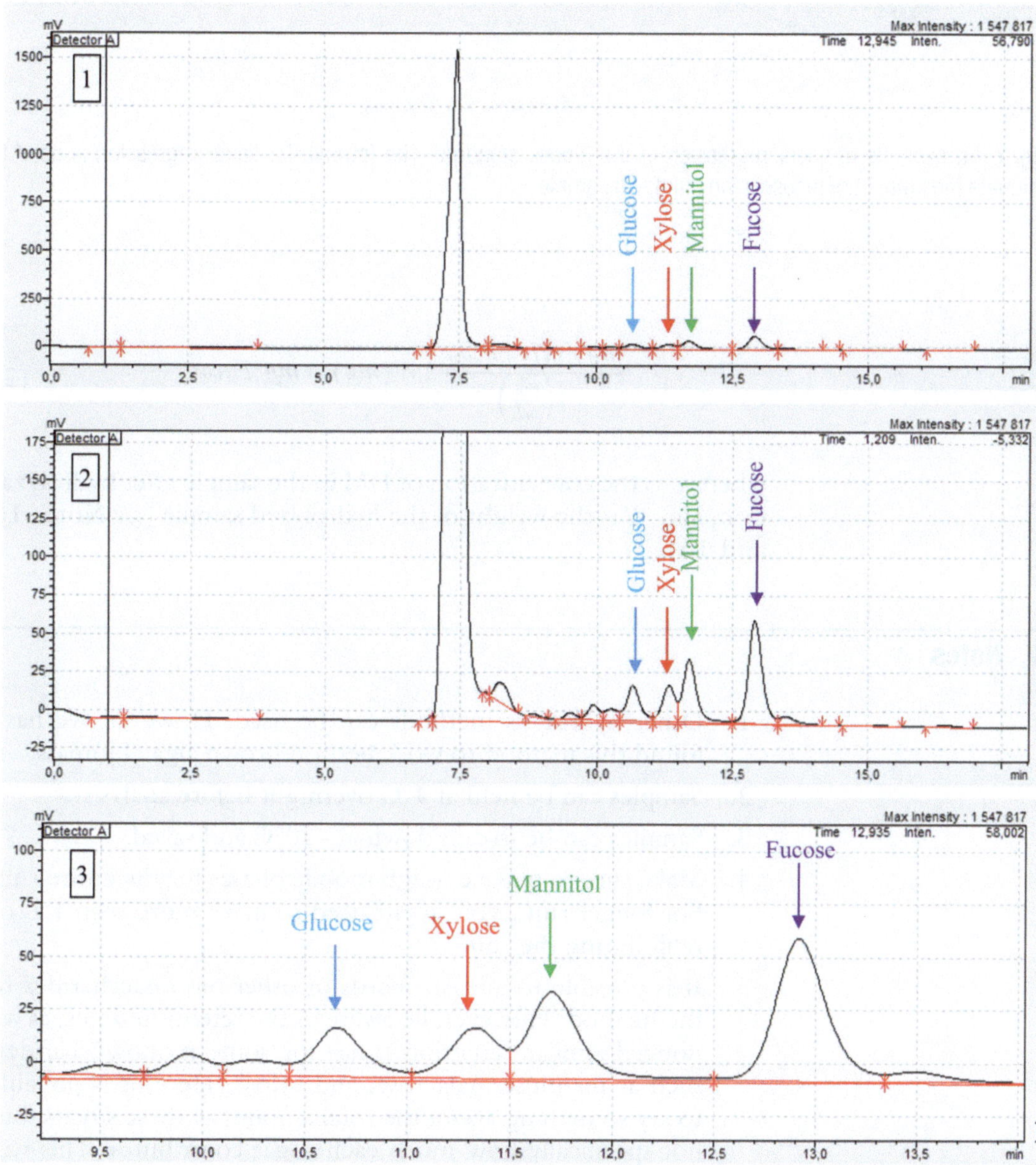

Fig. 1 Example of a chromatogram from H_2SO_4 hydrolyzed *Ascophyllum nodosum*. 1 is the initial chromatogram, and 2 and 3 are zoomed versions of the peaks. The different sugar peaks are indicated in the chromatograms

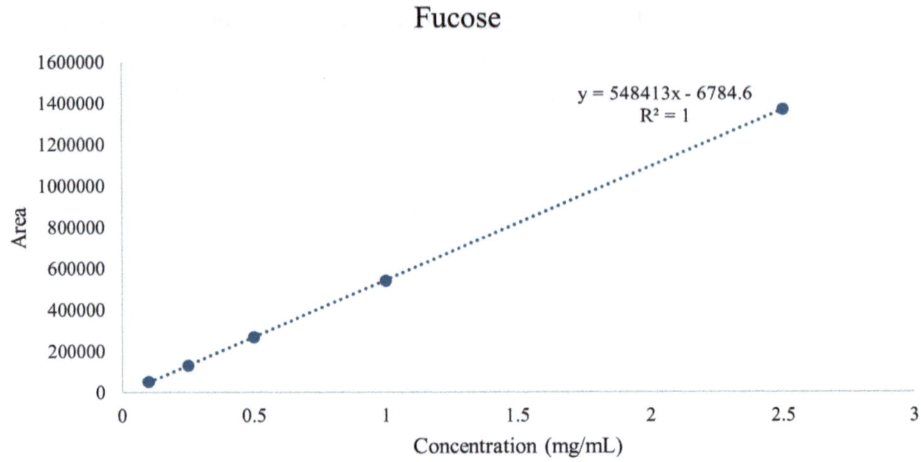

Fig. 2 An example of linear regression of the fucose standard. The formula for linear regression is used to estimate the amount of fucose in an unknown sample

$$\frac{mg/mL\ Fucose}{c\left(\frac{mg}{mL}\right)} * W(mg) = mg\ Fucose \tag{2}$$

where C is the concentration of DM in the sample after hydrolyzation, and W is the weight of the hydrolyzed sample (*see* **Notes 15 and 16**).

4 Notes

1. Other hydrolysis methods can be used. However, we have found this method to work best on brown algae biomass.
2. Samples can be held at 4°C overnight before analysis.
3. Samples can be frozen down at –20°C and saved.
4. Make sure to make enough mobile phases for the entire run. For longer runs, you might need to make more than 1 L or refill during the run.
5. It is possible to add standards of other monosaccharides to the method. However, be aware of the retention times, as we noticed similar retention times for some monosaccharides, such as mannose, galactose, and xylose. Here, it is possible to say something about the total amount of these sugars, but not specifically how much each sugar contributes. This will depend on the biomass. We have noticed that this method works very well for brown algae and not as well for red algae.

6. When starting up the analysis, it is always good to run the standards again, especially if it's been a long time since the last time, as the retention times might shift slightly.
7. Concentrations were calculated using $C_1V_1 = C_2V_2$.
8. While you wait for the instrument to reach the temperature, prepare your samples and batch file.
9. As the RID detector is very sensitive, the first runs tend to not be good. Although the oven temperature is correct, the column might not have reached the right temperature. Start the batch file with a few blank runs (mobile phase) as it can take a few runs for the instrument to stabilize. It can also be a good idea to add a few standards to the batch file as a positive control. For longer runs, one can add blanks and standards regularly during the runs for controlling the stability of the instrument and quality check the results. We have noticed that in very long runs the retention times might be shifted slightly, where regular standard runs will help compensate for potential shifts in retention times.
10. When the instrument has not been used for a while, we notice that the light bulb deteriorates, resulting in too low lamp voltage (total.en). For the analysis, this can be altered in the detector's settings. The light bulb must be changed when the total.en is too low after altering.
11. The linear regression can also be based on the height of the peaks.
12. Regular column and HPLC apparatus cleaning processes can limit potential peak errors. If a column has been used frequently and/or has not been cleaned for a long time, the peaks may deviate from what they are supposed to, and double peaks often occur.
13. The retention times of the sugars can deviate slightly between runs, especially if it has been a long time since the last run. We have had retention times between 10.61 and 10.92 for glucose, 11.29 and 11.69 for xylose, 11.67 and 11.99 for mannitol, and 12.87 and 13.38 for fucose. Therefore, always run new standard curves and add standards regularly in longer batch runs.
14. Always check the instrument's integration. We have noticed that sometimes it integrates areas that are not sugars, and other times it integrates the tops to include areas before and after, making it bigger than it is, resulting in an overestimation of the sugar.
15. Based on their linear regression, the formula will vary for each sugar and often between each standard run.

16. The calculations shown here calculate the total fucose content in the hydrolyzed sample. Depending on your needs, you can add calculation steps to calculate the sugar content in the entire raw material, at both dry weight and wet weight.

References

1. Galant AL, Kaufman RC, Wilson JD (2015) Glucose: detection and analysis. Food Chem 188:149–160. https://doi.org/10.1016/j.foodchem.2015.04.071
2. Swartz M (2010) HPLC detectors: a brief review. J Liquid Chromatogr Related Technol 33:1130–1150. https://doi.org/10.1080/10826076.2010.484356
3. Batool A, Menaa F (2020) Concentration and purification of seaweed components by chromatography methods. In: Torres MD, Kraan S, Dominguez H (eds) Sustainable seaweed technologies. Elsevier. pp. 315–370. https://doi.org/10.1016/B978-0-12-817943-7.00013-5
4. Manns D, Deutschle AL, Saake B, Meyer AS (2014) Methodology for quantitative determination of the carbohydrate composition of brown seaweeds (Laminariaceae). RSC Advances 4:25736–25746. https://doi.org/10.1039/C4RA03537B
5. Nguyen TT, Mikkelsen MD, Tran VHN et al (2020) Enzyme-assisted fucoidan extraction from brown macroalgae *Fucus distichus* subsp. *evanescens* and *Saccharina latissima*. Mar Drugs 18:296. https://doi.org/10.3390/md18060296
6. Sterner M, Ribeiro MS, Gröndahl F, Edlund U (2017) Cyclic fractionation process for *Saccharina latissima* using aqueous chelator and ion exchange resin. J Appl Phycol 29:3175–3189. https://doi.org/10.1007/s10811-017-1176-5
7. Sharma S, Neves L, Funderud J et al (2018) Seasonal and depth variations in the chemical composition of cultivated *Saccharina latissima*. Alg Res 32:107–112. https://doi.org/10.1016/j.algal.2018.03.012
8. Maribu I (2021) Chemical and enzyme-assisted extraction of fucoidan from two species of brown macroalgae (*Ascophyllum nodosum* and *Saccharina latissima*). MSc Thesis, UiT The Arctic University of Norway

Check for updates

Chapter 14

Chromatographic Purification of Fucoxanthin from Diatom Biomass

Mahammed Ilyas Khazi, Zeliha Demirel, Fakhra Liaqat, Elif Demirkan and Meltem Conk Dalay

Abstract

Fucoxanthin, a bioactive carotenoid, is well known for its antioxidant, anti-inflammatory, anti-obesity, and anticancer effects, making it a vital compound in nutraceuticals, pharmaceuticals, and cosmetics. Predominantly derived from diatoms such as *Phaeodactylum tricornutum*, fucoxanthin plays a pivotal role in photosynthesis and photoprotection. However, its commercial use is often hampered by extraction and purification issues owing to its limited natural abundance and sensitivity to deterioration under light and heat. This chapter presents a procedure for the extraction and purification of fucoxanthin from *Phaeodactylum* biomass using silica gel column chromatography. The methodology involves using solvent-based extraction to obtain fucoxanthin from microalgal biomass and silica gel column chromatography to purify fucoxanthin. This protocol provides a scalable and efficient method for producing high-quality fucoxanthin, making it suitable for use in a wide range of industrial and research applications.

Keywords Carotenoid, Extraction, Fucoxanthin, Purification, *Phaeodactylum*, Silica gel chromatography

1 Introduction

Fucoxanthin, a distinct xanthophyll carotenoid, is well known for its biological application and potential uses in a variety of sectors. Fucoxanthin, which is mostly found in diatoms and other brown-colored algae, is defined by its unique allene bond and epoxide group, which are responsible for its excellent light-harvesting abilities and photooxidative protection during photosynthesis [1, 2]. Besides its biological role, fucoxanthin has drawn interest due to its bioactive properties, including anti-inflammatory, anti-cancer, and antioxidant effects. These properties have prompted its use in nutraceuticals, cosmetics, and therapeutics [3].

The growing demand for fucoxanthin necessitates the development of efficient and scalable methods for its extraction and purification from microalgal biomass. *Phaeodactylum tricornutum* has emerged as

Kristian Spilling (ed.), *Biofuels from Algae*, Methods in Molecular Biology, vol. 2996,
https://doi.org/10.1007/978-1-0716-5031-8_14,

a promising source owing to its relatively high fucoxanthin content, ease of cultivation, and adaptability to controlled growth conditions [4]. However, the purification of fucoxanthin from complex microalgal matrices poses significant challenges. These include its limited solubility, thermal and photodegradability, as well as the co-extraction of structurally similar pigments and lipophilic compounds [5].

In the context of method development, silica gel column chromatography has been a widely employed method for the purification of fucoxanthin. This method offers several advantages, such as ease of setup, scalability for larger biomass processing, and efficient separation of fucoxanthin from other carotenoids. In addition to silica gel column chromatography, various chromatographic methods have been used for the purification of fucoxanthin, such as centrifugal partition chromatography, thin-layer chromatography, and preparative high-performance liquid chromatography [6]. This chapter provides a protocol for the extraction and purification of fucoxanthin from *Phaeodactylum tricornutum* biomass using silica gel column chromatography.

2 Materials

1. Conical flasks.
2. Glass columns (2 × 30 cm).
3. Test tubes for the collection of fractions.
4. Rotary evaporator.
5. UV-visible spectrophotometer.
6. High-performance liquid chromatography with diode array detection (HPLC-DAD).
7. Ultrasonic bath.
8. Vortex mixer.
9. Centrifuge.
10. 0.45 μm PTFE disk filters.
11. Pasteur pipettes for sample handling.
12. Solvents: Methanol, acetone, n-hexane (HPLC grade).
13. Silica Gel: Mesh size 0.063–0.200 mm for column packing.

3 Methods

3.1 Extraction of Fucoxanthin

1. Harvest *P. tricornutum* cells by centrifugation (6000 × g, 15 min).
2. Wash with deionized water to remove salt residues and lyophilize the biomass.

3. Weigh 1 g of dried biomass and transfer it to a conical flask.
4. Add 100 mL of methanol containing 0.010% (w/v) pyrogallol to prevent degradation and mix thoroughly (*see* **Note 1**).
5. Place the flasks containing mixture in an ultrasonic bath for 15 min at 27°C for extraction.
6. After ultrasonic treatment centrifuge the mixture at 6000 × g for 15 min at 4°C to separate cell debris.
7. Collect the supernatant and concentrate using a rotary evaporator under reduced pressure.

3.2 Column Chromatography for Purification

Perform all experiments at low light/dark to prevent fucoxanthin degradation, unless otherwise specified.

1. Suspend silica gel in n-hexane: acetone (90:10, v/v) and let it stand for 1 h to settle. Degas the slurry by gentle stirring to remove air bubbles (*see* **Note 2**).
2. Pour the silica gel slurry into the column, ensuring an even packing without voids or air pockets.
3. Rinse the packed column with 3–4 column volumes of pure n-hexane to equilibrate.
4. Dissolve the crude extract in a minimal volume of n-hexane.
5. Carefully load the sample onto the top of the column without disturbing the silica bed.
6. Start with a n-hexane for initial elution to remove non-polar impurities.
7. Gradually adjust the eluent composition to n-hexane: acetone (80:20, v/v) and then to (70:30, v/v) (*see* **Notes 3** and **4**). Collect fractions at each stage.
8. Finally, switch the eluent to n-hexane: acetone (65:35, v/v) to elute fucoxanthin (*see* **Note 5**).
9. Identify and collect the fucoxanthin-containing fractions by their distinct red-orange color.
10. Additional purification step:
11. Combine and concentrate the collected fucoxanthin fractions from the first round of purification using a rotary evaporator.
12. Redissolve the concentrated eluate in a minimal volume of n-hexane and apply it to the top of the freshly packed column.
13. Repeat the gradient elution sequence used previously, starting with n-hexane (100%) and gradually transitioning through n-hexane: acetone ratios (80:20, 70:30, and 65:35, v/v).
14. Collect the red-orange fucoxanthin fractions.
15. Assess purity using thin-layer chromatography (TLC) and high-performance liquid chromatography (HPLC) to confirm further impurity removal.

16. Combine the fractions showing high purity and concentrate them using a rotary evaporator under reduced pressure, then purge with nitrogen gas (*see* **Note 6**).

The thin-layer chromatography (TLC) of crude extract using different solvent systems of n-hexane and acetone (v/v) is shown in Fig. 1, and a silica gel column chromatography for fucoxanthin separation from crude extract is shown in Fig. 2.

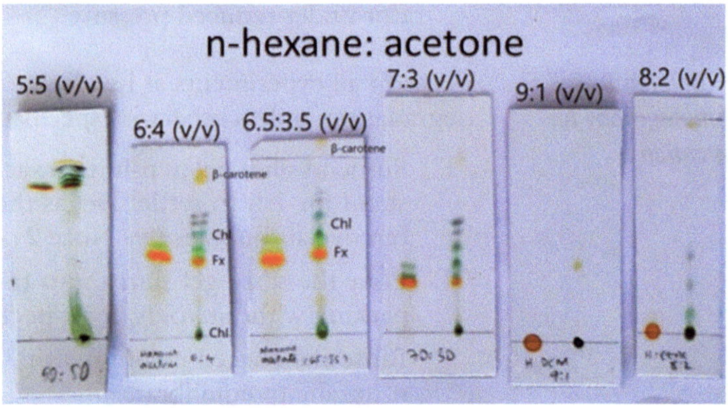

Fig. 1 Thin-layer chromatography (TLC) of crude extract using different solvent systems of n-hexane and acetone (v/v)

Fig. 2 Silica gel column chromatography for fucoxanthin separation from crude extract

3.3 Analysis and Validation

3.3.1 By HPLC Analysis

1. Filter the purified fucoxanthin fractions through 0.45 μm PTFE disk filters to remove residual particulates.
2. Analyze the filtered sample using high-performance liquid chromatography with diode array detection (HPLC-DAD) to confirm purity and quantify the fucoxanthin content.

3.3.2 By Thin-Layer Chromatography (TLC)

1. Spot samples alongside a fucoxanthin standard on a silica gel TLC plate.
2. Develop the plate in an appropriate solvent system (e.g., n-hexane: acetone, 65:35, v/v).
3. Visualize the bands to confirm the presence and purity of fucoxanthin.

4 Notes

1. Perform all steps in subdued light to prevent fucoxanthin degradation.
2. Use HPLC-grade solvents to avoid contamination and improve resolution.
3. Carefully adjust solvent ratios during elution to achieve optimal separation of fucoxanthin from other co-extracted pigments.
4. Ensure consistent flow rates during elution to avoid band broadening.
5. Perform all steps at low temperatures to prevent thermal degradation.
6. Storage: Encapsulation of fucoxanthin, and storage of purified fucoxanthin at –20°C in a dark, airtight container to maintain stability.

References

1. Anjana K, Arunkumar K (2024) Brown algae biomass for fucoxanthin, fucoidan and alginate; update review on structure, biosynthesis, biological activities and extraction valorisation. Int J Biol Macromol 280:135632. https://doi.org/10.1016/j.ijbiomac.2024.135632
2. Sahin MS, Khazi MI, Demirel Z, Dalay MC (2019) Variation in growth, fucoxanthin, fatty acids profile and lipid content of marine diatoms Nitzschia sp. and Nanofrustulum shiloi in response to nitrogen and iron. Biocatal Agric Biotechnol 17:390–398. https://doi.org/10.1016/j.bcab.2018.12.023
3. Doolaanea AA, Alfatama M, Alkhatib H, Mawazi SM (2023) Fucoxanthin. In: Jafari SM, Rashidinejad A, Simal-Gandara J (eds) Handbook of food bioactive ingredients: properties and applications. Springer International Publishing, Cham, pp 729–755

4. Pocha CKR et al (2022) Current advances in recovery and biorefinery of fucoxanthin from Phaeodactylum tricornutum. Algal Res 65:102735. https://doi.org/10.1016/j.algal.2022.102735
5. Seth K et al (2021) Bioprospecting of fucoxanthin from diatoms — challenges and perspectives. Algal Res 60:102475. https://doi.org/10.1016/j.algal.2021.102475
6. Sun J et al (2022) A simple and efficient strategy for fucoxanthin extraction from the microalga Phaeodactylum tricornutum. Algal Res 61:102610. https://doi.org/10.1016/j.algal.2021.102610

Chapter 15

Analytical Grade Purification of Phycocyanin from Cyanobacteria

Mahammed Ilyas Khazi, Zeliha Demirel, Fakhra Liaqat and Meltem Conk Dalay

Abstract

Phycocyanin is a blue-colored pigment–protein complex that exhibits numerous biofunctions such as anti-inflammation, antioxidation, antitumor, neuroprotective effect, and immunological enhancement. Purified phycocyanin has pharmaceutical and nutraceutical applications. In addition, as a nontoxic and non-carcinogenic natural coloring agent, phycocyanin has many applications in the food and cosmetic industries. This chapter describes a protocol for extraction and analytical grade purification of phycocyanin from cyanobacteria. The purification steps include (1) extraction of phycocyanin from biomass, (2) ammonium sulfate precipitation of phycocyanin and dialysis, and (3) purification of phycocyanin by gel filtration and ion-exchange chromatography.

Keywords Cyanobacteria, Extraction, Gel filtration chromatography, Ion-exchange chromatography, Phycocyanin, Purification

1 Introduction

Phycocyanin (PC) is one of the most prominent light-harvesting, water-soluble, protein–pigment complex that belongs to phycobiliprotein family. PC is an accessory photosynthetic pigment mainly present in Cyanobacteria, Rhodophyta, and Cryptomonads to increase the efficiency of photosynthesis by collecting light energy that is poorly utilized by chlorophyll [1–3]. The PC has absorption maximum at 620 nm and a fluorescence emission maximum at ~650 nm [4].

PC is composed of α- and β-subunits (heterodimer) which contain three linear tetrapyrrole chromophores (phycocyanobilins). Three αβ-units oligomerize as trimers $(\alpha\beta)_3$ which in turn form hexamers $(\alpha\beta)_6$. The exact number of α and β pairs may vary among the species [5]. PC has exhibited antioxidant, anti-inflammatory, antiplatelet, antitumor, nephroprotective, and hepatoprotective

Kristian Spilling (ed.), *Biofuels from Algae*, Methods in Molecular Biology, vol. 2996,
https://doi.org/10.1007/978-1-0716-5031-8_15,

properties [6]. Because of its bioactivities, PC not only has nutraceuticals and pharmaceuticals applications but also has obtained wide usage in the fields of molecular biology, immunology, cytology, and molecular diagnose due to its strong fluorescence. However, the industrial production of PC is limited due to challenging and unstable purification process [7]. Appropriate control of temperature, pH, and ionic strength during the extraction and purification of PC is important for the stability of PC. Consequently, an efficient, eco-friendly, and economical PC purification methods are essential to attain the desired yield, quality, and purity and also to preserve its properties [8, 9]. A number of methods for extraction and purification of PC from cyanobacteria have been reported in the literature [7, 9–12].

The selection of suitable cell disruption method and appropriate buffer for phycocyanin extraction are also very crucial factors for obtaining the high yield and purity. Several cell disruption methods including mechanical, chemical, and enzymatic disruption have been used for PC extraction from wet/dry biomass of cyanobacteria [10, 11]. The efficiency of cell disruption method depends on cell wall characteristics of species and operating conditions. Generally, mechanical disruption is the preferred method for PC extraction because of high cell disruption efficiency compared with other methods, and also it avoids chemical contamination. Methods reported for the purification of PC from cyanobacteria involve the combination of chromatographic and non-chromatographic methods such as ultrafiltration, aqueous two-phase extraction, expanded bed chromatography, gel filtration, and ion-exchange chromatography [3, 13–15]. Cost, complexity, and selectivity are the main factors that should be considered while choosing suitable purification method for PC. This chapter describes simple, economical, and effective procedure to extract and purify analytical grade PC from cyanobacteria using three-step sequential method consisting of extraction of PC, fractional ammonium sulfate precipitation, and chromatographic purification by using sephadex G-25 gel filtration and DEAE-sephadex ion-exchange column chromatography.

2 Materials

2.1 General Equipment

1. Ultrasonic homogenizer.
2. Refrigerated centrifuge.
3. UV-visible spectrophotometer.
4. Vortex.
5. Magnetic stirrer.

2.2 Extraction Buffer

Na-phosphate buffer (50 mM, pH 7.0): Weigh 3.29 g of $NaH_2PO_4{\cdot}2H_2O$ and 7.73 g of $Na_2HPO_4{\cdot}7H_2O$, and transfer to the 1 L glass beaker containing 250 mL deionized water. Mix well to dissolve salts completely. Further, add water to a volume of 700 mL, and adjust pH with HCl or NaOH if required, make up to 1 L with deionized water, and store at 4°C until use (*see* **Note 1**).

2.3 Ammonium Sulfate Precipitation

1. Ammonium sulfate, magnetic stirrer.
2. Glass beakers.

2.4 Dialysis

1. 50 mM Na-phosphate buffer (pH 7.0).
2. Dialysis cellulose membrane.

2.5 Gel Filtration

1. Sephadex G-25 media packed in glass column (3 × 30 cm).
2. Running buffer: 50 mM Na-phosphate buffer (pH 7.0).
3. Test tubes for the collection of fractions.

2.6 Ion-Exchange Chromatography

1. DEAE-sephadex in glass column (3 × 30 cm).
2. Column buffer: 50 mM Na-phosphate buffer (pH 7.0).
3. Running buffer: 0.1, 0.15, 0.2, and 0.25 M NaCl in column buffer (50 mM Na-phosphate).
4. Test tubes for the collection of fractions.

3 Methods

Perform all experiments at low light/dark and at 4°C to prevent PC denaturation, unless otherwise specified. A scheme of the overall purification process is shown in Fig. 1.

3.1 Crude Extract Preparation

1. Weigh 100 mg of lyophilized/dry cell mass of *A. platensis* into a centrifuge tube.
2. Add 10 mL of ice-cold Na-phosphate buffer to the biomass and vortex for 30 s.
3. Sonicate the suspension on the ice at a frequency of 20 kHz for 2 min (nine cycles, 50% power).
4. Centrifuge at 6000 × g for 15 min at 4°C to remove the cell debris.
5. Transfer the supernatant to a clean beaker (*see* **Notes 2 and 3**).

3.2 Ammonium Sulfate Precipitation and Dialysis

1. Place the beaker containing the crude phycocyanin extract on the ice (this can be done by placing crude extract containing beaker in an another larger beaker containing crushed ice) and stir with a magnetic stirrer.

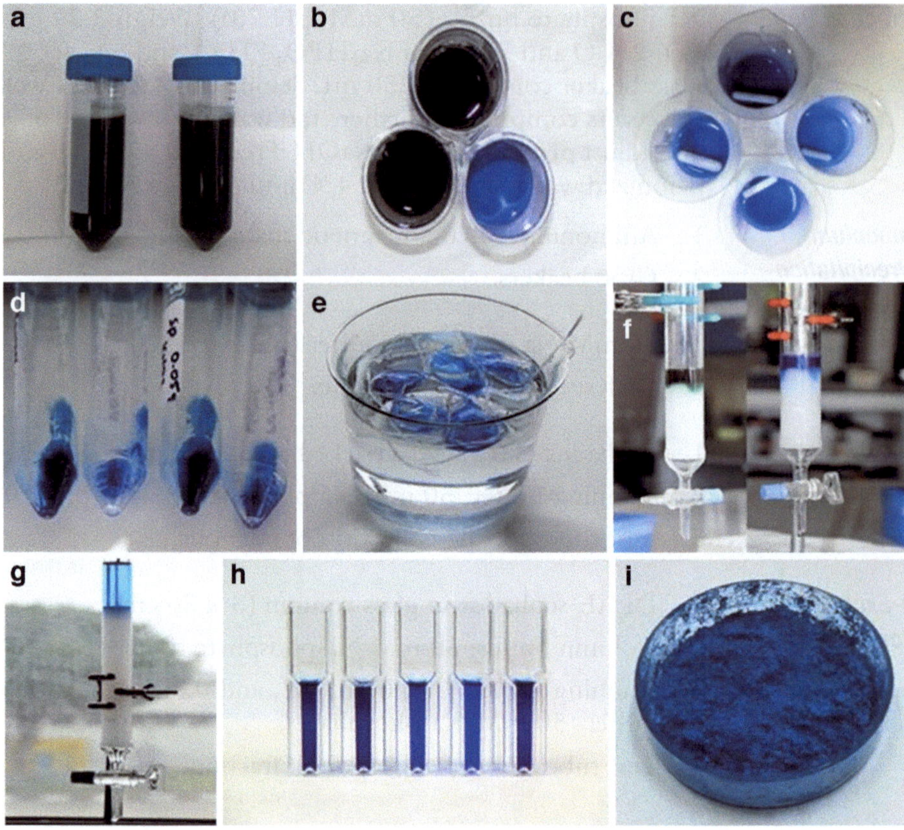

Fig. 1 The scheme of the overall purification process: (**a**) biomass suspended in Na-phosphate buffer, (**b**) crude PC extracts obtained with Na-phosphate buffer (green) and $CaCl_2$ (blue), (**c**) and (**d**) ammonium sulfate precipitates, (**e**) dialysis, (**f**) gel filtration chromatography, (**g**) ion-exchange chromatography, (**h**) purified C-PC fractions, and (**i**) lyophilized C-PC

2. While stirring slowly, add a pre-weighed quantity of ammonium sulfate to reach a saturation of 30%, and stir the mixture for nearly 2 h (*see* **Notes 4 and 5**).
3. Transfer the ammonium sulfate precipitated solution to conical centrifuge tubes and centrifuge at 6000 × g for 15 min at 4°C.
4. Collect the supernatant and discard the pellet.
5. Slowly add a pre-weighed quantity of ammonium sulfate to reach a saturation of 70%, and stir the mixture for >5 h.
6. Transfer the ammonium sulfate precipitated solution to conical centrifuge tubes and centrifuge at 6000 × g for 15 min 4°C.
7. Remove the supernatant carefully, and add a few milliliters of Na-phosphate buffer to the pellet to resuspend the phycocyanin.
8. Transfer the phycocyanin solution to dialysis tubing cellulose membrane using a Pasteur pipette and dialyze overnight against 2 L Na-phosphate buffer (*see* **Note 6**).

9. Decant phycocyanin solution into centrifuge tubes and centrifuge at 6000 × g for 10 min to remove any undissolved material.

3.3 Gel Filtration

1. Suspend Sephadex G-25 in Na-phosphate buffer and allow swelling (*see* **Note 7**).
2. Pour the swollen Sephadex G-25 slurry into the column (*see* **Notes 8 and 9**).
3. After the slurry has settled, rinse the column with three column volumes of Na-phosphate buffer.
4. Load the dialyzed phycocyanin solution on the column using Pasteur pipette.
5. Elute with the running buffer and collect the phycocyanin fractions.
6. Combine the collected fractions of high purity and yield.

3.4 Ion-Exchange Chromatography

1. Suspend DEAE-sephadex in Na-phosphate buffer and allow it to stand for 2–4 h.
2. Pour the DEAE-sephadex slurry into the column and allow settling down.
3. After the slurry has settled, rinse the column with three column volumes of Na-phosphate buffer in order to stabilize and equilibrate the bed.
4. Load the phycocyanin fractions obtained from gel filtration chromatography on the DEAE-sephadex column.
5. Rinse the column with Na-phosphate buffer.
6. Elute the phycocyanin with a linearly increasing ionic concentration gradient of NaCl from 0.1 to 0.25 M (C-phycocyanin (C-PC) usually elute between 0.15 and 0.25 M NaCl).
7. Collect the fractions.
8. Check the phycocyanin concentration and purity of the fractions collected after each step.

3.5 Spectrophotometric Estimation of Phycocyanin

1. The PC concentration can be calculated by using Eq. 1 [16]

$$\mathrm{PC}\left(\mathrm{mg\ mL^{-1}}\right) = \frac{A_{615} - 0.474A_{652}}{5.34} \tag{1}$$

where PC is the C-PC concentration (mg mL^{-1}), A_{615} is the absorbance of the sample at 615 nm, and A_{652} is the absorbance of the sample at 652 nm.

2. PC purity can be calculated by using Eq. 2 [16]

$$\text{Purity} = \frac{A_{620}}{A_{280}} \tag{2}$$

The absorbance at 620 and 280 nm correspond to PC and total protein, respectively. PC is considered food grade when A_{620}/A_{280} absorbance ratio is greater than 0.7, absorbance ratio 0.7 to 3.9 and above 4 have been considered as reagent and analytical grade, respectively [17].

4 Notes

1. All buffers should be prepared using deionized water and analytical grade reagents.
2. Generally, the color of the crude extract will depend on the type of the buffer used. Water and Na-phosphate buffer produce green-colored crude extracts, whereas $CaCl_2$ solution (1.5%) produces blue-colored crude extract (Fig. 2).
3. The crude C-PC purity and yield will depend on the type of buffer and cell disruption method used for the phycocyanin extraction. In our previous study, 12 h incubation of biomass with the $CaCl_2$ solution on rotary shaker gave food-grade purity, whereas using Na-phosphate buffer and sonication did not give food-grade purity [18].
4. If the ammonium sulfate contains lumps, then these should be broken up before adding.

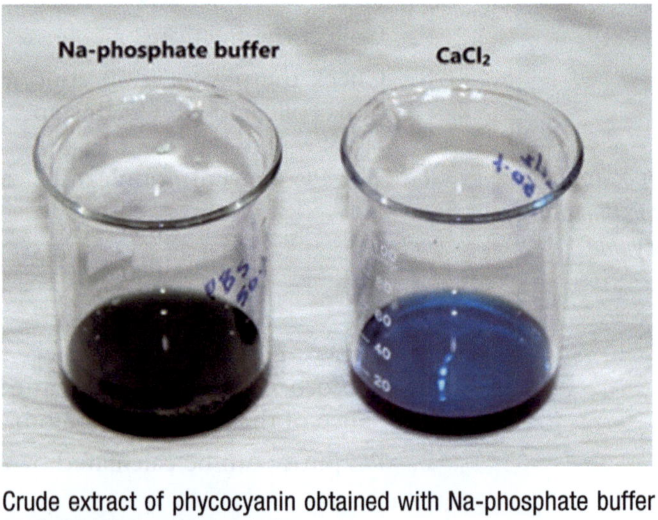

Fig. 2 Crude extract of phycocyanin obtained with Na-phosphate buffer (green color) and $CaCl_2$ solution (blue color)

5. Determine the volume of saturated ammonium sulfate solution necessary to give the desired percentage of saturation using web page (ammonium sulfate calculator http://www.proteinchemist.com/cgi-bin/s2.pl).
6. Dialysis should be performed with large external volumes and adequate stirring, as well as frequent exchanges of the buffer.
7. Swelling may take 24–48 h depending on matrix; the swelling step can be speeded up by heating in a boiling water bath for 30–60 min.
8. Do not agitate the matrix using magnetic stirrers or glass rods, since the gel beads will be pulverized.
9. Pour all gel in a single operation, and make sure that air bubbles do not become trapped into the matrix.

References

1. Glazer AN, Apell GS (1977) A common evolutionary origin for the biliproteins of cyanobacteria rhodophyta and cryptophyta. FEMS Microbiol Lett 1(2):113–116
2. Eriksen NT (2008) Production of phycocyanin—a pigment with applications in biology, biotechnology, foods and medicine. Appl Microbiol Biotechnol 80:1–14
3. Khazi MI et al (2021) Acclimation and characterization of marine cyanobacterial strains *Euryhalinema* and *Desertifilum* for C-phycocyanin production. Front Bioeng Biotechnol 9:752024
4. Dasgupta CN (2015) Algae as a source of phycocyanin and other industrially important pigments. Algal Biorefin:253–276
5. Kuddus M, Singh P, Thomas G, Al-Hazimi A (2013) Recent developments in production and biotechnological applications of C-phycocyanin. BioMed Res Int 2013(1):742859
6. Fernández-Rojas B, Hernández-Juárez J, Pedraza-Chaverri J (2014) Nutraceutical properties of phycocyanin. J Funct Foods 11: 375–392
7. Yu P et al (2017) Purification and bioactivities of phycocyanin. Crit Rev Food Sci Nutr 57(18):3840–3849
8. Yuan B et al (2022) A review of recent strategies to improve the physical stability of phycocyanin. Curr Res Food Sci 5:2329–2337
9. Manirafasha E et al (2016) Phycobiliprotein: potential microalgae derived pharmaceutical and biological reagent. Biochem Eng J 109:282–296
10. Usai L et al (2024) Recent advancements in production and extraction methods of phycobiliprotein c-phycocyanin by *Arthrospira* (*Spirulina*) *platensis*: a mini review. Curr Microbiol 81(12):428
11. Chamorro-Cevallos G (2016) Methods for extraction, isolation and purification of C-phycocyanin: 50 years of research in review. Int J Food Nutr Sci 3:1–10
12. Khazi MI et al (2024) Development of an efficient process to extract C-phycocyanin from marine cyanobacterial strains and evaluation of its bioactive potential. J Chem Technol Biotechnol 99(8):1897–1906
13. Moraes CC, Sala L, Cerveira GP, Kalil SJ (2011) C-phycocyanin extraction from *Spirulina platensis* wet biomass. Braz J Chem Eng 28:45–49
14. Jian-Feng N, Guang-Ce W, Lin X-z, Zhou B-C (2007) Large-scale recovery of C-phycocyanin from *Spirulina platensis* using expanded bed adsorption chromatography. J Chromatogr B 850(1–2):267–276
15. Singh NK, Parmar A, Madamwar D (2009) Optimization of medium components for increased production of C-phycocyanin from *Phormidium ceylanicum* and its purification by single step process. Bioresour Technol 100(4):1663–1669
16. Bennett A, Bogorad L (1973) Complementary chromatic adaptation in a filamentous blue-green alga. J Cell Biol 58(2):419–435
17. Sørensen L, Hantke A, Eriksen NT (2013) Purification of the photosynthetic pigment C-phycocyanin from heterotrophic *Galdieria sulphuraria*. J Sci Food Agric 93(12):2933–2938
18. Khazi MI, Demirel Z, Dalay MC (2018) Evaluation of growth and phycobiliprotein composition of cyanobacteria isolates cultivated in different nitrogen sources. J Appl Phycol 30:1513–1523

Check for updates

Chapter 16

Phytoplankton Pigment Analysis by High Performance Liquid Chromatography

Thomas Rohrlack

Abstract

Pigments form a chemotaxonomic system widely used to determine phytoplankton composition and reconstruct water body histories by using sedimentary pigments as biological archives. This chapter presents protocols for pigment extraction from water and sediment samples. It also describes a method to separate, identify, and quantify major phytoplankton pigments in a single run. The method is based on high performance liquid chromatography with diode array detection (HPLC-DAD).

Keywords Pigment analysis, Phytoplankton, Chlorophylls, Carotenoids, Extraction, HPLC

1 Introduction

Pigments represent the phytoplankton community of freshwater and marine ecosystems. Some pigments, such as chlorophyll a, are present in most types of phytoplankton and can serve as universal markers of phytoplankton biomass [1]. Others are specific to particular families, genera, or even species, forming the basis of a chemotaxonomic systems for phytoplankton analysis [2, 3]. Most pigments are hydrophobic. As a result, they are not released into the water after cell death. Instead, they sediment as a part of cell debris to form a sedimentary record of fossilized pigments, which can be used to reconstruct the historical development of lakes and marine ecosystems [4, 5]. Finally, some pigments are altered in a defined manner when phytoplankton cells are digested by other organisms such as zooplankton. The resulting derivatives can be quantified, allowing to reconstruct biological interactions involving phytoplankton [4].

Phytoplankton pigments are commonly classified into chlorophylls and carotenoids. While both chlorophylls and carotenoids are hydrophobic by nature, they differ in polarity, both among and within these groups of pigments. The here described method for pigment analysis utilizes these differences in polarity to chromatographically

Kristian Spilling (ed.), *Biofuels from Algae*, Methods in Molecular Biology, vol. 2996,
https://doi.org/10.1007/978-1-0716-5031-8_16,

separate major pigments found in living phytoplankton organisms and fossilized in the sediment. The method bases on protocols originally described by [6], although some modifications were made to improve separation of xanthophylls. The method should be seen as a compromise between the desire to have a single method for a rapid analysis of major phytoplankton pigments and the need to adjust extraction and separation protocols to the polarity of a given type of pigments.

This chapter describes protocols for extraction of pigments from phytoplankton biomass or sediments, and their quantitative analysis by high performance liquid chromatography (HPLC).

2 Materials

1. Glass fiber filters (preferably 47 mm diameter, GF/F grade)—for the analysis of water samples.
2. Filter equipment for glass fiber filters—for analysis of water samples.
3. Scale (precision ± 10 mg or better)—for the analysis of sediment samples.
4. 15 ml plastic tubes that can be tightly locked, rated for centrifugation at min 3000 × g.
5. 1 ml and 5 ml pipettes and tips for handling low viscosity organic solvents (tight opening).
6. Forceps.
7. Centrifuge with rotor for above-mentioned 15 ml plastic tubes (min 3000 × g).
8. Vortex mixer.
9. Refrigerator.
10. Freezer (−18°C)
11. Freeze dryer
12. High performance liquid chromatography with diode array detection (HPLC-DAD), equipped with four solvent channels, refrigerated autosampler, heated column compartment, degasser.
13. Dark HPLC vials.
14. C30 reverse phase column (2.1 × 250 mm, 3 μm particle size), can be replaced with C18 column if carotenes are not of interest.
15. Solvent A: Methanol (HPLC grade).
16. Solvent B: Freshly prepared 90% (v/v) acetonitrile in water (HPLC grade).

17. Solvent C: Ethyl acetate (HPLC grade).
18. Solvent D: Freshly prepared 400 mM solution of ammonium acetate in water (HPLC grade).
19. Pigment standards—selection depends on expected composition of samples. Most common: chlorophyll a (all algae), chlorophyll b (green algae), fucoxanthin (diatoms, dinoflagellates), neoxanthin (green algae, euglenophytes), lutein (green algae, euglenophytes), alloxanthin (cryptophytes), peridinin (dinoflagellates), diatoxanthin (diatoms, dinoflagellates, chrysophytes), zeaxanthin, canthaxanthin, myxoxanthophyll, echinenone (all four occurring in cyanobacteria), beta-carotene (most algae), alpha-carotene (cryptophytes, prochlorophytes, rhodophytes), pheophytin a (breakdown product of chlorophyll a—indicates senescent or dead algae), pheophorbide a (breakdown product of chlorophyll a—indicates grazing).

3 Methods

3.1 Pigment Extraction from Phytoplankton Samples (see Notes 1 and 2)

1. Filter a known volume of water samples through glass fiber filters.
2. Place the filters into 15 ml plastic tubes such that the cells on the filter face the center of the tube. The filters will be extracted inside these plastic tubes.
3. Freeze the tubes and store them at freezing temperatures until analysis.
4. Freeze-dry the filters immediately before extraction.
5. Extract cells (*see* **Note 3**). Add 5 mL of acetone to each sample. Close the tubes immediately, vortex them for 10 s, and transfer the samples to a refrigerator. Ensure that the filters are fully covered by acetone.
6. Keep the samples in the refrigerator overnight.
7. Recover the extract by vortexing the samples for 10 s. Then press the filters against the tube wall using forceps to collect as much of the extract as possible and remove and dispose of the filters.
8. Centrifuge the extracts at 3000 × g for 15 min to remove particles.
9. Proceed with HPLC analysis within 24 h.

3.2 Pigment Extraction from Sediment Samples (see Notes 1 and 2)

1. Freeze-dry samples.
2. Transfer 100–300 mg dry sediment into 15 ml plastic tubes.

3. Extract samples (*see* **Note 3**): Add 3 ml of acetone to each sample. Close the tubes immediately, vortex them for 10 s, and transfer the samples to a refrigerator.
4. Keep the samples in the refrigerator overnight.
5. Vortex for 10 s and centrifuge the extracts at 3000 × g for 15 min.
6. Proceed with HPLC analysis within 24 h.

3.3 HPLC Analysis

1. Prepare the HPLC for analysis: cool down the autosampler to 4°C and heat the column compartment to 25°C. Maintain these temperatures consistently throughout the analysis.
2. Transfer pigment extracts and pigment standards to HPLC vials. Add water to each vial to achieve a final mixture of 20% water and 80% extract (v/v). Vortex and transfer the vials to the autosampler, ensuring no particles are transferred. Do not add water to pigment standards if these are supplied in methanol, ethanol, or other solvents with similar polarity.
3. Program the HPLC pump to deliver the gradient detailed in Table 1.
4. Program the DAD detector to scan absorption from 350 to 700 nm and to separately monitor absorption at 436 nm throughout the gradient.
5. Set autosampler to injection volume of maximal 50 μl. Complete run.
6. Pigment identification: Use absorption chromatogram at 436 nm to find peaks with sample to noise ratio >3.
7. Use retention time and absorption spectrum of sample peaks and standards to identify pigments. Make use of library function of chromatography software if available.

Table 1
Solvent gradient for HPLC analysis

Time (min)	Flow (ml min^{-1})	%B	%C	%D	Curve[1]
0	0.5	0	0	20	
4	0.5	100	0	0	5
26	0.5	20	80	0	7
28	0.5	20	80	0	5
30	0.5	100	0	0	5
32	0.5	0	0	20	5
40	0.5	0	0	20	5

[1]Curve refers to shape of the gradient between two time steps, with 1 being the most convex gradient, 5 being a linear gradient, and 9 being the most concave gradient.

8. Spiking experiments may be necessary for the identification of pigments with similar retention times.
9. Additional information on spectrum characteristics of known pigments is given in [7].
10. Refer to [3], [7], or [4] to use pigment patterns to identify major phytoplankton groups.
11. Pigment quantification (*see* **Note 4**). Use absorption chromatogram at 436 nm to find peaks with sample to noise ratio > 10. Quantify using respective standards.

4 Notes

1. Pigments are highly sensitive to white light, especially after freeze-drying and in solution. To prevent artifacts, samples should preferably be handled in rooms with yellow or green lighting, as these wavelengths are weakly absorbed by most pigments. Alternatively, handling in low light conditions may be attempted. In all cases, light exposure should be minimized, and samples should be stored in complete darkness.
2. Some pigments are prone to oxidation, particularly in sediments, freeze-dried material, and solutions. Therefore, storage in a nitrogen atmosphere is recommended. If this is not feasible, the storage duration of phytoplankton and sediment samples should be kept to a minimum. Freeze-dried samples, pigment extracts, and pigment standards should be processed immediately or stored under nitrogen.
3. Despite careful handling, some acetone evaporation from pigment extracts is unavoidable, leading to an increase in pigment concentration. To correct this, an internal pigment standard should be added at its final concentration to the acetone before extraction begins. β-Apocarotenal is a cost-effective and readily available option, as it does not naturally occur in phytoplankton and does not coelute with major phytoplankton pigments.
4. Ideally, pigments should be quantified using standards at different concentrations to construct a calibration curve. However, standards are often available only in low concentrations. In such cases, quantification should rely on the molar absorption coefficient determined using the HPLC-DAD system employed for sample analysis. Theoretically, these coefficients can be used for subsequent analyses without rerunning standards, but this is only valid for HPLC-DAD systems with highly stable retention times. Stability should be tested regularly, and absorption coefficients should be remeasured as frequently as possible.

References

1. Wright SW, Jeffrey S (2006) Pigment markers for phytoplankton production. In: Volkman JK (ed) Marine organic matter: biomarkers, isotopes and DNA. Springer-Verlag Berlin Heidelberg 2005, pp 71–104
2. Hagman CHC, Rohrlack T, Uhlig S, Hostyeva V (2019) Heteroxanthin as a pigment biomarker for Gonyostomum semen (Raphidophyceae). PLoS One 14:e0226650
3. Peltomaa E, Asikainen H, Blomster J et al (2023) Phytoplankton group identification with chemotaxonomic biomarkers: in combination they do better. Phytochemistry 209:113624
4. Leavitt PR (1993) A review of factors that regulate carotenoid and chlorophyll deposition and fossil pigment abundance. J Paleolimnol 9:109–127
5. Sanger JE (1988) Fossil pigments in paleoecology and paleolimnology. Palaeogeogr Palaeoclimatol Palaeoecol 62:343–359
6. Wright S, Jeffrey S, Mantoura R et al (1991) Improved HPLC method for the analysis of chlorophylls and carotenoids from marine phytoplankton. Mar Ecol Prog Ser 77: 183–196
7. Roy S, Llewellyn CA, Egeland ES, Johnsen G (2011) Phytoplankton pigments: characterization, chemotaxonomy and applications in oceanography. Cambridge University Press, New York NY, USA

Chapter 17

Extracellular Enzyme Activity Measurements Using Fluorescent Substrate Derivatives

Mari Vanharanta

Abstract

Enzymes play an important role in biochemical processes by being responsible for hydrolysis of various biomolecules such as proteins, carbohydrates, and organic phosphorus compounds. Microbial extracellular enzyme activities can be measured using fluorescent model substrates as proxies for natural substrates. Here 4-methylumbelliferyl phosphate, 4-methylumbelliferyl β-D-glucopyranoside, and L-leucine-7-amido-4-methylcoumarin hydrochloride are utilized to measure potential hydrolytic activities of phosphatase, glucosidase, and aminopeptidase, respectively, in natural seawater samples.

Keywords Biomolecule hydrolysis, Enzymes, Marine degradation processes, Fluorometry

1 Introduction

Biomolecules are often larger than the proteins that form microbial cell membrane transport systems. Therefore, microorganisms utilize extracellular enzymes to degrade organic substrates outside the cell prior to uptake. Extracellular enzymes are synthesized inside the microbial cell and subsequently secreted outside, either as attached to the cell surface or in a dissolved cell-free form. Microbial degradation processes are commonly studied by quantifying rates of extracellular enzyme activities via fluorescent substrate derivatives. In this method, water samples are incubated with a synthetic substrate comprising of an artificial fluorescent molecule bound to a more natural molecule that upon enzymatic hydrolysis of a specific linkage (e.g., peptide or phosphomonoester bond) releases a soluble fluorescent molecule which can be detected by fluorometry [1]. This yields potential hydrolytic rates of extracellular enzymes when the substrate is provided at saturation concentration. Two commonly used fluorogenic compounds are 4-methylumbelliferone (MUF) and 7-amino-4-methylcoumarin (MCA) that remain nonfluorescent until the fluorophore–substrate bond is cleaved by enzymatic hydrolysis [2]. Organic solvents used to prepare the substrate stock solutions

Kristian Spilling (ed.), *Biofuels from Algae*, Methods in Molecular Biology, vol. 2996,
https://doi.org/10.1007/978-1-0716-5031-8_17,

are utilized to dissolve fluorogenic substrate derivatives and fluorescence standard compounds to solution. The effect of the solvent at its working solution concentration has been shown negligible [3, 4].

2 Materials

2.1 Fluorogenic Analogue Model Substrates

1. 4-Methylumbelliferyl phosphate (fluorogenic substrate for phosphatase)
2. 4-Methylumbelliferyl β-D-glucopyranoside (fluorogenic substrate for β-glucosidase)
3. L-Leucine-7-amido-4-methylcoumarin hydrochloride (fluorogenic substrate for leucine aminopeptidase)

2.2 Fluorescence Standard Compounds

1. 4-Methylumbelliferone sodium salt (reference fluorescence standard compound for phosphatase and glucosidase enzyme assays)
2. 7-Amino-4-methylcoumarin (reference fluorescence standard compound for amino- peptidase enzyme assay)

2.3 Solutions

1. 2-Methoxyethanol (MTXE)
2. Dimethyl sulfoxide (DMSO)
3. Ultrapure water

2.4 Supplies

1. Multichannel pipette and pipette tips
2. Low-protein binding 96-well microplates
3. Formulated aluminum foil sealing films (to keep the samples dark and avoiding evaporation)
4. Sterile 1–2 mL vials/tubes (e.g., Eppendorf)
5. Aluminum foil to keep stocks and working solutions in dark

2.5 Equipment

1. High precision weighing balance
2. Fluorometric microplate reader
3. Temperature controlled incubator
4. Vortex mixer
5. Freezer
6. Refrigerator
7. Autoclave

3 Methods

The fluorogenic substrates and their fluorophores are provided in the form of powder and should be weighed with a high precision weighing balance. Concentrations of the stock solutions depend on the saturation concentration of the environment and therefore the saturation concentration of samples should be defined for the study area and season before each experiment. For instance, in the Gulf of Finland (Baltic Sea), the saturation concentrations were 100 μM for 4-methylumbelliferyl phosphate (AP) and 4-methylumbelliferyl β-D-glucopyranoside (BG) and 500 μM for L-Leucine-7-amido-4-methylcoumarin hydrochloride (LAP) during summer [5]. Saturation concentrations are used for model substrates to compete with naturally occurring ones for enzyme active sites, thus the results reflect the potential rate of extracellular enzyme activity.

3.1 Preparation of Fluorophore Stock Solution

1. Prepare a 2% MTXE solution by diluting in ultrapure water.
2. Weigh the fluorophore powders (4-methylumbelliferone sodium salt and 7-amino-4-methylcoumarin) with a high precision weighing balance.
3. Transfer the weighed 4-methylumbelliferone sodium salt powder into 2% MTXE solution, utilize vortex mixer to dissolve the powder, and prepare the standard curve for calibration according to the instructions in Sect. 3.4. Keep the solution in dark with, e.g., aluminum foil.
4. Transfer the weighed 7-amino-4-methylcoumarin powder into 100% DMSO solution, utilize vortex mixer to dissolve the powder, and prepare the standard curve for calibration according to the instructions in Sect. 3.4. Keep the solution in dark with, e.g., aluminum foil.

3.2 Preparation of Fluorogenic Substrate Stock Solution

1. Prepare a 2% MTXE solution and a 2% DMSO solution by diluting in ultrapure water.
2. Weigh the fluorogenic substrate powders with a high precision weighing balance.
3. Transfer weighed 4-methylumbelliferyl phosphate substrate and 4-methylumbelliferyl β-D-glucopyranoside substrate analogue powders into the 2% MTXE solution in separate vials. Stir the solutions with a vortex mixer (*see* **Note 1**). Keep the solution in dark with, e.g., aluminum foil.
4. Transfer L-leucine-7-amido-4-methylcoumarin hydrochloride substrate analogue powder into the 2% DMSO solution. Stir the solution with a vortex mixer until the powder has dissolved. Keep the solution in dark with, e.g., aluminum foil.
5. Transfer the stock solutions in 2 mL sterile vials and store at −20°C in dark until analysis.

3.3 Defining the Saturation Concentration

The concentrations used to determine the saturation concentration should be based on previously published work, with testing conducted across different concentration ranges if no prior measurements are available for the study area. Prepare a dilution series of different concentrations of fluorogenic substrate stock solutions to natural sample to yield final concentrations of, e.g., 0, 20, 60, 100, 150, 200, 250, 300, 350, 400, 450, 500, and 550 μM for LAP enzyme assay (Fig. 1). Based on literature from the same area [6], lower concentrations were utilized for defining the saturation concentrations of AP and BG (10, 20, 40, 60, 80, 100, 120, 140, 160, 180, and 200 μM).

1. Prepare a dilution series of desired final concentrations of fluorogenic substrate from stock solutions in ultrapure water.
2. Add 330 μl of sample water in each well in the microplate.
3. Add 20 μl of fluorogenic substrate on sample.
4. Measure fluorescence at excitation wavelength of 360 nm and emission wavelength of 450 nm immediately after substrate addition and after a 4-h incubation period.
5. Define the saturation concentration based on the saturation curve as plotted in Fig. 1.

3.4 Standard Curve for Calibration

Fluorescence is converted to microbial extracellular enzyme activity (EEA nmol L^{-1} h^{-1}) by standard curves established with a range of different concentrations of 4-methylumbelliferone sodium salt (MUF) and 7-amino-4-methylcoumarin (MCA) added in 0.2 μm filtered sample water (*see* **Note 2**). Fluorescence of the fluorophore is measured with same settings as samples. Here eight different

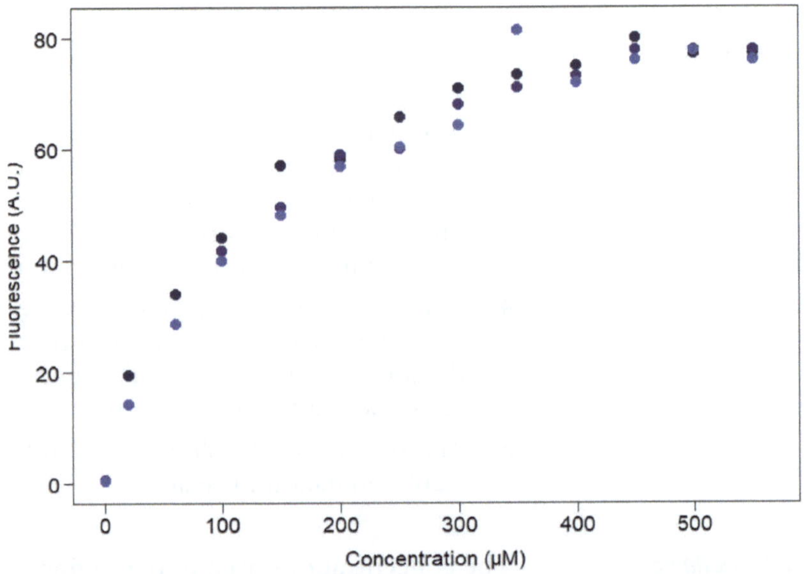

Fig. 1 Determination of the saturation concentration of the fluorogenic peptidyl substrate for leucine aminopeptidase (L-leucine-7-amido-4-methylcoumarin hydrochloride) activity measured in triplicates after a 4-h incubation.

concentrations of methylcoumarin fluorophore (MCA) were utilized in triplicate (0, 100, 400, 600, 800, 1000, 1200, and 1400 nM) to obtain a standard curve for analyzing the intensity of fluorescence of the reference fluorescence standard compound in different concentrations (Fig. 2).

1. Prepare a dilution series of desired final concentrations of fluorophore from the fluorophore stock solution in triplicate by diluting in ultrapure water.
2. Add 330 μl of 0.2 μm filtered sample water in each well of the low protein binding 96-well microplate.
3. Add 20 μl of fluorophore solution.
4. Measure fluorescence at excitation wavelength of 360 nm and emission wavelength of 450 nm.
5. Determine the standard curve for calibration according to Fig. 2.

3.5 Extracellular Enzyme Activity Measurements

Hydrolytic rates of alkaline phosphatase (APase), β-glucosidase (BGase), and leucine aminopeptidase (LAPase) are measured as an increase in fluorescence of sample incubated with fluorescently labelled model substrates over a 4 h incubation. Fluorescence of methyl-umbelliferone (MUF) and 7-amino-4-methyl-coumarine (MCA) caused by the enzyme cleavage of the substrate analogues is measured with a spectrofluorometer with excitation wavelength of 360 nm and emission wavelength of 450 nm. For studying extracellular enzyme activity

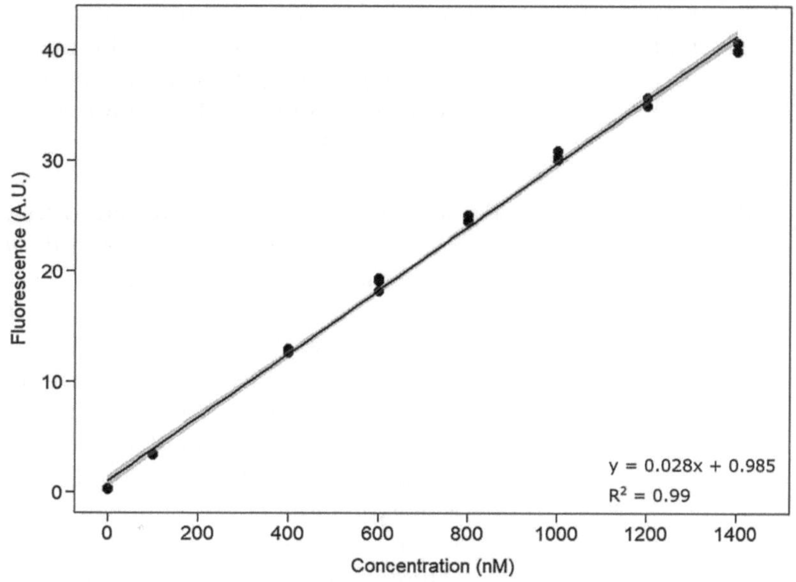

Fig. 2 A standard curve of fluorescence (arbitrary units (A.U.)) versus different concentrations of methylcoumarin fluorophore (MCA) prepared in triplicate.

in different size fractions (e.g., cell-free and cell-attached activities) sample water is filtered through low protein binding syringe filters (*see* **Note 2**). Extracellular enzyme assays should be conducted as soon as possible after seawater sampling to minimize any alterations of the microbial community after sampling [4].

3.5.1 Fluorescence Measurements

1. Prepare working solutions by thawing frozen stock solutions, stirring with a vortex mixer, and diluting stocks to target concentration with ultrapure water. Keep the stocks and working solutions in dark.
2. Add 330 μl of sample water in each well of the low protein binding 96-well microplate.
3. Utilize the multichannel pipette to add 20 μl of substrate working solution in quadruplicates using four wells in a microplate (*see* **Notes 3 and 4**).
4. Measure fluorescence of the samples in the microplate right after addition of model substrate at excitation wavelength of 360 nm and emission wavelength of 450 nm.
5. Cover the microplates with formulated aluminum foil sealing films to keep samples in dark and to avoid evaporation (*see* **Note 5**).
6. Place in dark at in situ temperature for incubation.
7. Measure fluorescence at five timepoints, every ~1 h over a ~4 h incubation period.
8. It is recommended not to reuse the microplate (*see* **Note 6**).

3.5.2 Control Measurements

A possible increase in abiotic substrate decay over time is ruled out with controls prepared by diluting model substrates in autoclaved sample water. Prepare a quadruplicate of autoclaved sample water control in four wells in a microplate according to the Steps 2–7 in Sect. 3.5.1.

3.5.3 Calculations

Fluorescence values are converted to concentrations of hydrolyzed substrate using a standard curve of fluorescence versus different concentrations of methylcoumarin (MCA) or methylumbelliferyl (MUF) fluorophores (see an example of MCA in Fig. 2). The increase in sample relative fluorescence (arbitrary units (A.U.)) is converted to concentration of hydrolyzed substrate based on a linear model that is fitted to the fluorescence time series of the four replicates measured over the 4 h incubation period and comparing to the model fitted to the calibration standard in Fig. 2. This calculation is made for each model substrate based on its reference fluorescence calibration standard curve.

Here, based on the increase in sample fluorescence of L-leucine-7-amido-4-methylcoumarin hydrochloride (A.U. h^{-1}) (Fig. 3) and the standard curve for 7-amino-4-methylcoumarin fluorophore

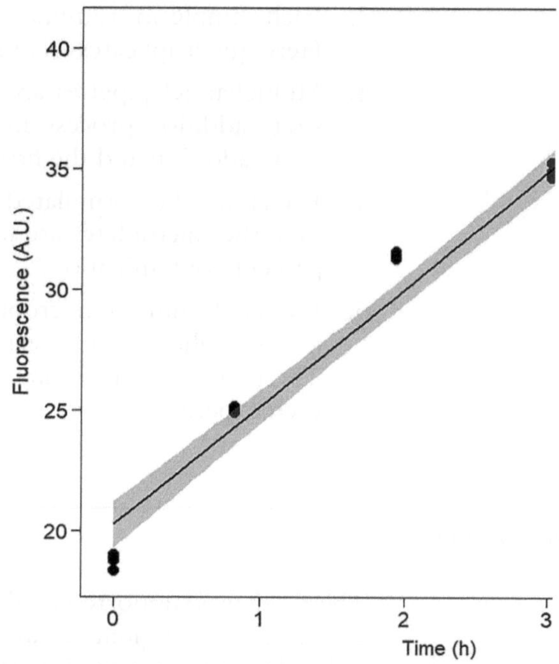

Fig. 3 Scatter plot showing fluorescence in arbitrary units (A.U.) over time in hours (h). Increasing fluorescence of L-leucine-7-amido-4-methylcoumarin hydrochloride demonstrating the activity of leucine-aminopeptidase in natural seawater sample over a 4-h incubation period.

(MCA) (Fig. 2) the hydrolytic rate of leucine aminopeptidase was calculated according to the formula:

$$\text{Hydrolytic rate (nmol L}^{-1}\text{ h}^{-1}) = \text{Fluorescence (A.U.) increase per hour} / 0.028 - 0.985$$

4 Notes

1. The 4-methylumbelliferyl phosphate (AP) powder dissolves rapidly in 2% MTXE but the 4-methylumbelliferyl β-D-glucopyranoside (BG) powder dissolves slower and requires to be stirred with a vortex mixer for 15 min to ensure all powder has dissolved.
2. To study enzyme activities in different size classes, water samples can be divided by gently filtering through membrane filters of various pore sizes [7]. Low protein binding syringe filters (e.g., Acrodisc® with Supor® Membrane, PALL Life Sciences) are recommended due to underestimation of the enzyme activity in the filtrates resulting from adsorption effects to the filter [8]. A low protein binding syringe filter with 0.2 μm pore size is also utilized in preparing the standard curve.

3. Each sample measurement should be performed in replicate. Here quadruplicates were used.
4. Multichannel pipettes are convenient as they accelerate substrate addition process and minimize the time between substrate addition and the first fluorescence measurement.
5. Exchange the formulated aluminum foil sealing films every time the microplates are taken from incubation to reading to prevent contamination.
6. Proper cleaning of microplates is challenging, and any residues of fluorophores can interfere with subsequent measurements. Therefore, it is recommended to use a fresh microplate for each experiment.

Acknowledgments

This work was supported by Walter and Andrée de Nottbeck foundation, Kone Foundation, and Finnish Cultural Foundation, and utilized marine research infrastructure at the Finnish Environment Institute as a part of the national FINMARI RI consortium.

References

1. Hoppe H-G (1983) Significance of exoenzyme activities in the ecology of brackish water: measurements by means of methylumbelliferyl-substrates. Mar Ecol Prog Ser 11:299–308
2. Hoppe H-G (1993) Use of fluorogenic model substrates for extracellular enzyme activity (EEA) measurement of bacteria. In: Kemp PF, Cole JJ, Sherr BF, Sherr EB (eds) Handbook of methods in aquatic microbial ecology. CRC Press, pp 423–431
3. Kim S-J, Hoppe H-G (1986) Microbial extracellular enzyme detection on agar plates by means of fluorogenic methylumbelliferyl substrates. GERBAM — Deuxième Colloque International de Bactériologie marine — CNRS, Brest, 1-5 octobre 1984. 1FREMER, Actes de Colloques, 3, pp 175–183
4. Obayashi Y, Wei Bong C, Suzuki S (2017) Methodological considerations and comparisons of measurement results for extracellular proteolytic enzyme activities in seawater. Front Microbiol 8:1952
5. Vanharanta M, Santoro M, Villena-Alemany C et al (2024) Microbial remineralization processes during post-spring-bloom with excess phosphate available in the northern Baltic Sea. FEMS Microbiol Ecol 100:fiae103
6. Baltar F, Legrand C, Pinhassi J (2016) Cell-free activity is linked to seasonal temperature changes: a case study in the Baltic Sea. Biogeosciences 13:2815–2821
7. Baltar F, De Corte D, Thomson B, Yokokawa T (2019) Teasing apart the different size pools of extracellular enzymatic activity in the ocean. Sci Total Environ 660:690–696
8. Obayashi Y, Suzuki S (2008) Adsorption of extracellular proteases in seawater onto filters during size fractionation. J Oceanogr 64:367–372

Chapter 18

Determining Inorganic and Organic Carbon

Jaana Koistinen, Mervi Sjöblom and Kristian Spilling

Abstract

Carbon is the element that makes up the major fraction of lipids and carbohydrates, which can be used for making biofuel. It is therefore important to provide enough carbon and also to follow the flow into particulate organic carbon and potential loss to dissolved organic forms of carbon. Here, we present methods for determining dissolved inorganic carbon, dissolved organic carbon, and particulate organic carbon.

Keywords Dissolved inorganic carbon, Dissolved organic carbon, Particulate organic carbon, pH

1 Introduction

Carbon (C) is a fundamental component of all life. It is fixed from carbon dioxide (CO_2) into biomass during photosynthesis, and microalgae account for approximately half the carbon fixation globally [1]. In culture, being able to measure different carbon pools is useful for understanding the conditions the algae have for carbon fixation (e.g., ensuring that they are not carbon-limited) and follow the flow into particulate organic carbon and potential loss to dissolved organic forms of carbon.

Particulate organic carbon (POC) is fundamental for understanding biologically relevant stoichiometric relationships, most notably the carbon:nitrogen:phosphorus (C:N:P) ratio, but other ratios can be useful, such as carbon-to-lipids ratio, which reveals information about carbon allocation of the algae. The Redfield ratio is often used as an indicator of algal growth requirements [2]. This ratio was originally derived from the average nutrient need for marine microalgae and the availability of these nutrients in the open ocean. The Redfield ratio is 106 carbon atoms to 16 nitrogen atoms to 1 phosphorus atom, but there is a large variability in this ratio depending on the species cultivated and available resources (C, N, P). For example, increasing light leads normally to higher C fixation and an increase in the C:N:P ratio; in addition, temperature and pH might also affect the stoichiometric ratio of the harvested biomass [2, 3]. For biofuel as the end product of microalgal cultivation, getting the

Kristian Spilling (ed.), *Biofuels from Algae*, Methods in Molecular Biology, vol. 2996,
https://doi.org/10.1007/978-1-0716-5031-8_18,

C:N:P ratio as high as possible is beneficial as it is the carbon which forms the main component of lipids and carbohydrates.

In this chapter, we present methods for determining dissolved inorganic carbon (DIC), dissolved organic carbon (DOC), and particulate organic carbon (POC). DIC consists of dissolved CO_2 and bicarbonate and carbonate ions (HCO_3^- and CO_3^{2-}), and is a direct measure of inorganic carbon availability for photosynthesis, but how much of the DIC is in the form of CO_2 is depending on the pH [4]. As such, pH is a good indicator of carbon availability, as the removal of CO_2 during photosynthesis increases pH (dissolved CO_2 is a weak acid). Measurements of DIC are needed for most methods used to determine carbon fixation (e.g., converting measurements of radiolabeled CO_2 fixation into total carbon fixation). Some of the carbon fixed during photosynthesis may be released into the water as DOC. High concentrations of DOC can inhibit growth and would in most cases increase bacterial production and biomass.

During DIC measurements, the carbonates within the sample are converted with an acid to CO_2, which is detected with an infrared gas analyzer [5]. The DOC method is based on high-temperature catalytic oxidation (~700°C) of a sample from which particulate carbon has been removed by filtration and all inorganic carbon has been removed by acidification and sparging [4]. DOC can be determined together with total dissolved nitrogen (TDN) using Total Organic Carbon (TOC) instrument equipped with nitrogen oxide analyzer and POC with particulate organic nitrogen (PON). The method for determining POC is based on filtration and is consequently a measure of all particulate carbon in the culture, but unless there are other particles in the cultivation water, it would be equal to the carbon found in the algal biomass. The sample on a filter is combusted at a very high temperature (~1000°C; catalytic oxidation) and POC is determined by a CN analyzer or a mass spectrometer [4]. Determination of TDN and PON are presented in a separate chapter.

All these methods are the same as those used for samples from the natural environment [e.g., 6, 7], but the possible higher concentration in algal cultures must be considered.

2 Materials

Use only analytical grade reagents.

2.1 Dissolved Inorganic Carbon

1. 25–50 ml glass BOD bottles with stoppers.
2. 24 ml glass autosampler vials with screw caps.
3. 50 ml graduated cylinder.
4. 100–500 ml volumetric flasks.
5. 100–500 ml storage bottles.
6. Sodium bicarbonate ($NaHCO_3$).

7. Sodium carbonate (Na_2CO_3).
8. Spatulas.
9. Analytical balance.
10. Small aluminum cups (approximately 5 mm Ø, 5 mm high).
11. Ultrapure water (e.g., Milli-Q).
12. Plastic film (e.g., Parafilm).
13. Dissolved inorganic carbon analyzer with instrument-specific carrier gas and oxidizing agents.
14. Concentrated phosphoric acid (85% H_3PO_4).
15. Refrigerator.

2.2 Dissolved Organic Carbon

1. 24 ml glass autosampler vials with screw caps.
2. Oven (450°C).
3. Concentrated hydrochloric acid (37% HCl).
4. Ultrapure water (e.g., Milli-Q).
5. 0.7 μm glass fiber filters (GF/F; 25 mm Ø) and filter holders (e.g., Swinnex) or 0,2 μm syringe filters.
6. Disposable 20 ml sterile syringes.
7. Potassium hydrogen phthalate ($C_8H_5KO_4$).
8. Spatulas.
9. Analytical balance.
10. 50 ml graduated cylinder.
11. 100–1000 ml volumetric flasks.
12. 100–1000 ml storage bottles.
13. Adjustable pipettes and pipette tips.
14. Ultrasonicator bath.
15. Plastic film (e.g., Parafilm).
16. Refrigerator and freezer.
17. Carbon analyzer with autosampler and instrument-specific carrier gas and oxidizing agents.

2.3 Particulate Organic Carbon

1. 0.7 μm glass fiber filters (e.g., 25 mm Ø).
2. Concentrated hydrochloric acid (37% HCl).
3. Forceps.
4. Ultrapure water (e.g., Milli-Q).
5. Oven (450°C).
6. Vacuum filtration device with a vacuum pump.
7. Small (30 mm Ø) disposable Petri dishes.
8. Aluminum foil circles (30 mm).

9. Handheld pliers suitable for making small balls out of aluminum.
10. 96-well plates.
11. Bunsen burner.
12. Ethanol (96%).
13. Glycine ($C_2H_5NO_2$).
14. Spatulas.
15. Analytical balance.
16. 100 ml volumetric flask and storage bottle.
17. Adjustable pipettes and pipette tips.
18. Refrigerator.
19. CN elemental analyzer with autosampler and instrument-specific carrier gas and oxidizing agents.

3 Methods

Use acid-washed glassware (soaked for 4 h in 6% HCl and rinsed with deionized water).

3.1 Dissolved Inorganic Carbon

1. Combust the sample bottles and aluminum cups for 4 h at 450°C and cool them to room temperature before use.
2. Prepare a stock standard solution (or weigh smaller amounts of standards for working standards in aluminum cups for transfer to volumetric flasks; *see* **Note 1**). For a 1000 mg/l stock solution, dissolve 1.75 g sodium bicarbonate and 2.205 g sodium carbonate in ultrapure water, and add ultrapure water to a final volume of 500 ml (stable for 4 months)
3. Prepare a 25% phosphoric acid solution by carefully dissolving 25 ml concentrated phosphoric acid in about 50 ml ultrapure water, fill up to a volume of 100 ml, and store in a glass bottle (*see* **Note 2**).
4. Take samples and put it in BOD bottles or autosampler vials, filling the whole volume, and store in a refrigerator (*see* **Note 3**). Collect in the BOD bottle if the sample needs to be prepared by filtering before analysis (*see* **Note 4**). The samples can be stored in a refrigerator for about 1 day (*see* **Note 3**).
5. Prepare calibration standards (e.g., 0–30 mg/l; *see* Step 2 and **Note 1**). Calibration standards should be made on the day of analysis.
6. Prepare the analyzer for analysis according to the operation manual (*see* **Note 5**).
7. Place standards in the autosampler of the analyzer for a calibration curve and linearity test (*see* **Note 6**).

8. Replace autosampler vial's screw cap with a plastic film.
9. Run the samples with the analyzer by measuring 3–5 replicates per sample. Verify proper operation of the instrument with reference material and include ultrapure water blanks in the run (*see* **Note 7**).

3.2 Dissolved Organic Carbon

1. Combust glass vials and GF/F filters for 4 h at 450°C and cool them to room temperature before use.
2. Prepare 2 mol/l HCl by dissolving 33.2 ml concentrated hydrochloric acid in ultrapure water and fill up to a volume of 200 ml. Store in a glass bottle.
3. Filter the sample with the syringe and filter. Rinse the syringe with the sample before filtering. Discard the first 5 ml of the filtrate (or more depending on the filter) and collect 20 ml into the glass vial. Prepare a filter blank by filtering ultrapure water.
4. Add 80 μl, 2 mol/l HCl to the glass vial.
5. Put on the lid and place the glass vial in a refrigerator or freezer until it can be run in the carbon analyzer (*see* **Note 8**).
6. Prepare a stock standard solution containing 1000 mg C/l by dissolving 2.125 g potassium hydrogen phthalate (pre-dried at 105°C for 1 h) in ultrapure water, and then add ultrapure water to a final volume of 1 l. Store the stock solution in a glass flask in the refrigerator (stable for 1 month).
7. At the time of measurement, prepare suitable working standards (e.g., 0–10 mg/l) for the calibration of the analyzer (*see* **Note 9**). Pipette potassium hydrogen phthalate stock solution and 2 mol/l HCl (200 μl/100 ml) into volumetric flasks and fill with ultrapure water (for blanks, use only ultrapure water and HCl). Calibration standards should be made weekly.
8. Fill each calibration vial about 1/2 to 2/3 full, put on a plastic film (*see* **Note 10**), and make holes in the film with a clean syringe needle.
9. Prepare the analyzer for analysis according to the operation manual (*see* **Note 5**).
10. Place standards in the autosampler of the analyzer and run the standards for a calibration curve and linearity test (*see* **Note 6**).
11. Take the samples out of the refrigerator/freezer and let them come to room temperature.
12. Place the samples in an ultrasonic bath for 10 min.
13. Remove the lids and put on plastic film (*see* **Note 10**) and make holes in the film with a clean syringe needle.
14. Run samples with the analyzer (*see* **Note 11**). Verify proper operation of the instrument with reference material (Consensus

Reference Water or equivalent) and include ultrapure water blanks in the run (*see* **Note 7**).

3.3 Particulate Organic Carbon

1. Prepare glass fiber filters by placing them in an acid bath (6% HCl) for 4 h, and then rinse thoroughly with plenty of ultrapure water (*see* **Note 12**).
2. Combust the filters and aluminum circles for 4 h at 450°C, and cool to room temperature.
3. Place a filter in the filtration device and apply suction with the vacuum pump.
4. Pipette a known volume of culture through the filter (*see* **Note 13**). Rinse with 5 ml ultrapure water to make sure all POC is on the filter. For blanks use ultrapure water.
5. Remove the filter with forceps while the suction is on (*see* **Note 14**). Rinse the filtration device with ultrapure water between samples.
6. Place the filter in a small Petri dish and dry it (*see* **Note 15**). For blanks, place pure glass fiber filters in the Petri dishes.
7. Prepare a stock standard solution containing 4290 mg C/l for POC by dissolving 1.340 g glycine in ultrapure water, and add ultrapure water to a final volume of 100 ml. Store the stock solution in a glass flask in a refrigerator (stable for 1 month).
8. Prepare standards for the calibration of the analyzer by adding a known volume (e.g., 40 μl) glycine stock solution onto glass fiber filters and dry the filters (*see* **Note 15**).
9. Place dried filters on the aluminum circles, one filter per circle.
10. Fold with two clean forceps (cleaned with ethanol and burned).
11. Make into a small ball with a handheld plier.
12. Put the small balls into a 96-well plate (*see* **Note 16**).
13. At the time of measurement, prepare the analyzer for analysis according to the operation manual (*see* **Note 5**).
14. Place the aluminum balls in the autosampler of the analyzer (*see* **Note 17**). Verify proper operation of the instrument with reference material and include blanks in the run (*see* **Note 7**).

4 Notes

1. Prepare 3–5 calibration standards according to the level of DIC in the water you are measuring.
2. Specifics of acid and carrier gas vary between instruments. We have used 25% phosphoric acid for the Shimadzu TOC-VCPH

instrument with ASI-V autosampler. TOC analyzers acidify a sample which causes inorganic carbon present to evolve into CO_2, which is detected by a non-dispersive infrared detector.

3. The point is that no headspace exists so that there is no gas exchange. The bottles should preferably be kept in a refrigerator as temperature determines gas solubility, i.e., keeping them warm would create bubbles inside, which would lead to underestimation of the concentration of inorganic carbon.
4. Samples must be pre-filtered in cases where calcium-containing organisms are cultivated in order to measure only the dissolved inorganic carbon. Filter straightly to autosampler vials.
5. It is critical to have proper training before running the analyzer. This will depend on the system that you are using, and it is beyond the scope of this chapter to go into the details of the operation of the instrument.
6. The calibration curve is prepared by plotting the measured area of standards versus standard concentrations (linear regression).
7. Start each sample run with ultrapure water blank(s) and reference material to verify proper operation of the instrument. Analyze samples in sequences of a maximum of 20 samples followed by a blank and reference material. Also, insert standards in the sequence to test accuracy and drift during the analytical runs (calibration standards at the beginning and end of the sample set).
8. These can be stored in the refrigerator for up to 3 weeks, or in the freezer for up to 6 months.
9. We have used ca. 5–10 different dilution steps for calibration depending on the concentration span of the samples.
10. This is to prevent contamination of other samples in case of bubbles forming during measurements.
11. We have a Shimadzu TOC-VCPH with ASI-V autosampler. The sample is first purged with CO2 free air to remove inorganic carbon, and then an aliquot is injected onto a combustion tube filled with platinum-coated alumina beads. The sample is purged through the combustion column where non-purgeable organic carbon compounds are converted to CO_2, which is detected by a non-dispersive infrared detector.
12. Using a large filtration device (e.g., 50 mm Ø), you may place several filters on top of each other and suck ultrapure water through the filters.
13. There should be a clear color on the filter but not forming a "cake" of biomass. The detection limits of the instrument used to determine the POC should be considered when making the filtration, and measuring the dry weight first will give you some idea of how much water you need to filter.

14. This is to minimize the water content as much as possible.
15. Make sure to use lids not closing the Petri dish completely, or alternatively use some aluminum foil, to let the filter dry completely. The filter can be stored dry at room temperature in a desiccator.
16. Marking the sample name with the well plate name makes it easy to follow the samples.
17. We have used Europa Scientific ANCA 20–20 stable isotope analyzer with a Sercon GSL sample preparation system. The combustion tube is filled with Cr_2O_3 (oxidation catalyst), CuO wire, and Ag wool. The balls are combusted at 1000°C to oxidize hydrocarbons, and the gases produced (CO_2, etc.) are transferred through a reduction tube for measurement in the detector of the analyzer.

Acknowledgments

This updated chapter was supported by the AlgaeProBANOS project, co-funded by the European Union through the Horizon Europe program (project number 101112943), and the equipment used were part of the Finnish Marine Research Infrastructure (FINMARI) consortium.

References

1. Falkowski PG, Raven JA (2013) Aquatic photosynthesis, 2nd edn. Princeton University Press, Princeton
2. Sterner RW, Elser JJ (2002) Ecological stoichiometry. Princeton University Press, Princeton
3. Spilling K, Ylöstalo P, Simis S, Seppälä J (2015) Interaction effects of light, temperature and nutrient limitations (N, P and Si) on growth, stoichiometry and photosynthetic parameters of the cold-water diatom Chaetoceros wighamii. PLoS One 10:e0126308
4. Grasshoff K, Ehrhardt M, Kremling K (1999) Methods of seawater analysis. Wiley VCH Verlag Gmbh, Weinheim
5. Salonen K, Kotimaa A-L (1975) The determination of dissolved inorganic carbon, a possible source of error in determining the primary production of lake water phytoplankton. Ann Bot Fennici 12:187–189
6. Spilling K, Asmala A, Haavisto N et al (2022) Brownification affects phytoplankton community composition but not primary productivity in eutrophic coastal waters: a mesocosm experiment in the Baltic Sea. Sci Total Environ 841:156510
7. Spilling K, Asmala A, Haavisto N et al (2022) Dataset from a mesocosm experiment on brownification in the Baltic Sea. Data Brief 45:108755

Check for updates

Chapter 19

Determining Inorganic and Organic Nitrogen

Jaana Koistinen, Mervi Sjöblom and Kristian Spilling

Abstract

Nitrogen (N) is one of the key nutrients for algal growth and is an integral part of many cellular components, for example, in proteins. Being able to determine the inorganic and organic pools of N is consequently critical for algal cultivation. In this chapter, we present the methods we use for determining dissolved inorganic nitrogen (DIN), dissolved organic nitrogen (DON), and particulate organic nitrogen (PON). The methods presented here for DIN rely on colorimetric methods, and those for DON and PON on filtration and high-temperature catalytic oxidation.

Keywords Ammonium, Dissolved organic nitrogen, Nitrate, Nutrient uptake, Particulate organic nitrogen

1 Introduction

Nitrogen (N) is one of the key nutrients for algal growth and is an integral part of many cellular components, for example, in proteins. There is plenty of nitrogen around, but most of it is in the form of nitrogen gas—N_2 (~78% of the atmosphere is nitrogen gas). This is not biologically available except for a few nitrogen fixers that can use N_2 gas, for example, some cyanobacteria. The basic inorganic forms of nitrogen that are used for algal growth are nitrate (NO_3), nitrite (NO_2), and ammonium (NH_4). In addition, some organic forms of nitrogen can be used by some algae, but large pools of dissolved organic nitrogen can yield high biomass of bacteria.

When cultivating algae, it is therefore important to have some idea of the availability of inorganic nitrogen, as this is a prerequisite for growth. Inorganic nitrogen is taken up and forms organic forms (e.g., protein), and being able to determine organic forms (particulate and dissolved forms) might be important for any downstream processing of the biomass and cultivation water after biomass removal.

In this chapter we present the methods we use for determining dissolved inorganic, dissolved organic, and particulate organic forms of N. The dissolved inorganic nitrogen (DIN; nitrite, nitrate, and

Kristian Spilling (ed.), *Biofuels from Algae*, Methods in Molecular Biology, vol. 2996,
https://doi.org/10.1007/978-1-0716-5031-8_19,

ammonium) methods are based on colorimetric principles, measuring the absorption of specific wavelengths after addition of chemicals reacting with the N-containing molecule [1]. In (NO_3+ NO_2) analysis, nitrate is reduced to nitrite, and all nitrite in the sample is reacted with an aromatic amine, which is then coupled with another aromatic amine to form an azo dye. The reduction of nitrate to nitrite here is performed with an acidic vanadium chloride reagent [2]. Ammonium analysis is based on the formation of a blue compound by salicylate and hypochlorite in the presence of sodium nitroprusside. The method also records unionized ammonia (NH_3). Dissolved organic nitrogen (DON) can be derived from high-temperature (~700°C) catalytic oxidation of a filtered and acidified sample [1]. The sample is sparged with synthetic air, and non-purgeable dissolved nitrogen compounds are then combusted and converted to nitrogen monoxide (NO), which is cooled and detected by gas-phase chemiluminescence [3], resulting in total dissolved nitrogen (TDN). DON is calculated as the difference between TDN and DIN, where DIN is the sum of all dissolved inorganic nitrogen species. TDN can be determined together with dissolved organic carbon (DOC) using Total Organic Carbon (TOC) instrument equipped with nitrogen oxide analyzer. The particulate organic nitrogen (PON) is measured simultaneously with particulate organic carbon (POC) after combustion. In this technique, the sample placed on a filter is combusted at a very high temperature (~1000°C) in a combustion tube filled with an oxidation catalyst, and oxides of N are reduced to N_2 in a reduction tube and determined by a CN analyzer or a mass spectrometer [1]. Determination of DOC and POC are presented in a separate chapter.

All these methods are the same as those used for samples from the natural environment [e.g., 4, 5], but the possible higher concentration in algal cultures must be considered.

2 Materials

Use only analytical grade reagents.

2.1 Dissolved Inorganic Nitrogen

1. Ammonium chloride (NH_4Cl).
2. Concentrated hydrochloric acid (37% HCl).
3. Concentrated nitric acid (65% HNO_3).
4. Disodium nitroprusside dihydrate ($Na_2Fe(CN)5NO{\cdot}2H_2O$).
5. N-(1-naphthyl)-ethylenediamine dihydrochloride ($C_{10}H_7NH\ CH_2{\cdot}CH_2{\cdot}NH_2{\cdot}2HCl$).
6. Sodium salicylate (HOC_6H_4COONa).
7. Potassium nitrate (KNO_3).
8. Sodium hydroxide (NaOH).
9. Sulfanilamide ($NH_2{\cdot}C_6H_6{\cdot}SO_2{\cdot}NH_2$).

10. Sodium dichloroisocyanurate dihydrate ($C_3Cl_2N_3NaO_3 \cdot 2H_2O$).
11. Tri-sodium citrate dihydrate ($C_6H_5Na3O_7 \cdot 2H_2O$).
12. Vanadium chloride (VCl_3).
13. Ultrapure water (e.g., Milli-Q).
14. 50 ml reaction flasks (for manual ammonium analysis).
15. 25 ml reaction flasks with screw caps (e.g., Pyrex containers for manual nitrate analysis).
16. 50 ml graduated cylinders.
17. 100–1000 ml volumetric flasks.
18. 100–1000 ml storage bottles.
19. Spatulas.
20. Adjustable pipettes and pipette tips.
21. Analytical balance.
22. Oven capable of being maintained at 45°C (for manual nitrate analysis).
23. Refrigerator.
24. 1 cm quartz glass cuvette or flow through cuvette (for manual nitrate analysis).
25. 5 cm quartz glass cuvette (for manual ammonium analysis).
26. Spectrophotometer or automatic analyzer.

2.2 Dissolved Organic Nitrogen (Total Dissolved Nitrogen)

1. 24 ml glass autosampler vials with screw caps.
2. Oven (450°C).
3. Concentrated hydrochloric acid (37% HCl).
4. Ultrapure water (e.g., Milli-Q).
5. 0.7 μm glass fiber filters (GF/F; 25 mm Ø) and syringe filter holders (e.g., Swinnex) or 0,2 μm syringe filters.
6. Disposable 20 ml sterile syringes.
7. Ammonium sulfate (NH_4SO_4).
8. Spatulas.
9. Analytical balance.
10. 100–1000 ml volumetric flasks and storage bottles.
11. Adjustable pipettes, pipette tips.
12. Ultrasonicator bath.
13. Plastic film (e.g., Parafilm).
14. Refrigerator and freezer.
15. Nitrogen analyzer with autosampler and instrument-specific carrier gas and oxidizing agents.

2.3 Particulate Organic Nitrogen

1. 0.7 μm glass fiber filters (25 mm Ø)
2. Concentrated hydrochloric acid (37% HCl).
3. Forceps.
4. Ultrapure water (e.g., Milli-Q).
5. Oven (450°C).
6. Vacuum filtration device with a vacuum pump.
7. Small (30 mm Ø) disposable petri dishes.
8. Aluminum foil circles (30 mm).
9. Handheld plier suitable for making small balls of the aluminum foil with filter.
10. 96-well plate.
11. Bunsen burner.
12. Ethanol (96%).
13. Glycine ($C_2H_5NO_2$).
14. Spatulas.
15. Analytical balance.
16. 100 ml volumetric flask and storage bottle.
17. Adjustable pipette and pipette tips.
18. Refrigerator.
19. CN elemental analyzer with autosampler and instrument-specific carrier gas and oxidizing agents.

3 Methods

Use acid-washed glassware (soaked for 4 h in 6% HCl and rinsed with deionized water).

3.1 Nitrate and Nitrite

This method will provide the combined nitrate plus nitrite concentration. The procedure follows that of Koroleff in the first edition of [1] and [6] (excluding the reduction and reaction to the azo dye that is performed here with a combined reagent, containing VCl_3 reduction solution as in [2]).

1. Prepare sulfanilamide solution by dissolving 1.0 g sulfanilamide in a mixture of 85 ml ultrapure water and 14.5 ml concentrated HCl. Store in a glass bottle at room temperature.
2. Prepare N-(1-naphthyl)-ethylenediamine dihydrochloride (NED) solution by dissolving 0.07 g NED in 100 ml ultrapure water. Store in a brown glass bottle in a refrigerator (renew if the color is brown; stable for about 1 month).

3. Prepare vanadium chloride solution by dissolving 0.8 g vanadium chloride (preferably in a fume hood due to possible corrosive fumes) in a mixture of 85 ml ultrapure water and 8.4 ml concentrated HCl; then add ultrapure water to a final total volume of 100 ml. Store in a glass bottle in a refrigerator (stable for about 1 month).
4. Prepare NO_3 mix reagent by mixing 100 ml vanadium chloride solution with 20 ml sulfanilamide solution and 20 ml amine solution. Store in plastic bottle in a refrigerator (stable about 1 week) or freeze (prolongs stability).
5. Prepare nitrate standard 1000 μg/ml N (71 μmol/ml N) by dissolving 0.7218 g potassium nitrate (pre-dried for 2 h at 110°C) in ultrapure water; then add ultrapure water to a final total volume of 100 ml. Store in a glass bottle in a refrigerator (stable for about 1 month).
6. On the day of analysis, prepare working solutions of nitrate (0–100 μg/l) for calibration by diluting the standard stock solution (see Step 5) with ultrapure water (e.g., 1 ml to 100 ml for 10 000 μg/l solution and then, e.g., 1 ml of this dilution to 100 ml ultrapure water for a 100 μg/l solution).
7. Prepare the analyzer for analysis according to the operation manual (*see* **Note 1**).
8. Place standards in the autosampler of an analyzer or measure the standards manually by a spectrophotometer for a calibration curve and linearity test (*see* **Note 2**).
9. For manual analysis, pipette three, e.g., 2 ml portions of each working standard and three 2 ml aliquots of ultrapure water as blanks into reaction flasks and add 2 ml mix reagent to standards and blanks. Close the flasks, mix well by swirling, and incubate for 1 h at 45°C.
10. Measure the absorbance of blanks and standards with a spectrophotometer in a 1 cm cuvette at 540 nm for a calibration curve (*see* **Note 2**).
11. Collect the samples in glass bottles or polyethylene bottles and analyze the samples immediately after collection (within 1 h). The sample can be stored in a glass bottle in a refrigerator for about 5 h.
12. Place samples in the autosampler of an analyzer or perform manual analysis with a spectrophotometer. Verify proper operation of the instrument with certified reference material and include ultrapure water blanks in the run (*see* **Note 3**).
13. For manual analysis, pipette 2 ml aliquot of each sample with a pipette to a reaction flask, add 2 ml mix reagent to the flask, close, mix well by swirling, and incubate 1 h at 45°C.

14. Measure the absorbance of the sample in a 1 cm cuvette at 540 nm (if the absorbance is greater than that of the highest calibration standard, prepare a diluted sample), and calculate the nitrate concentration by the calibration curve (*see* **Notes 4 and 5**).

3.2 Ammonium

Perform analyses in a well-ventilated room where no ammoniacal solutions are stored. The procedure follows that of SFS-EN ISO 15923-1 [7], which is based on ISO 7150-1 [8].

1. Prepare 0.06 mol/l nitric acid solution by carefully adding 1 ml concentrated nitric acid to 200 ml ultrapure water, mix, and add ultrapure water to a final volume of 250 ml. Store in a glass bottle at room temperature (stable for 1 year).
2. Prepare reagent I (sodium nitroprusside reagent) by dissolving 32.5 g sodium salicylate and 32.5 g tri-sodium citrate dihydrate in about 200 ml ultrapure water. Ensure that pH is <8.0 (if necessary, acidify with nitric acid solution). Dissolve 243 mg disodium nitroprusside dihydrate in this solution and add ultrapure water to a final volume of 250 ml. Store in a brown glass bottle in a refrigerator (stable for 1 week).
3. Prepare reagent II (sodium dichloroisocyanurate reagent, DIC reagent) by dissolving 8.0 g of sodium hydroxide in about 200 ml ultrapure water and cool down. Dissolve 500 mg sodium dichloroisocyanurate dihydrate in this solution and add ultrapure water to a final volume of 250 ml. Store in a brown glass bottle in a refrigerator (stable for 1 week).
4. Prepare ammonium standard 300 μg/ml N (21 μmol/ml) by dissolving 0.1146 g ammonium chloride (pre-dried at 100°C for 1 h) in ultrapure water and then adding ultrapure water to a final volume of 100 ml. Store in a glass bottle in a refrigerator (stable for approximately 1 month).
5. On the day of analysis, prepare suitable working solutions of ammonium (e.g., 0–90 μg N/l) for calibration by diluting the standard stock solution (see Step 4) with ultrapure water (e.g., 1 ml to 100 ml for a 3000 μg/l solution and then, e.g., 3 ml of this dilution to 100 ml ultrapure water for a 90 μg/l solution).
6. Prepare the analyzer for analysis according to the operation manual (*see* **Note 1**).
7. Place standards in the autosampler of an analyzer or measure the standards manually by a spectrophotometer for a calibration curve and linearity test (*see* **Note 2**).
8. For manual analysis, measure 20 ml of working standards and 20 ml ultrapure water for blanks with graduated cylinder and pour into reaction flasks.

9. Add 2 ml reagent I and 2 ml reagent II to standards and blanks (*see* **Note 6**). Swirl between additions to mix and close the flasks. Let stand for 1 h at room temperature (shorter for higher temperatures; *see* **Note 7**).
10. Measure the absorbance of blanks and standards in a 5 cm cuvette at 660 nm for the calibration curve (*see* **Note 2**).
11. Collect the samples in glass bottles or polyethylene bottles and analyze the samples immediately after collection. The sample can be stored in a glass bottle in a refrigerator for about 3 h.
12. Place samples in the autosampler of an analyzer or perform manual analysis with a spectrophotometer. In the case of saline samples, interference should be removed before analysis (*see* **Note 8**). Verify proper operation of the instrument with certified reference material and include ultrapure water blanks in the run (*see* **Note 3**).
13. For manual analysis, measure two 20 ml aliquots of sample with a graduated cylinder into reaction flasks (sample and reference).
14. Add 2 ml reagent I and 2 ml reagent II to the sample flask (*see* **Note 6**). Add 2 ml reagent I and 2 ml ultrapure water to the reference sample for turbidity. Swirl between additions to mix and close the flask. Let stand for 1 h at room temperature (shorter for higher temperatures; *see* **Note 7**). Remove possible precipitate by centrifugation (*see* **Note 8**).
15. Measure the absorbance of samples in a 5 cm cuvette at 660 nm (if the absorbance is greater than that of the highest calibration standard, prepare a diluted sample), and calculate the ammonium concentration by the calibration curve (*see* **Note 4**).

3.3 Dissolved Organic Nitrogen (Total Dissolved Nitrogen, TDN)

1. Combust the vials and GF/F filters for 4 h at 450°C and cool them to room temperature before use.
2. Prepare 2 mol/l HCl by dissolving 33.2 ml concentrated hydrochloric acid in ultrapure water and fill up to a volume of 200 ml. Store in a glass bottle.
3. Filter the sample with the syringe and filter. Rinse the syringe with the sample before filtering. Discard the first 5 ml of the filtrate (or more depending on the filter) and collect 20 ml into the glass vial. Prepare filter blank by filtering ultrapure water.
4. Add 80 μl 2 mol/l HCl to the vial.
5. Put on the lid and place the vial in the refrigerator or freezer until it can be run in the nitrogen analyzer (*see* **Note 9**).
6. Prepare a stock standard solution containing 1000 mg N/l by dissolving 4.716 g ammonium sulfate (pre-dried at 105°C for 1 h) in ultrapure water and fill up to a volume of 1 l. Store the stock solution in a glass flask in the refrigerator (stable for 1 month).

7. At the time of measurement, prepare suitable working standards (e.g., 0–1 mg/l) for the calibration of the analyzer (*see* **Note 10**). Prepare an intermediate standard solution 100 mg/l from the stock solution and pipette appropriate amounts of this solution and 2 mol/l HCl (200 μl/100 ml) for working standards into volumetric flasks and fill with ultrapure water (for blanks only ultrapure water and HCl). Calibration standards should be made weekly.
8. Fill each calibration vial about 1/2 to 2/3 full, put on a plastic film (*see* **Note 11**), and make holes on the film with a clean syringe needle.
9. Prepare the TDN analyzer for analysis according to the operation manual (*see* **Note 1**).
10. Place standards in the autosampler of an analyzer and run the standards for the TDN calibration curve and linearity test.
11. On the day of analysis, take samples out of the refrigerator/freezer and let them get room temperature before further processing.
12. Place the samples in an ultrasonication bath for 10 min.
13. Remove the lids, put on a plastic film (*see* **Note 11**), and make holes in the film with a clean syringe needle.
14. Run samples with the analyzer (*see* **Note 12**). Verify proper operation of the instrument with reference material (Consensus Reference Water or equivalent) and include ultrapure water blanks in the run (*see* **Note 3**).
15. Calculate DON by subtracting DIN from TDN.

3.4 Particulate Organic Nitrogen

1. Prepare glass fiber filters by placing them in an acid bath (6% HCl) for 4 h and rinse them thoroughly with plenty of ultrapure water (*see* **Note 13**).
2. Combust the filters and aluminum circles for 4 h at 450°C and cool to room temperature.
3. Place a filter in the filtration device and apply suction with the vacuum pump.
4. Pipette a known volume of culture through the filter (*see* **Note 14**). Rinse with 5 ml ultrapure water to make sure all PON is on the filter. For blanks use ultrapure water.
5. Remove the filter with forceps while the suction is on (*see* **Note 15**). Rinse the filtration device with ultrapure water between samples.
6. Place the filter in the Petri dish, and dry it (*see* **Note 16**). For blanks, place pure glass fiber filters in the Petri dishes.
7. Prepare a stock standard solution containing 2500 mg N/l by dissolving 1.340 g of glycine in ultrapure water and fill up

to a volume of 100 ml. Store the stock solution in a glass flask in the refrigerator (stable for 1 month).

8. Prepare standards for calibration of the analyzer by adding a known volume (e.g., 40 μl) glycine stock solution onto pure glass fiber filters and dry the filters (*see* **Note 16**).
9. Place dried filters on the aluminum circles, one filter per circle.
10. Fold with two clean forceps (cleaned with ethanol and quickly put into the flame of the Bunsen burner).
11. Make into a small ball with the handheld pliers.
12. Put the small balls into a 96-well plate (*see* **Note 17**).
13. At the time of the measurement, prepare the analyzer for analysis according to the operation manual (*see* **Note 1**).
14. Place the aluminum balls in the autosampler of the analyzer (*see* **Note 18**). Verify proper operation of the instrument with certified reference material and include blanks in the run (*see* **Note 3**).

4 Notes

1. It is critical to have proper training before running the analyzer. This will depend on the system that you are using, and it is beyond the scope of this chapter to go into the details of the operation of the instrument. We have used the Aquakem 250 analyzer for nitrate/nitrite and ammonium, the Shimadzu TOC-VCPH with ASI-V autosampler, and TNM-1 Total Nitrogen detector for TDN, and the Europa Scientific ANCA 20-20 Stable Isotope Analyser with a Sercon GSL sample preparation system for PON.
2. A calibration curve is prepared by plotting the measured absorbance of standards versus standard concentrations (linear regression).
3. Start each sample run with blank(s) and reference material to verify proper operation of the instrument. Analyze samples in sequences of a maximum of 20 samples, followed by a blank and reference material. Insert also standards in the sequence to test accuracy and drift during the analytical runs.
4. Determine the concentration in the sample by the slope of the calibration curve. Reduce the absorbance of the blank/reference from the absorbance of the sample.
5. The reduction capacity of NO_3 mix-reagent is not enough for concentrations >100 μg/l.

6. The maximum sample:reagent ratio is 10:1.
7. At incubation temperatures between 30 and 40°C, the recommended incubation time is 8 min.
8. The complexing capacity of citrate is exceeded in saline samples due to cations. This interference can be prevented by distillation [8]. Optionally, pretreatment with NaOH followed by centrifugation can be used to remove interferences. In manual analyses, precipitate formed after adding reagents can be removed by centrifugation.
9. These can be stored short term (2–3 weeks) in a refrigerator or alternatively in a freezer for 6 months.
10. Use ca. 5–10 different dilution steps for calibration depending on the concentration span of the samples.
11. This is to prevent contamination of other samples in case of bubbles forming during measurements.
12. In the Shimadzu TOC-VCPH system, the sample is purged with CO2 free air to remove inorganic purgeable species, then an aliquot is injected into a combustion tube filled with platinum-coated alumina beads, and combustion products are measured by the analyzer. Non-purgeable dissolved nitrogen compounds are converted to NO, which is mixed with ozone chemiluminescence and detected by a photomultiplier.
13. Using a large filtration device (e.g., 50 mm Ø) you may place several filters on top of each other and suck ultrapure water through the filters.
14. There should be a clear color on the filter, but not forming a "cake" of biomass.
15. This is to minimize the water content as much as possible.
16. Make sure to use lids not closing the Petri dish completely, or alternatively use some aluminum foil, to let the filter dry completely. The dried filter can be stored at room temperature in a desiccator.
17. Note, marking the sample name with the well plate name makes it easy to follow the samples.
18. In our instrument, the combustion tube is filled with an oxidation catalyst (Cr_2O_3), CuO wire, and Ag wool, and the reduction tube with Cu wires. The balls are combusted at 1000°C to oxidase hydrocarbons and combustion products (N_2, NO_x, etc.) are transferred through a reduction tube (to reduce oxides of N to N_2) for measurement in the detector of the analyzer.

Acknowledgments

This updated chapter was supported by the AlgaeProBANOS project, co-funded by the European Union through the Horizon Europe program (project number 101112943), and the authors used equipment part of the Finnish Marine Research Infrastructure (FINMARI) consortium.

References

1. Grasshoff K, Ehrhardt M, Kremling K (1999) Methods of seawater analysis. Wiley-VCH Verlag Gmbh, Weinheim
2. Schnetger B, Lehners C (2014) Determination of nitrate plus nitrite in small volume marine water samples using vanadium(III)chloride as a reduction agent. Mar Chem 160:91–98
3. Hutte RS, Sievers RE, Birks JW (1986) Gas chromatography detectors based on chemiluminescence. J Chromatogr Sci 24:499–505
4. Spilling K, Asmala A, Haavisto N et al (2022) Brownification affects phytoplankton community composition but not primary productivity in eutrophic coastal waters: a mesocosm experiment in the Baltic Sea. Sci Total Environ 841:156510
5. Spilling K, Asmala A, Haavisto N et al (2022) Dataset from a mesocosm experiment on brownification in the Baltic Sea. Data Brief 45:108755
6. Koroleff F (1979) Methods for the chemical analysis for seawater. Meri 7:1–60. (In Finnish)
7. SFS-EN ISO 15923-1 (2024) Water quality. Determination of selected parameters by discrete analysis systems. Part 1: Ammonium, nitrate, nitrite, chloride, orthophosphate, sulfate and silicate with photometric detection (ISO 15923-1:2013). https://www.iso.org/standard/55559.html
8. ISO 7150-1 (1984) Water quality – Determination of ammonium – Part 1: Manual spectrometric method. https://www.iso.org/standard/13742.html

Chapter 20

Total Nitrogen Determination by a Spectrophotometric Method

Jaana Koistinen, Mervi Sjöblom and Kristian Spilling

Abstract

Being able to measure total nitrogen (TN) is important for following the nitrogen budget. In this chapter, we present the spectrophotometric method we use for determining TN. The method relies on oxidation and reduction steps, involving persulfate digestion of nitrogen compounds into nitrate followed by spectrophotometric determination.

Keywords Nutrient uptake, Total nitrogen, Inorganic nitrogen, Organic nitrogen

1 Introduction

Total nitrogen (TN) includes all forms of nitrogen (organic and inorganic) that is found in water, with the exception of the gas phase (N_2), and is defined as the nitrogen that gives rise to nitrite/ nitrate ions. In practice, TN is the sum of nitrite, nitrate, ammonia, and dissolved and particulate organic nitrogen. TN is an important variable to measure for keeping track of the nitrogen budget and is commonly used for monitoring the state of the environment in lakes and the ocean and also for monitoring waste streams and cultivations in aquatic environments.

In this chapter, we present the method we use for determining total nitrogen. The method is based on colorimetric principles, measuring nitrogen after a persulfate oxidation of nitrogen compounds in alkaline conditions to nitrate, reduction step, and addition of chemicals reacting with the N-containing molecule [1]. Nitrate formed during oxidation is reduced to nitrite here with an acidic vanadium chloride reagent [2], which is simultaneously added with amines forming a colored compound, an azo dye, that is detectable spectrophotometrically. Total dissolved nitrogen (TDN) can be determined after filtering the sample, which excludes the particulate organic carbon (PON). TDN can be measured together with dissolved organic

Kristian Spilling (ed.), *Biofuels from Algae*, Methods in Molecular Biology, vol. 2996,
https://doi.org/10.1007/978-1-0716-5031-8_20,

carbon (DOC) using high-temperature (~700°C) catalytic oxidation [1]. PON can be determined using combustion at higher temperature (~1000°C) with an oxidation catalyst followed by reduction to N_2 (or determined as the difference between TN in unfiltered and filtered samples) [1]. Dissolved organic nitrogen (DON) can be determined as the difference between TDN and dissolved inorganic nitrogen (DIN; the sum of all dissolved inorganic nitrogen species). The method is versatile and have been used in a number of studies including algae, bacteria, and higher plants [e.g., 3–5]

2 Materials

Use only analytical grade reagents.

2.1 Total Nitrogen

1. Boric acid (H_3BO_3).
2. Concentrated hydrochloric acid (37% HCl).
3. Disodium ethylenediaminetetraacetate dihydrate (Na_2-EDTA; $C_{10}H_{14}N_2Na_2O_8 \cdot 2H_2O$; e.g., Titriplex® III).
4. N-(1-naphtyl)-ethylenediamine dihydrochloride ($C_{10}H_7NH\ CH_2 \cdot CH_2 \cdot NH_2 \cdot 2HCl$).
5. Potassium peroxodisulfate ($K_2S_2O_8$) with low nitrogen content (e.g., Merck 1.05092).
6. Sodium hydroxide (NaOH).
7. Sulfanilamide ($NH_2 \cdot C_6H_6 \cdot SO_2 \cdot NH_2$).
8. Vanadium chloride (VCl_3).
9. Ultrapure water (e.g., Milli-Q).
10. 25 ml reaction flasks with screw caps (e.g., Pyrex containers).
11. 10 ml reaction tubes with screw caps (e.g., Pyrex tubes for manual analysis).
12. 50–200 ml graduated cylinders.
13. 50–1000 ml volumetric flasks.
14. 100–1000 ml storage bottles.
15. Spatulas.
16. Adjustable pipettes and pipette tips.
17. 1 cm plastic cuvettes or flow-through cuvette (for manual analysis).
18. Analytical balance.
19. Autoclave.
20. Oven capable of being maintained at 45°C (for manual analysis).
21. Refrigerator.
22. Spectrophotometer or automatic analyzer.

3 Methods

Use acid-washed glassware (soaked for 4 h in 6% HCl and rinsed with deionized water).

3.1 Total Nitrogen

The procedure follows that of Koroleff in the first edition of [1, 6] (excluding the reduction and reaction to the azo dye that is performed here with a combined reagent, containing VCl_3 reduction solution as in [2]).

1. Prepare 0.075 mol/l sodium hydroxide solution by dissolving 3 g sodium hydroxide in 1000 ml of ultrapure water (*see* **Note 1**).
2. Prepare oxidizing reagent by dissolving 10 g potassium peroxodisulfate and 6 g boric acid in 1 l 0.075 mol/l sodium hydroxide solution (*see* **Notes 2 and 3**). Store in a brown glass bottle in a refrigerator (stable for several weeks).
3. Prepare sulfanilamide solution (nitrite reagent I) by dissolving 1.0 g sulfanilamide in a mixture of 85 ml ultrapure water and 14.5 ml concentrated HCl. Store in a glass bottle at room temperature.
4. Prepare N-(1-naphtyl)-ethylenediamine dihydrochloride (NED) solution (nitrite reagent II) by dissolving 0.07 g NED in 100 ml ultrapure water. Store in a brown glass bottle in a refrigerator (renew if the color is brown; stable about 1 month).
5. Prepare vanadium chloride solution by dissolving 1.6 g vanadium chloride (preferably in a fume hood due to possible corrosive fumes) in a mixture of 170 ml ultrapure water and 16.8 ml concentrated HCl, and then add ultrapure water to a final total volume of 200 ml. Store in glass bottle in a refrigerator (stable about 1 month).
6. Prepare TN mix reagent by mixing 200 ml vanadium chloride solution with 40 ml sulfanilamide solution and 40 ml amine solution. Store in plastic bottle in a refrigerator (stable about 1 week) or freeze (prolongs stability).
7. Prepare total nitrogen standard 140 µg/ml N (10 µmol/ml N) by dissolving 0.1862 g Na_2-EDTA in ultrapure water, and then add ultrapure water to a final total volume of 100 ml. Store in a glass bottle in a refrigerator (stable for several months).
8. On the day of analyses, prepare working solutions of total nitrogen (0–600 µg/l) for calibration by diluting the standard stock solution (see Step 7) with ultrapure water (e.g., 1 ml to 50 ml for 2800 µg/l solution and then, e.g., 10 ml of this dilution to 50 ml ultrapure water for 560 µg/l solution).

9. Pipette 5 ml of each calibration standard and 5 ml of ultrapure water for blanks to reaction flasks (*see* **Note 4**).
10. Add 5 ml of oxidizing reagent solution to each flask and close the flask. Mix by swirling, autoclave for 30 min, and allow to cool to room temperature.
11. Prepare analyzer for analyses according to the operation manual (*see* **Note 5**)
12. Place standards in the autosampler of an analyzer or measure the standards manually by a spectrophotometer for a calibration curve and linearity test (*see* **Note 6**).
13. For manual analyses, pipette 0.5 ml of the oxidized standards and blanks to reaction tubes. Add 3 ml TN mix reagent to each tube and close the tubes. Mix well by swirling, incubate for 1 h at 45°C, and allow to cool to room temperature.
14. Measure the absorbance of blanks and standards with a spectrophotometer in a 1 cm cuvette (or flow-through cuvette) at 545 nm for a calibration curve (*see* **Note 6**).
15. Collect the samples in glass bottles or polyethylene bottles and analyze the samples immediately after collection. The sample can be stored in a glass bottle in a refrigerator for a day.
16. Pipette 5 ml of sample (or diluted sample; *see* **Note 7**) to reaction flasks.
17. Add 5 ml of oxidizing reagent to each flask and close the flasks. Add 5 ml of oxidizing reagent to three empty flasks for reagent blanks (*see* **Note 8**). Mix by swirling, autoclave for 30 min, and allow to cool to room temperature.
18. Add 5 ml of ultrapure water to the reagent blanks, close the flasks, and mix well.
19. Place the samples and reagent blanks in the autosampler of an analyzer or perform manual analyses with a spectrophotometer. Verify proper operation of the instrument with certified reference material, and include ultrapure water blanks in the run (*see* **Note 9**).
20. For manual analyses, pipette 0.5 ml of the oxidized samples and reagent blanks to reaction tubes. Add 3 ml of TN mix reagent to each tube and close the tubes. Mix well by swirling, incubate 1 h at 45°C, and allow to cool to room temperature.
21. Measure the absorbance of the reagent blanks and samples with a spectrophotometer in a 1 cm cuvette (or flow-through cuvette) at 545 nm, and calculate the total nitrogen concentration by the calibration curve (*see* **Note 10**).

4 Notes

1. The strength of the NaOH solution should ensure that excess OH^- ions are present after the decomposition of the oxidant. The amount of NaOH in the solution should be adequate for neutralization of the 2 mol of H^+ formed per mol of persulfate.
2. For complete oxidization of organic material, oxygen available from the persulfate oxidation should be ten times higher than that of organic material in the sample.
3. Total nitrogen and total phosphorus can be analyzed after simultaneous oxidation with persulfate using more concentrated oxidizing reagent.
4. Use five different dilution steps for calibration and determine the blank with ultrapure water since ultrapure water may contain nitrogen traces.
5. It is critical to have proper training before running the analyzer. This will depend on the system that you are using, and it is beyond the scope of this chapter to go into the details of the operation of the instrument. We have used Thermo Scientific Aquakem 250 analyzer for TN analysis.
6. Calibration curve is prepared by plotting measured absorbance of standards versus standard concentrations (linear regression). Reduce the absorbance of the blank from the absorbance of the standards.
7. If the total nitrogen content is higher than 600 μg/l, the sample should first be diluted with ultrapure water. The total carbon content of the sample should not exceed 20 mg/l when using 5 ml of the oxidizing solution for 5 ml of sample.
8. Determine the blank caused by the oxidant since persulfate contains traces of nitrogen.
9. Start each sample run with ultrapure water blank(s) and reference material to verify proper operation of the instrument. Analyze samples in sequences of maximum 20 samples followed by a blank and reference material. Insert also standards in the sequence to test accuracy and drift during the analytical runs.
10. Determine concentration in the sample by the slope of the calibration curve. Reduce the absorbance of the measured reagent blank and the reagent blank value taken from the intercept of the calibration line with the ordinate from the absorbance of the sample.

Acknowledgments

This updated chapter was supported by the AlgaeProBANOS project, co-funded by the European Union through the Horizon Europe program (project number 101112943), and the authors used equipment part of the Finnish marine research infrastructure (FINMARI) consortium.

References

1. Grasshoff K, Ehrhardt M, Kremling K (1999) Methods of seawater analysis. Wiley-VCH Verlag Gmbh, Weinheim
2. Schnetger B, Lehners C (2014) Determination of nitrate plus nitrite in small volume marine-water amples using vanadium(III)chloride as a reduction agent. Mar Chem 160:91–98
3. Spilling K, Asmala A, Haavisto N et al (2022) Dataset from a mesocosm experiment on brownification in the Baltic Sea. Data Brief 45:108755
4. Fang J, Adams JM, Liu Z, Deng Y (2025) Soil salinity and moisture have contrasting effects on deterministic versus stochastic assembly of bacterial communities in alpine lake shores. Appl Soil Ecol 211:106115
5. Rosli KA, Misran A, Yazan LS, Wahab PEM (2025) High-light and nutrient interactions drive carbohydrate and proton pump dynamics in Agastache rugosa (Fisch. & CA Mey.) Kuntze. Plant Physiol Biochem 219:109374
6. Koroleff F (1979) Methods for the chemical analysis for seawater. Meri 7:1–60. (in Finnish)

Check for updates

Chapter 21

Determining Inorganic and Organic Phosphorus

Jaana Koistinen, Mervi Sjöblom and Kristian Spilling

Abstract

Phosphorus (P) is a macronutrient for all microalgal species, and the main form of uptake is orthophosphate (PO_4). In this chapter we present a colorimetric method for determining the PO_4 concentration and dissolved organic phosphorus (DOP) based on total phosphorus (TP) measurements. We also describe a method for determining particulate organic phosphorus (POP) based on the same principles.

Keywords Orthophosphate, Particulate organic phosphorus, Total phosphorus

1 Introduction

In addition to nitrogen (see separate chapter), phosphorus (P) is the other main nutrient that often limits microalgae growth in natural environments. Algae are in general only able to use orthophosphate (PO_4), but some species are known to be able to use organic forms to some extent [1].

Phosphorus inside the cells can be found, e.g., in the DNA and RNA, and is as such critical for the growth machinery of algal cells. Comparing different algae with different growth characteristics has revealed that the internal ratio of carbon (C):nitrogen (N):phosphorus (P) is lower for species growing rapidly compared with slow-growing species [2], and as such providing more phosphorus than the much used Redfield ratio (106 C:16 N:1 P) might be required to get maximum growth rates.

In this chapter, we present a colorimetric method for determining the PO_4 and dissolved organic phosphorus (DOP) based on total phosphorus (TP) measurements [3, 4] and a method for determining particulate organic phosphorus (POP) based on the same principles [5]. The determination of TP is based on acid oxidation of organic components with peroxodisulfate in an autoclave, releasing phosphorus as phosphate [3]. The POP method involves baking at a high temperature (to decompose organic phosphorus compounds) and hydrolysis of polyphosphate to orthophosphate. Phosphate analysis is based on the antimony-molybdate method. Orthophosphate

Kristian Spilling (ed.), *Biofuels from Algae*, Methods in Molecular Biology, vol. 2996,
https://doi.org/10.1007/978-1-0716-5031-8_21,

reacts with molybdate and antimony in an acidic solution to form an antimonyphosphomolybdate complex, which is then reduced by ascorbic acid, forming a blue color. For determining the DOP, there is no direct way as with dissolved organic carbon or nitrogen. Total phosphorus can be used to calculate DOP by subtracting the other P pools (POP could also be removed by first filtering the sample).

All these methods are the same as what is used for samples from the natural environment [e.g., 6, 7], but does also have other uses including higher plants and algal cultures [8], but then the possible higher concentrations must be taken into consideration.

2 Materials

Use only analytical grade reagents.

2.1 Phosphate and Dissolved Organic Phosphorus from Total Phosphorus Measurements

1. Ammonium heptamolybdate tetrahydrate ($(NH_4)6Mo_7O_{24}\cdot 4H_2O$).
2. Ascorbic acid ($C_6H_8O_6$).
3. Potassium persulfate ($K_2S_2O_8$).
4. Potassium antimony tartrate ($K(SbO)C_4H_4O_6$).
5. Potassium dihydrogen phosphate (KH_2PO_4).
6. Concentrated sulfuric acid (96% H_2SO_4).
7. Concentrated hydrochloric acid (37% HCl).
8. Ultrapure water (e.g., Milli-Q).
9. 50 ml reaction flasks (e.g., Erlenmeyer flasks for manual phosphate analysis).
10. 25–50 ml reaction flasks with screw caps (e.g., Pyrex containers for total phosphorus).
11. 50–500 ml graduated cylinders.
12. 50–500 ml volumetric flasks.
13. 100–500 ml storage bottles.
14. Adjustable pipettes and pipette tips.
15. Spatulas.
16. Analytical balance.
17. Refrigerator.
18. Autoclave.
19. Spectrophotometer or automatic analyzer.
20. 1–5 cm quartz glass cuvette (for manual analysis).
21. 0.7 μm glass fiber filters (25 mm Ø) and filter holders (e.g., Swinnex).
22. 50 ml disposable sterile syringes.
23. Oven (450°C).

2.2 Particulate Organic Phosphorus

1. 0.7 μm glass fiber filters (25 mm Ø).
2. Concentrated sulfuric acid (96% H_2SO_4).
3. Concentrated hydrochloric acid (37% HCl).
4. Sodium sulfate (Na_2SO_4).
5. Magnesium sulfate heptahydrate ($MgSO_4 \cdot 7H_2O$).
6. Ammonium heptamolybdate tetrahydrate ($(NH_4)6Mo_7O_{24} \cdot 4H_2O$).
7. Ascorbic acid ($C_6H_8O_6$).
8. Potassium antimony tartrate ($K(SbO)C_4H_4O_6$).
9. Potassium dihydrogen phosphate (KH_2PO_4).
10. Ultrapure water (e.g., Milli-Q).
11. 50–500 ml graduated cylinders
12. 50–1000 ml volumetric flasks.
13. 100–1000 ml storage bottles.
14. 20 glass vials with screw caps (e.g., scintillation vials).
15. Forceps.
16. Spatulas.
17. Adjustable pipette and pipette tips.
18. Oven (450°C).
19. Vacuum filtration device with a vacuum pump.
20. Analytical balance.
21. Refrigerator.
22. Water bath or dry block heater.
23. Spectrophotometer or automatic analyzer.
24. 2 cm quartz glass cuvette (for manual analyses).

3 Methods

Use acid-washed glassware (soaked for 4 h in 6% HCl and rinsed with deionized water).

3.1 Phosphate

The procedure follows that of Koroleff [4].

1. Prepare 4.5 mol/l sulfuric acid solution. Carefully add 125 ml concentrated acid to 375 ml ultrapure water, cool, and make up to 500 ml with ultrapure water. Store in an amber glass at room temperature.
2. Prepare molybdate solution dissolving 4.75 g ammonium heptamolybdate tetrahydrate in 45 ml ultrapure water and make up to 50 ml with ultrapure water. Store in a glass or

polyethylene bottle at room temperature (usable as long as clear).

3. Prepare tartrate solution by dissolving 1.67 g potassium antimony tartrate in 50 ml ultrapure water. Store in a glass bottle at room temperature (usable as long as clear).
4. Prepare reagent I (acidified ascorbic acid solution) by dissolving 7.0 g ascorbic acid in 65 ml ultrapure water, and then add 35 ml 4.5 mol/l H_2SO_4. Store in an amber glass bottle in a refrigerator (usable as long as colorless).
5. Prepare reagent II (mix reagent) by mixing 22.5 ml molybdate solution to 60 ml 4.5 mol/l H_2SO_4. Add 2.5 ml tartrate solution and 35 ml ultrapure water. Store in an amber glass bottle at room temperature (stable for months).
6. Prepare phosphate standard 310 μg P/ml by dissolving 0.1361 g potassium dihydrogen phosphate (dried for 2 h at 110°C) in ultrapure water containing 1 ml 4.5 mol/l H_2SO_4, and make up to 100 ml with ultrapure water. Store in a glass bottle in a refrigerator (stable for months).
7. On the day of analysis, prepare suitable working solutions of phosphate (e.g., 0–310 μg P/l) for calibration by diluting the standard stock solution (see Step 6) with ultrapure water (e.g., 0.1 ml in 100 ml ultrapure water for 310 μg P/l solution).
8. Prepare analyzer for analyses according to the operation manual (*see* **Note 1**). Dilute the reagents 1:10 for analyzer.
9. Place standards in the autosampler of an analyzer, or measure the standards manually by a spectrophotometer for a calibration curve and linearity test (*see* **Note 2**).
10. For manual analyses, measure 35 ml of each calibration standard and 35 ml of ultrapure water for blanks with graduated cylinder to reaction flasks.
11. Add 1 ml of reagent I to standards and the blanks. Mix well by swirling.
12. Add 1 ml of reagent II to standards and the blanks. Mix well by swirling and allow to react for 10 min.
13. Measure the absorbance of standards and blanks in a 5 cm cuvette at 880 nm for a calibration curve (*see* **Note 2**).
14. Collect the samples in glass bottles or polyethylene bottles and analyze the samples immediately after collection (within 2 h; store protected from light in a refrigerator).
15. Place samples in the autosampler of an analyzer or perform manual analyses with a spectrophotometer. Verify proper operation of the instrument with certified reference material and include ultrapure water blanks in the run (*see* **Note 3**).

16. For manual analyses, measure two 35 ml portions of the sample with a graduated cylinder (sample and reference).
17. Add 1 ml of reagent I to both flasks. Mix well by swirling.
18. Add 1 ml of reagent II to the first flask and 1 ml ultrapure water to the second flask (the reference). Mix well by swirling and allow to react for 10 min.
19. Measure the absorbance of samples in a 5 cm cuvette at 880 nm (if the absorbance is greater than that of the highest calibration standard, prepare diluted sample or use smaller cuvette), and calculate the phosphate concentration by the calibration curve (*see* **Note 4**).

3.2 Dissolved Organic Phosphorus from Total Phosphorus Measurements

The procedure follows that of Koroleff [4].

1. Prepare 4.5 mol/l sulfuric acid solution (see Sect. 3.1).
2. Prepare oxidizing reagent by diluting 15 ml of 4.5 mol/l sulfuric acid to 100 ml with ultrapure water. Dissolve 5 g potassium peroxodisulfate in the solution. Store in a polyethylene bottle protected from light at room temperature (stable for about 1 week).
3. Prepare molybdate solution (see Sect. 3.1).
4. Prepare tartrate solution (see Sect. 3.1).
5. Prepare reagent I (ascorbic acid solution) by dissolving 7.0 g ascorbic acid in 100 ml ultrapure water. Store in an amber glass bottle in a refrigerator (stable for at least 1 month, usable as long as almost colorless).
6. Prepare reagent II (mix reagent) (see Sect. 3.1).
7. Prepare phosphate standard 310 μg P/ml (see Sect. 3.1).
8. On the day of analysis, prepare suitable working solutions of phosphate (e.g., 0–600 μg P/l) for calibration by diluting the standard stock solution (see Sect. 3.1) with ultrapure water (e.g., 0.2 ml in 100 ml ultrapure water for 620 μg P/l solution).
9. Measure 35 ml of each calibration standard and 35 ml of ultrapure water for blanks with graduated cylinder to reaction flasks.
10. Add 2 ml of oxidizing reagent solution to the flasks and dissolve by swirling. Close the flasks and place in the autoclave. Autoclave for 30 min and allow to cool to room temperature.
11. Prepare analyzer for analyses according to the operation manual (*see* **Note 1**). Dilute the reagents 1:10 for analyzer.
12. Place standards in the autosampler of an analyzer, or measure the standards manually by a spectrophotometer for a calibration curve and linearity test (*see* **Note 2**).

13. For manual analyses, add 1 ml of reagent I (see Step 5) to standards and blanks. Mix well by swirling.
14. Add 1 ml of reagent II (see Step 6) to standards and blanks. Mix well by swirling and allow to react for 10 min.
15. Measure the absorbance of blanks and standards in a 5 cm cuvette at 880 nm for a calibration curve (*see* **Note 2**).
16. Filter the sample with the syringe and filter. Rinse the syringe with the sample before filtering. Prepare filter blank by filtering ultrapure water.
17. Collect the samples in polyethylene bottles and analyze the samples immediately after collection (within 2 h; store protected from light in a refrigerator).
18. Measure 35 ml of the sample with a graduated cylinder to reaction flasks.
19. Add 2 ml of oxidizing reagent solution to the flasks and dissolve by swirling. Close the flasks and place in the autoclave. Autoclave for 30 min and allow to cool to room temperature.
20. Place samples in the autosampler of an analyzer or perform manual analyses with a spectrophotometer. Verify proper operation of the instrument with certified reference material and include ultrapure water blanks in the run (*see* **Note 3**).
21. For manual analyses, add 1 ml of reagent I (see Step 5) to the flasks. Mix well by swirling.
22. Add 1 ml of reagent II (see Step 6) to the flasks. Mix well by swirling and allow to react for 10 min.
23. Measure the absorbance of samples in a 5 cm cuvette at 880 nm (if the absorbance is greater than that of the highest calibration standard, prepare diluted sample or use 1 cm cuvette), and calculate the total phosphorus concentration by the calibration curve (*see* **Note 4**).
24. The results provide the total phosphorus for the filtered sample. For determining the dissolved organic phosphorus, subtract the PO_4 concentration (see Sect. 3.1).

3.3 Particulate Organic Phosphorus

The method is a modified version of Solórzano and Sharp [5].

1. Prepare glass fiber filters and glass scintillation vials by acid wash (soak in 6% HCl for 4 h), rinse with ultrapure water, and combust for at 450°C for 4 h.
2. Prepare sodium sulfate solution (0.17 mol/l) by dissolving 12 g anhydrous Na_2SO_4 in 500 ml ultrapure water. Store in a glass bottle at room temperature (stable for several months).
3. Prepare magnesium sulfate solution (0.17 mol/l) by dissolving 42 g $MgSO_4{\cdot}7H_2O$ in 1 l ultrapure water. Store in a glass bottle at room temperature (stable for several months).

4. Place the filter in the filtration device and apply suction with the vacuum pump.
5. Pipette a known volume of culture through the filter (*see* **Note 5**), and for blanks use ultrapure water.
6. Rinse with 4 ml 0.17 mol/l Na_2SO_4.
7. Remove the filter with forceps while the suction is on (*see* **Note 6**). Rinse the device with ultrapure water between samples.
8. Place the filter in the glass scintillation vial and dry it (*see* **Note 7**). For blanks, place pure filters in scintillation vials.
9. Prepare 4.5 mol/l sulfuric acid solution (see Sect. 3.1).
10. Prepare molybdate solution (see Sect. 3.1).
11. Prepare tartrate solution (see Sect. 3.1).
12. Prepare POP reagent I by dissolving 7.0 g ascorbic acid in 100 ml ultrapure water (as in Sect. 3.2 reagent I). Store in amber glass bottle in a refrigerator (usable as long as colorless).
13. Prepare POP reagent II (mix reagent; as in Sect. 3.1 reagent II, except 100 ml 4.5 mol/l H_2SO_4 and no water).
14. Prepare phosphate standard 310 μg P/ml (see Sect. 3.1).
15. Prepare 0.2 mol/l HCl solution. Carefully add 16.6 ml of concentrated HCl to 500 ml of ultrapure water, cool, and make up to 1000 ml with ultrapure water. Store in a glass bottle at room temperature (stable for months).
16. Prepare working standards of phosphate (e.g., 0–600 μg P/l) for calibration by diluting the standard stock solution with ultrapure water (e.g., 0.2 ml in 100 ml for 620 μg P/l solution).
17. Pipette 10 ml calibration solution and 10 ml ultrapure water for blanks to scintillation vials and add 200 μl of 0.17 mol/l $MgSO_4$.
18. Standards and blanks are put in warm cabinet at 95°C until solution has dried completely.
19. Put in oven at 450°C for 2 h.
20. Add 5 ml 0.2 mol/l HCl and add caps. Mix by swirling.
21. Place in water bath (or dry block heater), 80°C for 30 min, and cool to room temperature.
22. Add 5 ml ultrapure water, add caps, and mix by swirling.
23. Prepare analyzer for analyses according to the operation manual (*see* **Note 1**). Dilute the reagents 1:10 for analyzer.
24. Place standards in the autosampler of an analyzer or measure the standards manually by a spectrophotometer for a calibration curve and linearity test (*see* **Note 2**).

25. For manual analyses, add 300 μl of POP reagent I and mix by swirling. Add 300 μl POP reagent II. Mix well by swirling and allow to react for 10 min.
26. Measure the absorbance of standards and blanks in a 2 cm cuvette at 880 nm for a calibration curve and linearity test (*see* **Note 2**).
27. When measuring samples, start by adding 2 ml 0.017 mol/l $MgSO_4$ to samples and filter blanks. Prepare 0.017 mol/l $MgSO_4$ solution by diluting 0.17 mol/l solution with ultrapure water.
28. Put in warm cabinet at 95°C until solution has dried.
29. Put in oven at 450°C for 2 h (*see* **Note 8**).
30. Add 5 ml 0.2 mol/l HCl and add caps. Mix by swirling.
31. Place in water bath (or dry block heater), 80°C for 30 min, and cool to room temperature.
32. Add 5 ml ultrapure water, add caps, and mix by swirling.
33. Place samples in the autosampler of an analyzer or perform manual analyses with a spectrophotometer. Verify proper operation of the instrument with certified reference material and include ultrapure water blanks in the run (*see* **Note 3**).
34. For manual analyses, add 300 μl of POP reagent I and mix by swirling. Add 300 μl of POP reagent II and mix well by swirling, and allow to react for 10 min.
35. Measure absorption of the samples at 880 nm with 2 cm cuvette (if the absorbance is greater than that of the highest calibration standard, prepare diluted sample), and calculate the particulate organic phosphate concentration by the calibration curve (*see* **Note 4**).

4 Notes

1. It is critical to have proper training before running the analyzer. This will depend on the system that you are using, and it is beyond the scope of this chapter to go into the details of the operation of the instrument. We have used Thermo Scientific Aquakem 250 analyzer for phosphate.
2. Calibration curve is prepared by plotting measured absorbance of standards versus standard concentrations (linear regression).
3. Start each sample run with ultrapure water blank(s) and reference material to verify proper operation of the instrument. Analyze samples in sequences of max 20 samples followed by

a blank and reference material. Also insert standards in the sequence to test accuracy and drift during the analytical runs.

4. Determine concentration in the sample by the slope of the calibration curve. Reduce the absorbance of the blank/reference from the absorbance of the sample.
5. There should be a clear color on the filter, but not forming a "cake" of biomass.
6. This is to minimize the water content as much as possible.
7. Make sure to loosely cover the scintillation vials, e.g., with aluminum foil; once the filter is completely dry, close the cap. The dried filter can be stored at room temperature.
8. Any marking will burn off, so they need to be placed in a way that you can keep track of sample numbers; mark again after they have cooled down.

Acknowledgments

This updated chapter was supported by the AlgaeProBANOS project, co-funded by the European Union through the Horizon Europe program (project number 101112943), and the authors used equipment part of the Finnish marine research infrastructure (FINMARI) consortium.

References

1. Dyhrman ST (2016) Nutrients and their acquisition: phosphorus physiology in microalgae. In: Borowitzka M, Beardall J (eds) The physiology of microalgae. Springer International Publishing, New York
2. Hillebrand H, Steinert G, Boersma M et al (2013) Goldman revisited: faster-growing phytoplankton has lower N:P and lower stoichiometric flexibility. Limnol Oceanogr 58:2076–2088
3. Grasshoff K, Ehrhardt M, Kremling K (1999) Methods of seawater analysis. Wiley-VCH Verlag Gmbh, Weinheim
4. Koroleff F (1979) Methods for the chemical analysis for seawater. Meri 7:1–60. (in Finnish)
5. Solórzano L, Sharp JH (1980) Determination of total dissolved phosphorus and particulate phosphorus in natural waters. Limnol Oceanogr 25:754–758
6. Spilling K, Asmala A, Haavisto N et al (2022) Brownification affects phytoplankton community composition but not primary productivity in eutrophic coastal waters: a mesocosm experiment in the Baltic Sea. Sci Total Environ 841:156510
7. Spilling K, Asmala A, Haavisto N et al (2022) Dataset from a mesocosm experiment on brownification in the Baltic Sea. Data Brief 45:108755
8. Rosli KA, Misran A, Yazan LS, Wahab PEM (2025) High-light and nutrient interactions drive carbohydrate and proton pump dynamics in *Agastache rugosa* (Fisch. & CA Mey.) Kuntze. Plant Physiol Biochem 219:109374

Chapter 22

Determining Dissolved and Biogenic Silica

Jaana Koistinen, Mervi Sjöblom and Kristian Spilling

Abstract

Most algae do not use silicon in any form with one notable exception, diatoms. Silicon is a major constituent of diatoms. Diatoms are characterized by high growth rates and are often one of the key groups in forming algal blooms in natural waters, and as such it is an interesting group for cultivation. In this chapter we present methods for determining dissolved silica (DSi) and biogenic silica (BSi), oxide forms of silicon, based on colorimetric methods. BSi is determined after filtration and alkaline digestion.

Keywords Biogenic silica, Diatoms, Dissolved silica, Silicon

1 Introduction

Most algae do not use silicon (Si) in any form with one notable exception, diatoms, in addition to some smaller groups such as silicoflagellates. Silicon, one of the most abundant elements on the surface of the earth, is a major constituent of diatoms and may become limiting for diatom growth. Silicon appears as the oxide (silica) or silicates, most of the silicon being bound to rocks as silicate minerals. Chemical weathering brings dissolved silica (DSi), the form that is available for diatoms. Diatoms are characterized by high growth rates and are often one of the key groups in forming algal blooms in natural waters, and as such it is an interesting group for cultivation [1, 2]. Part of the success of the diatoms has been attributed to the fact that they have silicified cell walls, hydrated amorphous silica forming the outer cell wall, termed frustules (Fig. 1). The metabolic cost of producing silicified walls has been hypothesized to the lower than for building carbon based (e.g., cellulose) walls and provide the diatoms with the potentially very high (>1/day) maximum growth rate [3, 4].

The diatom frustules consist of two halves that are joined in the middle similar to a box with a lid, where both the box and lid part are of equal size in most species. During cell division, both of the two

Kristian Spilling (ed.), *Biofuels from Algae*, Methods in Molecular Biology, vol. 2996,
https://doi.org/10.1007/978-1-0716-5031-8_22,

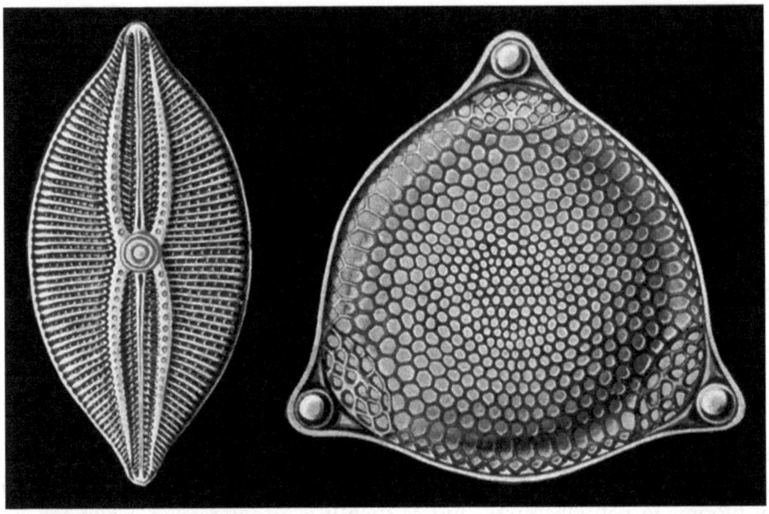

Fig. 1 Two examples of diatom frustules: *Navicula bullata* left and *Triceratium robertsianum* right [11]. This work is in the public domain

halves become the lid part of the new cell. The old lid forms a new cell of equal size whereas the old box part (now the new lid) forms a slightly smaller cell. This leads to smaller average size of the population over time (several generations). When cells reach the smallest possible size, a spore known as an auxiospore forms. The auxiospore increases its size to the maximum for the species, and the process starts form the beginning again [5].

Measuring silicon is only necessary when cultivating diatoms, but then it is critical to keep track of the silicon concentration. The need for silicon is often considered as 1:1 ratio with nitrogen (Redfield-Brzezinski ratio), but this is somewhat depending on the species in question. Many species are able to grow well also in water with N:Si ratio >1, and might have thinner frustules, which could be beneficial as silica frustules are hard to break. In order to determine the optimal nutrient ratio, it is important to also know the organic forms in the algal biomass, termed biogenic silica (BSi).

In this chapter we present a colorimetric method for determining dissolved silica (DSi), reactive silica (dissolved silicate, mono and polymeric silica, silicic acid) [6], and a method for biogenic silica (BSi), based on the same principles. Silicon analysis is based on the molybdate-blue method [6]. Silicomolybdic acid is formed when sample in acid solution is treated with a molybdate solution; then the complex is reduced by ascorbic acid to blue complex. The determination of BSi includes filtration and wet-alkaline procedure, which converts BSi to silicic acid ($Si(OH)_4$) [7].

These methods are traditionally used for natural environment samples [e.g., 8, 9] but could also be used for algal cultures. Then, the possible higher concentration in algal cultures must be considered.

2 Materials

2.1 Dissolved Silica

Use only plasticware (*see* **Note 1**) and analytical grade reagents.

1. Concentrated sulfuric acid (96% H_2SO_4).
2. Ammonium heptamolybdate tetrahydrate ($(NH_4)6Mo_7O_{24} \cdot 4H_2O$).
3. Oxalic acid dihydrate ($(COOH)_2 \cdot 2H_2O$).
4. Ascorbic acid ($C_6H_8O_6$).
5. Silicon standard 1000 mg Si in 14% sodium hydroxide solution.
6. Ultrapure water (e.g., Milli-Q).
7. 50–100 ml polypropylene graduated cylinders.
8. 50–1000 ml polypropylene volumetric flasks.
9. 100–1000 ml polyethylene storage bottles.
10. Spatulas.
11. Adjustable pipettes and pipette tips.
12. 50 ml polypropylene reaction flasks (for manual analysis).
13. Analytical balance.
14. Refrigerator.
15. Spectrophotometer or automatic analyzer.
16. 1 cm plastic cuvette (for manual analysis).

2.2 Biogenic Silica

1. Polycarbonate filters (e.g., 0.8 μm Nuclepore Ø 2.5 cm).
2. Vacuum filtration device (*see* **Note 1**) with a vacuum pump.
3. Forceps.
4. 50–100 ml polypropylene graduated cylinders.
5. 50–1000 ml polypropylene volumetric flasks.
6. 100–1000 ml polyethylene storage bottles.
7. Spatulas.
8. Adjustable pipettes and pipette tips.
9. 50 ml polypropylene bottles or centrifuge tubes with screw caps.
10. Silicon standard (see Sect. 2.1) and sodium hydroxide (NaOH).
11. Concentrated hydrochloric acid (37% HCl).
12. Concentrated sulfuric acid (96% H_2SO_4).
13. Ammonium heptamolybdate tetrahydrate ($(NH_4)6Mo_7O_{24} \cdot 4H_2O$).
14. Oxalic acid dihydrate ($(COOH)_2 \cdot 2H_2O$).
15. Ascorbic acid ($C_6H_8O_6$).

16. Ultra-pure water (e.g., Milli-Q water).
17. Analytical balance.
18. Refrigerator.
19. Water bath (100°C).
20. Spectrophotometer or automatic analyzer.
21. 1 cm plastic cuvette (for manual analysis).

3 Methods

Use acid-washed plasticware (soaked for 4 h in 6% HCl and rinsed with deionized water).

3.1 Dissolved Silica

The procedure follows that of Koroleff in the first edition of [6] and [10].

1. Prepare 3.6 mol/l sulfuric acid solution. Carefully add 10 ml concentrated acid to 37 ml ultrapure water in a plastic beaker (*see* **Note 2**), cool and make up to 50 ml with ultrapure water in a volumetric flask.
2. Prepare molybdate solution by dissolving 10 g ammonium molybdate tetrahydrate in 40 ml ultrapure water slightly heating, and then add ultrapure water to a final volume of 50 ml.
3. Prepare Reagent I by adding 50 ml of the ammonium molybdate solution (see Step 2) to 50 ml 3.6 mol/l sulfuric acid solution (see Step 1; *see* **Note 3**). Store in a polyethylene bottle at room temperature protected from light (stable for months).
4. Prepare Reagent II by dissolving 8.8 g oxalic acid in 90 ml ultrapure water and then add ultrapure water to a final volume of 100 ml. Store in a polyethylene bottle at room temperature.
5. Prepare Reagent III by dissolving 1.8 g ascorbic acid in 100 ml ultrapure water. Store in a polyethylene bottle in a refrigerator (usable as long as colorless) protected from light.
6. Prepare standard solution 1000 mg Si/l by diluting Silicon standard with ultrapure water to a final volume of 1000 ml. Store in a polyethylene bottle at room temperature.
7. On the day of analysis, prepare suitable working solutions of silicon (0–1000 μg Si/l) for calibration by diluting the standard stock solution (see Step 6) with ultrapure water.
8. Prepare analyzer for analyses according to the operation manual (*see* **Note 4**). Dilute the reagents 1:10 for analyzer.
9. Place standards in the autosampler of an analyzer or measure the standards manually by a spectrophotometer for a calibration curve and linearity test (*see* **Note 5**).

10. For manual analyses, measure 35 ml of each calibration standard and 35 ml of ultrapure water for blanks with graduated cylinder to reaction flasks.
11. Add 1 ml of Reagent I to standards and blanks, wait for 10 min, and then add 1 ml each of Reagents II and III and wait for 30 min (mix gently after each addition).
12. Measure the absorbance of standards and blanks in a 1 cm cuvette at 810 nm for a calibration curve (*see* **Note 5**).
13. Collect the samples in polyethylene bottles and analyze the samples immediately after collection. The sample can be stored in a plastic bottle in a refrigerator for about 1 day.
14. Place samples in the autosampler of an analyzer or perform manual analyses with a spectrophotometer. Verify proper operation of the instrument with certified reference material and include ultrapure water blanks in the run (*see* **Note 6**).
15. For manual analyses, measure 35 ml of the sample with a graduated cylinder.
16. Add 1 ml of Reagent I, wait for 10 min, and then add 1 ml each of Reagents II and III and wait for 30 min (mix gently after each addition).
17. Measure the absorbance of samples in a 1 cm cuvette at 810 nm (if the absorbance is greater than that of the highest standard, prepare diluted sample) and calculate the silicon concentration by the calibration curve (*see* **Note 7**).

3.2 Biogenic Silica

1. Place the polycarbonate filter in the filtration device and apply suction with the vacuum pump.
2. Pipette a known volume of culture through the filter (*see* **Note 8**).
3. Remove filter with forceps while the suction is on and place in a plastic screw-capped plastic bottle. Rinse the device with ultrapure water between samples.
4. Let the filter dry and store at room temperature in a desiccator until determination of BSi.
5. Place a clean filter into a vial for laboratory blank (*see* **Note 9**).
6. Prepare 3.6 mol/l sulfuric acid solution (see Sect. 3.1, Step 1).
7. Prepare molybdate solution (see Sect. 3.1, Step 2).
8. Prepare Reagent I (see Sect. 3.1, Step 3).
9. Prepare Reagent II (see Sect. 3.1, Step 4).
10. Prepare Reagent III (see Sect. 3.1, Step 5).
11. Prepare standard solution 1000 mg Si/l (see Sect. 3.1, Step 6).

12. Prepare 0.2 mol/l sodium hydroxide solution by dissolving 8 g NaOH in 1000 ml ultrapure water. Store in polyethylene bottle at room temperature.
13. Prepare 0.2 mol/l hydrochloric acid solution. Carefully add 16 ml of concentrated HCl to 500 ml of ultrapure water (*see* **Note 2**), cool and make up to 1000 ml with ultrapure water. Store in a polyethylene bottle at room temperature (stable for months).
14. At the time of measurements, prepare suitable working solutions of silicon (e.g., 0–50 mg Si/l) for calibration by diluting the standard stock solution (see Step 11) with ultrapure water.
15. Pipette 0.5 ml calibration solution in a plastic screw-capped plastic bottle.
16. Pipette 10 ml 0.2 mol/l NaOH to the vial and close the cap loosely.
17. Place in water bath at 100°C for 15 min and then cool to room temperature.
18. Add 10 ml 0.2 mol/l HCl, close the cap, and mix by swirling.
19. Prepare analyzer for analyses according to the operation manual (*see* **Note 4**). Dilute the reagents 1:10 for analyzer.
20. Place standards in the autosampler of an analyzer or measure the standards manually by a spectrophotometer for a calibration curve and linearity test (*see* **Note 5**).
21. For manual analyses, add 600 μl Reagent I, wait for 10 min, and then add 600 μl each of Reagents II and III and wait for 30 min (mix gently after each addition).
22. Measure the absorbance of standards and blanks in a 1 cm cuvette at 810 nm for a calibration curve and linearity test (*see* **Note 5**).
23. When measuring samples, start by adding 10 ml 0.2 mol/l NaOH and close the cap loosely.
24. Place in water bath at 100°C for 15 min and then cool to room temperature.
25. Add 10 ml 0.2 mol/l HCl, close the cap, and mix by swirling.
26. Place samples in the autosampler of an analyzer or perform manual analyses with a spectrophotometer. Verify proper operation of the instrument with certified reference material and include ultrapure water blanks in the run (*see* **Note 6**).
27. For manual analyses, add 600 μl Reagent I, wait for 10 min, and then add 600 μl each of Reagents II and III and wait for 30 min (mix gently after each addition).

28. Measure the absorbance of samples in a 1 cm cuvette at 810 nm (if the absorbance is greater than that of the highest standard, prepare diluted sample) and calculate the silicon concentration by the calibration curve (*see* **Note 7**).

4 Notes

1. Glass containing silica and samples/standards should not be allowed to come into contact with glass to avoid contamination.
2. Do this very carefully as a lot of heat is produced in this process.
3. Pour carefully the ammonium molybdate tetrahydrate solution to the acid solution, otherwise the solution turns white, milky looking, and cannot be used.
4. It is critical to have proper training before running the analyzer. This will depend on the system that you are using, and it is beyond the scope of this chapter to go into the details of the operation of the instrument. We have used Thermo Scientific Aquakem 250 analyzer for silica.
5. Calibration curve is prepared by plotting measured absorbance of standards versus standard concentrations (linear regression).
6. Start each sample run with ultrapure water blank(s) and reference material to verify proper operation of the instrument. Analyze samples in sequences of max 20 samples followed by a blank and reference material. Insert also standards in the sequence to test accuracy and drift during the analytical runs.
7. Determine concentration in the sample by the slope of the calibration curve. Reduce the absorbance of the blank from the absorbance of the sample.
8. The method is very sensitive; so for diatom cultures you should use a relatively little volume, just enough to see some color on the filter.
9. At least one for each batch of samples to assess and correct for laboratory contamination.

Acknowledgments

This updated chapter was supported by the AlgaeProBANOS project, co-funded by the European Union through the Horizon Europe program (project number 101112943), and the authors used equipment part of the Finnish marine research infrastructure (FINMARI) consortium.

References

1. Hildebrand M, Davis AK, Smith SR, Traller JC, Abbriano R (2012) The place of diatoms in the biofuels industry. Biofuels 3(2):221–240
2. Bozarth A, Maier UG, Zauner S (2009) Diatoms in biotechnology: modern tools and applications. Appl Microbiol Biotechnol 82(2):195–201
3. Armbrust EV (2009) The life of diatoms in the world's oceans. Nature 459:185–192
4. Martin-Jézéquel V, Hildebrand M, Brzezinski MA (2000) Silicon metabolism in diatoms: implications for growth. J Phycol 36:821–840
5. Round FE, Crawford RM, Mann DG (2007) Diatoms: biology and morphology of the genera. Cambridge University Press, Cambridge
6. Grasshoff K, Ehrhardt M, Kremling K (1999) Methods of seawater analysis. Wiley-VCH Verlag Gmbh, Weinheim
7. Krausse GL, Schelske CL, Davis CO (1983) Comparisons of three wet-alkaline methods of digestion of biogenic silica in water. Freshwater Biol 13:73–81
8. Spilling K, Asmala A, Haavisto N et al (2022) Brownification affects phytoplankton community composition but not primary productivity in eutrophic coastal waters: a mesocosm experiment in the Baltic Sea. Sci Total Environ 841:156510
9. Spilling K, Asmala A, Haavisto N et al (2022) Dataset from a mesocosm experiment on brownification in the Baltic Sea. Data Brief 45:108755
10. Koroleff F (1979) Methods for the chemical analysis for seawater. Meri 7:1–60 (in Finnish)
11. Haeckel EHPA (1904) Kunstformen der Natur. Bibliographisches Institut, Leipzig

Chapter 23

Microalgae Harvesting Using Fungal Mycelia in Suspension

Danielle Bansfield

Abstract

Fungi can be used to flocculate and harvest microalgae at a high efficiency in a more resource- and cost-efficient, and sustainable manner than traditional microalgal harvesting methods. However, the two most common fungal-assisted microalgal harvesting methods are lengthy. Here, we describe a more rapid fungal-assisted microalgal harvesting method using *Pleurotus ostreatus* or *Penicillium corylophilum* mycelia in suspension as bioflocculants. The fungal mycelia can be cultivated in any suitable growth medium, and the technique is applicable across microalgal species.

Keywords Microalgae, Filamentous fungi, Microalgal harvesting, Fungal assisted

1 Introduction

The high production costs of microalgal harvesting limit its commercial application to high-value markets such as supplements, cosmetics, and nutraceuticals, making lower-value uses like biofuels and animal feed economically unfeasible [1, 2]. A significant portion of these costs, around 20–30%, comes from energy- and chemical-intensive harvesting processes [2–4]. Bioflocculation has thus gained interest as a more sustainable and economical alternative. By using biodegradable flocculants and enabling medium recycling, bioflocculation can reduce both environmental impacts and production costs [5, 6].

Although other approaches, such as autoflocculation and extracellular polymer (EPS) use, are available, these methods can be unreliable or expensive [7, 8]. Fungal bioflocculants, however, provide distinct advantages, enhancing the nutrient and biochemical profile of the biomass without the contamination risks typical of chemical methods, making them suitable for downstream applications [9, 10]. Common fungal-assisted harvesting techniques include (1) co-cultivating the fungal and algal species for simultaneous harvesting and (2) adding pre-pelleted fungi to microalgal suspensions for subsequent harvesting. However, these methods are often time-intensive, taking more than a day to achieve the desired results [10–12].

Kristian Spilling (ed.), *Biofuels from Algae*, Methods in Molecular Biology, vol. 2996,
https://doi.org/10.1007/978-1-0716-5031-8_23,

The method presented here offers a faster alternative, directly applying fungal mycelium to microalgal suspensions for immediate bioflocculation and harvesting, thereby eliminating many preparatory steps needed in traditional methods. *Pleurotus ostreatus* or *Penicillium corylophilum* mycelium was used, as these species proved particularly effective in harvesting microalgae and are adaptable across various microalgal species [13, 14].

2 Materials

1. *Pleurotus ostreatus* or *Penicillium corylophilum* mycelia cultivated in a suitable medium (*see* **Note 1**).
2. Microalgae culture (species of choice).
3. 250 ml cell culture flasks.
4. Glass fiber filters with a pore size of 0.7 μm (pre-combusted and pre-weighed).
5. 100 Ø glass dishes with lids.
6. Erlenmeyer flasks (250 ml).
7. Orbital shaker (set at 90 rpm).
8. pH strips (range 2.0–9.0).
9. Fluorometer capable of measuring variable (Fv) and maximum (Fm) fluorescence (for Fv/Fm measurements).
10. Acid Lugol's solution.
11. Gridded Sedgwick–Rafter counting chamber or a flow imaging microscope.

3 Methods

3.1 Preparation of Fungal and Algal Cultures

1. Cultivate *Pleurotus ostreatus* and *Penicillium corylophilum* mycelia in an appropriate medium until sufficient biomass is produced (*see* **Notes 1–4**).
2. Grow the selected microalgal species to the desired concentration, typically in the exponential growth phase (*see* **Note 5**).
3. Place each glass fiber filter in its own glass dish with a lid.
4. Pre-dry the filters at 60°C over night or at 80°C for 30–60 min.
5. Weigh the filters in their glass dishes by first tarring the dish, placing the filter on the glass dish, weighing it, and recording the weight on the bottom of the glass dish with a permanent marker.

6. Filter 2–5 ml of both fungal and algal cultures separately onto pre-weighed filters in triplicate.
7. Dry the filters again as described above.
8. Reweigh the filters as described above and calculate the difference in weights to determine the dry weight (g DW l^{-1}) of the fungal and algal cultures.

3.2 Fungi and Algae Ratio Adjustment

1. Microalgal harvesting should be tried at a few different fungi-to-algae ratios to identify the optimal combination for harvesting efficiency. Suggested ratios include 1:1, 1:2, 1:3, 2:1, and 3:1, based on dry weight (*see* **Note 6**).
2. Adjust the fungal suspensions by diluting them in their respective cultivation mediums to match the dry weight concentration (g DW l^{-1}) of the algal culture.
3. Combine the fungal mycelia with the microalgae in the selected fungi-to-algae ratio (e.g., 1:1 in a total volume of 50 ml would be 25 ml of fungal filaments in suspension plus 25 ml of microalgae in suspension).

3.3 Incubation Setup

1. Transfer the fungal–algal mixtures into 250 ml Erlenmeyer flasks for a total working volume of 50 ml.
2. Place the flasks on an orbital shaker set to 90 rpm (*see* **Note 7**).
3. Incubate at 23–25°C under ambient laboratory light conditions (*see* **Note 8** and Fig. 1 for an example).

3.4 Harvesting Assessment

1. At designated time points (e.g., immediately after the addition of the fungal suspension and at regular intervals), sample the suspension for microalgal cell count and ratio of variable to maximum fluorescence (Fv/Fm) measurements to assess the health of the algae (*see* **Notes 9–12**).
2. Conduct the Fv/Fm measurements immediately with fresh aliquots.
3. Aliquots for cell counts can be preserved with one to two drops of acid Lugol's solution and stored at +4°C and counted later.
4. A gridded Sedgwick–Rafter counting chamber under the microscope or a flow imaging microscope can be used to determine cell counts. NB: If using a flow imaging microscope, the sample should first be filtered using an 80 μm filter to prevent clogging of the instrument.
5. Calculate the harvesting efficiency of the microalgae using the following equation:

$$\frac{c_1 - c_2}{c_1} \times 100$$

Fig. 1 (a) Microalgal harvesting set up with fungi and microalgae added in the desired proportions in Erlenmyer flasks, situated on an orbital shaker set to 90 RPM at ambient room temperature (24°C) and light conditions, (b) fungal–algal aggregates obtained at the end of harvesting (approx. 1–2 h)

where C1 is the initial cell concentration (cells ml^{-1}) and C2 is the concentration at the measured time point.

3.5 Batch Settling Tests

1. These tests are done to determine how readily the fungi–algae flocs settle and, therefore, whether the biomass should be collected by settling or sieve.
2. Gently resuspend the flocs and carefully pour the sample into a 250 ml glass measuring cylinder, taking care not to break or rupture the flocs during the process.
3. Record the changing height of the floc-liquid interface at 0-, 5-, 10-, 15-, 20-, and 30-min intervals (see Fig. 2 for an example).
4. Plot the data (height on the y-axis vs time on the x-axis) to produce a batch settling curve.

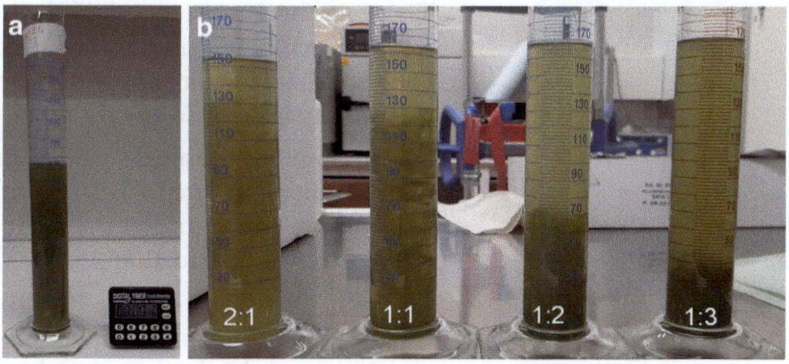

Fig. 2 (a) Batch settling tests set-up and (b) the end of batch settling tests at different fungi:algae ratios showing increasing settling with a decreasing proportion of fungi

4 Notes

1. Suitable mediums for filamentous fungi are Yeast Malt broth, Malt Extract Broth, Potato Dextrose Broth, or a nutrient-rich food waste stream such as whey.
2. Sufficient fungal biomass means equal to or more biomass concentration than the microalgal suspension.
3. Ideally, the fungal cultures should be grown long enough to have an abundance of filaments in suspension below a thin mat of fungus on the surface. If it grows much longer than this, all the filaments coalesce into a dense mat on the surface, and there will not be enough filaments in suspension.
4. You can check if the amount of fungal filamentous material is sufficient by checking the concentration of fungal and algal species in g of DW as described above before deciding to start the experiments.
5. Ensure microalgal cultures have access to air by purchasing cell culture flasks with vented caps or by screwing non-vented caps loosely.
6. According to our studies [13] the higher the proportion of fungi used, the more buoyant the fungi–algae flocs will be. Therefore, if you wish to collect the biomass via sedimentation, it is recommended to use as low a proportion of fungi as possible to gain the desired harvesting results.
7. If you wish to scale up and are using larger total volumes shaking velocity should be adjusted upward. This should be experimented with to determine the best-shaking velocity for your total volume and flask size.

8. These methods were developed with species with an optimal temperature range of 22–25°C. However, if one or both of your species have an optimal temperature range outside of this, you may need to use different temperature conditions.
9. When sampling the algal suspension, take 100–500 μl of algal suspension and dilute in 5–10 ml of ultrapure water (depending on how concentrated the algae in the suspension is) and immediately add acid Lugol's solution to prevent rupture of the cells.
10. When sampling for cell counts, it is better to err on the side of caution and take larger rather than smaller suspension aliquots.
11. Sampling for Fv/Fm measurements: take 100–300 μl of suspension and dilute in 3–10 ml of algal growth media before measuring.
12. Make sure the dilution is not too concentrated and that the Fv/Fm curve is not cut off at the top, as this will give an inaccurate reading. If this is the case, then repeat with a more dilute sample.

Acknowledgments

Special thanks to Drs Jonna Piiparinen, Kristian Spilling, and Anna Mikola for their support and input during this project. This study used equipment which is part of the Finnish marine research infrastructure (FINMARI) consortium.

References

1. Mathimani T, Mallick N (2018) A comprehensive review on harvesting of microalgae for biodiesel – key challenges and future directions. Renew Sustain Energy Rev 91:1103–1120. https://doi.org/10.1016/j.rser.2018.04.083
2. Muhammad G, Alam MA, Mofijur M et al (2021) Modern developmental aspects in the field of economical harvesting and biodiesel production from microalgae biomass. Renew Sustain Energy Rev 135:110209. https://doi.org/10.1016/j.rser.2020.110209
3. Fasaei F, Bitter JH, Slegers PM, Van Boxtel AJB (2018) Techno-economic evaluation of microalgae harvesting and dewatering systems. Algal Res 31:347–362. https://doi.org/10.1016/j.algal.2017.11.038
4. Li S, Hu T, Xu Y et al (2020) A review on flocculation as an efficient method to harvest energy microalgae: mechanisms, performances, influencing factors and perspectives. Renew Sustain Energy Rev 131:110005. https://doi.org/10.1016/j.rser.2020.110005
5. Chen J, Leng L, Ye C et al (2018) A comparative study between fungal pellet- and spore-assisted microalgae harvesting methods for algae bioflocculation. Bioresour Technol 259:181–190. https://doi.org/10.1016/j.biortech.2018.03.040
6. Yin Z, Zhu L, Li S et al (2020) A comprehensive review on cultivation and harvesting of microalgae for biodiesel production: Environmental pollution control and future directions. Bioresour Technol 301:122804. https://doi.org/10.1016/j.biortech.2020.122804
7. Pahl SL, Lee AK, Kalaitzidis T et al (2013) Harvesting, thickening and dewatering microalgae biomass. In: Borowitzka MA, Moheimani

NR (eds). Algae for biofuels and energy. Springer Netherlands, Dordrecht, pp 165–185
8. Spilling K, Seppälä J, Tamminen T (2011) Inducing autoflocculation in the diatom Phaeodactylum tricornutum through CO_2 regulation. J Appl Phycol 23:959–966. https://doi.org/10.1007/s10811-010-9616-5
9. Purchase D (ed) (2016) Fungal applications in sustainable environmental biotechnology. Springer International Publishing, Cham
10. Chu R, Li S, Zhu L et al (2021) A review on co-cultivation of microalgae with filamentous fungi: efficient harvesting, wastewater treatment and biofuel production. Renew Sustain Energy Rev 139:110689. https://doi.org/10.1016/j.rser.2020.110689
11. Bhattacharya A, Mathur M, Kumar P et al (2017) A rapid method for fungal assisted algal flocculation: critical parameters & mechanism insights. Algal Res 21:42–51. https://doi.org/10.1016/j.algal.2016.10.022
12. Wrede D, Taha M, Miranda AF et al (2014) Co-cultivation of fungal and microalgal cells as an efficient system for harvesting microalgal cells, lipid production and wastewater treatment. PLoS One 9:e113497. https://doi.org/10.1371/journal.pone.0113497
13. Bansfield D, Spilling K, Mikola A, Piiparinen J (2022) Bioflocculation of Euglena gracilis via direct application of fungal filaments: a rapid harvesting method. J Appl Phycol 34:321–334. https://doi.org/10.1007/s10811-021-02651-5
14. Bansfield D, Mikola A, Spilling K, Jonna P (2024) Fungal cultivation in a dairy side stream: valorisation potential of whey into added-value biomass and byproducts. Available at SSRN 4914896

Check for updates

Chapter 24

RNA Extraction Strategy for Algae Rich in Polysaccharides and Polyphenols

Timo Jensen

Abstract

RNA extraction in algal research is particularly challenging due to the high levels of polysaccharides and polyphenols that interfere with the purification process. This chapter presents a streamlined RNA extraction strategy, covering all steps from sample preparation to final purification. The method utilizes a CTAB-based lysis buffer and targeted separation steps to effectively eliminate proteins, polysaccharides, and polyphenols, employing lithium chloride for RNA precipitation. Designed for broad applicability across algal species, this strategy can be easily adjusted and performed with basic laboratory equipment, providing an accessible and effective tool for algae research.

Keywords RNA extraction, Algae, CTAB, Polyphenols, Polysaccharides

1 Introduction

High-quality RNA extracts are crucial for a range of biological research tasks, including hybridization experiments, RT-PCR, cDNA library construction, and transcriptomics. While commercially available kits and standardized protocols have become well established for animal tissues and bacteria, extracting RNA from plants with high levels of polysaccharides and polyphenols remains challenging [1–5]. Marine macroalgae, in particular, are often rich in polyphenolic compounds within their cell walls, which can interfere with or co-precipitate during RNA purification [6–8]. Since commercial kits frequently prove ineffective in such cases, numerous protocols have been developed for RNA extraction from specific macroalgal species [6–12]. However, these protocols often lack cross-species applicability, as an effective extraction strategy for one species may not be suitable for another.

In this chapter, an RNA extraction strategy suitable for a diverse array of algae is presented. This strategy builds upon previously published protocols, employing a hexadecyltrimethylammonium

Kristian Spilling (ed.), *Biofuels from Algae*, Methods in Molecular Biology, vol. 2996,
https://doi.org/10.1007/978-1-0716-5031-8_24,

bromide (CTAB)-based buffer for cell lysis and organic solvents for protein removal. This basic approach has been continuously refined by multiple groups to enhance extraction efficiency and simplify the process for various algal species [7, 9–13]. In a key study, Pearson et al. (2006) reported remarkable achievements in extracting RNA from the challenging brown alga *Fucus vesiculosus* and other species by optimizing the buffer composition, while also reducing buffer consumption and eliminating the need for toxic chaotropic agents. This protocol was then further modified by our group and established for a broader taxonomic range of algae [14]. The essence of this methodological development is presented here in an RNA extraction strategy that is both adjustable and executable with basic laboratory equipment, providing a practical and versatile tool for algal research.

2 Materials

All preparation steps should be performed using RNase-free ultrapure water as RNase activity will strongly diminish RNA integrity. Purchased reagents should be molecular biology grade and RNase-free if available (*see* **Note 1**).

2.1 RNA Extraction

1. CTAB lysis buffer: 2% (w/v) hexadecyltrimethylammonium bromide (CTAB), 100 mM trishydroxymethylaminomethane (Tris; pH 8.0), 50 mM EDTA (pH 8.0), 50 mM dithiothreitol (DTT) and 2 M NaCl (*see* **Note 2**) in RNase-free water. Optional: Add 2% (w/v) polyvinylpyrrolidone (PVP), e.g., with a M_W of 40 kDa, or polyvinylpolypyrrolidone (PVPP) for better separation of phenolic inhibitors (*see* **Note 3**). DTT should always be added freshly prior to extraction (*see* **Note 4**).
2. Organic phase: Always follow the safety guidelines when handling phenol or chloroform (*see* **Note 5**)! Two alternative organic phase options can either be purchased commercially (recommended) or prepared manually (*see* **Note 6**):
   (a) Phenol-chloroform-isoamylalcohol [25:24:1; PCI]
   (b) Chloroform-isoamylalcohol [24:1; CI]
3. An acidic pH of 4.5–5 is recommended (*see* **Note 7**).
4. Ice cold ethanol: 80% (v/v) ethanol in RNase-free water, store at −20°C until use.
5. Lithium chloride (LiCl): Use pre-made RNase-free LiCl solution (7.5 M) or prepare LiCl solution in RNase-free water.
6. RNase-inhibitor solution for surface cleaning.
7. Liquid nitrogen.
8. Ethanol, ≥96% (v/v).
9. β-mercaptoethanol.

10. Polyethylene glycol (PEG) 300 or 400 (*see* **Note 5**).
11. Grinding equipment (e.g., mortar and pestle).
12. Lab consumables such as suitable tubes and pipette tips (*see* **Note 8**).
13. Optional: diethyl pyrocarbonate (DEPC) for RNase inactivation (*see* **Note 1**).

3 Methods

3.1 Sampling and Sample Preparation

1. Algae samples should be immediately frozen on dry ice or liquid nitrogen after sampling to avoid RNA degradation. Samples can be stored at −20°C (or lower for long-term storage) or freeze-dried (*see* **Note 9**). To avoid influence from epibionts, clean leaf surfaces of macroalgae with 70% ethanol.
2. The equipment used for grinding should be RNase-free. Grinding equipment such as mortar and pestle can be dry-heated at 230°C for 2–4 h or autoclaved (*see* **Note 10**). Surfaces should be treated with RNase inhibitor solution before starting to work.
3. Sample tissues can be homogenized by grinding with mortar and pestle under liquid nitrogen until a very fine powder is achieved. Alternative grinding techniques may be used. Ensure that the samples remain frozen at all times during the process and do not thaw under any circumstances (*see* **Note 11**). After grinding, the powder can be weighed in the quantities chosen for extraction (*see* **Notes 12 and 13**).
4. Quick-freeze ground powder in liquid nitrogen and store at −20°C or lower until use.

4 Extraction

The following RNA extraction procedure can be adapted to different scales, depending on the desired RNA yield and the laboratory equipment available (*see* **Note 13**). Volume indications are provided relative to the recovered supernatant during extraction; for instance, if 800 μL of supernatant is recovered, "1 vol." refers to the addition of an equal 800 μL.

1. Add CTAB extraction buffer in a volume equivalent to four times the weight of the powder (e.g., for 200 mg of powder, add 800 μL of CTAB lysis buffer).

2. Incubate at RT for 15 min and vortex thoroughly every 2–3 min to properly lyse cells (*see* **Note 14**).
3. Add 1 vol. of the organic phase and vortex thoroughly.
4. Centrifuge at ≥4000 × *g* for 30 min (*see* **Note 15**) and transfer aqueous supernatant to a fresh tube without disrupting the phase underneath.
5. Add 0.3–0.6 vol. of ethanol (*see* **Note 16**).
6. Mix by hand and incubate at RT for 5 min until a white precipitate occurs.
7. Centrifuge at 4000 × *g* for 10 min and transfer supernatant to a fresh tube without disrupting the precipitate.
8. Add 1 vol. of the organic phase and vortex thoroughly.
9. Centrifuge at ≥4000 × *g* for 30 min (*see* **Note 15**) and transfer aqueous supernatant into fresh tube without disrupting the phase underneath.
10. Add LiCl to a final concentration of 2.5 M.
11. Add β-mercaptoethanol to a final concentration of 1%.
12. Precipitate RNA by incubation (*see* **Note 17**) at 4°C overnight or −20°C or lower for 30 min. Maintain samples on ice throughout the following steps, except during the drying process.
13. Centrifuge at 12 000 × *g* at 4°C for 30 min to pellet RNA.
14. Carefully remove liquid from the pellet.
15. Add 500 μL of pre-cooled 80% (v/v, −20°C) ethanol and gently invert tube to wash pellet.
16. Carefully remove ethanol from the pellet.
17. Centrifuge at 12 000 × *g* at 4°C for 2 min and remove residual ethanol with pipette.
18. Centrifuge at 12 000 × *g* at 4°C for 1 min and remove residual ethanol with pipette.
19. Let pellet dry for 10–15 min (*see* **Note 18**).
20. Resuspend pellet in 30–100 μL of RNase-free water or buffer of choice.

4.1 Quality Control and RNA Polishing

Photometric measurements offer valuable insights into RNA quality. High-quality RNA is characterized by an A_{260nm}/A_{280nm} ratio close to 2.0 and an A_{260nm}/A_{230nm} ratio ranging from 2.0 to 2.2. For first-time RNA extractions, it is advisable to examine the full UV/Vis spectrum to detect peak shifts or potential contaminants. Be aware that the measured UV/Vis ratios are influenced by the pH and ionic strength of the buffer used for resuspension (*see* **Note 19**). If RNA quality is insufficient, using an RNA cleanup kit is a practical approach

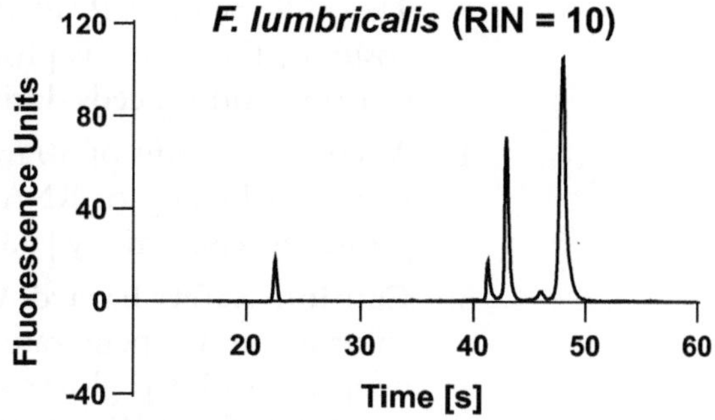

Fig. 1 Electropherogram of RNA sample from the red macroalga *Furcellaria lumbricalis* extracted by the presented method and analyzed by capillary electrophoresis with a 2100 Bioanalyzer using the RNA 6000 Nano Kit (Agilent Technologies Inc., USA). Major peaks at approx. 43 s and 48 s correspond to the 18S and 25S ribosomal subunits in plant RNA samples. Reproduced from [14] with permission from Springer

for polishing and concentration. RNA integrity can be evaluated through agarose gel electrophoresis or capillary electrophoresis. For macroalgae, the presence of distinct 25S and 18S ribosomal bands or peaks indicates high RNA integrity (*see* **Note 20**; Fig. 1).

5 Notes

1. Autoclaving ultra-pure water or prepared solutions at 121°C for 20 min will reduce, but not fully deactivate RNase activity. A DEPC treatment can be used to improve RNase inactivation. It is important to note that solutions containing Tris are incompatible with DEPC treatment. Tris reacts with DEPC, reducing its effectiveness and necessitating the use of a larger excess to achieve adequate treatment. However, residual DEPC left in the solution after treatment may interfere with biochemical reactions in downstream applications. Therefore, it is recommended to purchase pre-made nuclease-free solutions or dissolve molecular biology-grade powdered reagents in DEPC-treated water. Also, do not autoclave DTT or DTT-containing solutions.
2. High NaCl concentrations, such as 2 M, will increase the ionic strength of the buffer, thereby facilitating polysaccharide removal [5, 13, 15, 16].
3. PVP/PVPP have been shown to effectively bind phenolic inhibitors and prevent oxidation [17, 18]. Note that soluble PVP and insoluble PVPP with different molecular weights and degrees of crosslinking are available and that the choice of the polymer may have an impact on the extraction result.

The addition of PVP/PVPP may also alter the density and viscosity of the aqueous phase, potentially necessitating higher centrifugation speeds during phase separation steps.

4. At concentrations of 50 mM or higher, DTT prevents oxidative cross-linking of RNA by phenolic compounds and suppresses RNase activity [13].
5. **Caution: Safety hazard!** Work under fume hood. Phenol and chloroform will penetrate nitrile gloves quickly. Always wear neoprene or butyl gloves with high safety level against phenol and chloroform. Wear safety goggles. Phenol and chloroform cannot be easily removed with water. Keep polyethylene glycol (PEG) 300 or 400 readily available as a precaution and for removing potential spills. After centrifugation, open tubes and containers with caution as aerosols might have formed during the process. Use tubes with safety lids and remove spills immediately. Dispose of the organic phase and contaminated lab consumables as hazardous chemical waste.
6. In general, PCI is considered more effective for removing proteins [19] and may lead to higher RNA yields. However, when insufficient phase separation and phenolic carryover is observed (usually seen by a shifted maximum in the UV/Vis spectrum [20]), CI may be a suitable alternative for both and particularly the second separation step, as it is considerably denser than water and capable of removing residual phenol from the aqueous phase.
7. An acidic pH of 4.5–5 of the organic phase facilitates the separation of DNA into the interphase. However, traces of DNA will still be present in the sample and incorporation of a DNase treatment may be required, contingent on the intended downstream application.
8. Reaction tubes and pipette tips are often overlooked as potential sources of RNase contamination. If RNA degradation occurs, consider switching to RNase-free consumables.
9. Freeze-drying can increase sample stability during storage and decrease yield fluctuations caused by varying water contents of the samples [13].
10. Autoclaving at 121°C for 20 min will not completely inactivate RNase activity but is usually sufficient if grinding is performed strictly under liquid nitrogen. Alternatively, a DEPC-treatment could be considered for RNase inactivation of equipment. Always let mortar and pestle cool to room temperature before adding liquid nitrogen.
11. Insufficient grinding reduces RNA yield and quality. Samples must be kept frozen during the entire grinding process as thawing will immediately decrease RNA integrity.

12. It is advisable to pre-cool the tubes and equipment used for weighing (e.g., in liquid nitrogen) to prevent thawing of the powder during the weighing process. Prioritize speed over precision.
13. Using 200 mg of powder or less will enable the protocol to be performed in 2 mL tubes, thereby strongly reducing the consumption of chemicals and simplifying centrifugation procedures, while usually yielding enough RNA for most common downstream applications. However, the protocol can be scaled up to 1 g of algae powder and performed in 15 or 50 mL tubes. This is especially recommended if the used algae tissues contain lots of water and/or low yields are expected. For larger-scale extractions, the samples can be aliquoted into 1.5 or 2 mL tubes after Step 11 and subsequently centrifuged at high speed in a benchtop centrifuge during Step 13 if the available large-volume centrifuges operate at insufficient speed.
14. Often, the CTAB lysis buffer takes on a milky appearance during lysis.
15. Higher centrifugation speeds, e.g., up to 12 000 × *g*, will improve phase separation and are recommended.
16. Higher concentrations of ethanol can improve polysaccharide precipitation. For most samples, 0.3 volumes are sufficient.
17. In contrast to alcohols such as ethanol or isopropanol, LiCl will efficiently precipitate RNA while leaving DNA, carbohydrates, and proteins in solution. However, traces of DNA will still be present in the sample and incorporation of a DNase treatment may be required, contingent on the intended downstream application. The RNA will appear as a white pellet after centrifugation.
18. Do not let the pellet dry out completely as the solubility of the RNA will decrease. Tubes can be positioned upside down at an angle to prevent contaminants from entering while keeping the opening clear for proper drying.
19. With increasing pH, absorbance at 280 nm decreases, while absorbance at 260 nm remains unchanged. Consequently, the A_{260nm}/A_{280nm} ratio increases in alkaline buffers compared to more acidic or neutral buffers and water [21].
20. Clearly visible ribosomal RNA bands or peaks indicate RNA integrity (Fig. 1). In capillary gel electrophoresis, this integrity is reflected by RNA Integrity Numbers ≥7. However, some RNA samples from macroalgae display additional signals originating from chloroplast or mitochondrial RNA [22].

Acknowledgments

This work was conducted as part of the "Redox Enzymes from Algae" project, funded by the German Federal Ministry of Education and Research. Special thanks are extended to Livia Saleh, Verena Sandow, Dominik Bents, Steffen Krohn, Yu-Chen Wu, Maria Mucke, Ammelie Svea Boje, Stefan Veltel, Steffen Hennig, Levent Piker, Matthias Peipp, and Antje Labes for their invaluable contributions to the development of the presented method.

References

1. Loomis WD (1974) Overcoming problems of phenolics and quinones in the isolation of plant enzymes and organelles. Methods Enzymol 31:528–544. https://doi.org/10.1016/0076-6879(74)31057-9
2. Schneiderbauer A, Sandermann H, Ernst D (1991) Isolation of functional RNA from plant tissues rich in phenolic compounds. Anal Biochem 197:91–95. https://doi.org/10.1016/0003-2697(91)90360-6
3. Salzman RA, Fujita T, Zhu-Salzman K et al (1999) An improved RNA isolation method for plant tissues containing high levels of phenolic compounds or carbohydrates. Plant Mol Biol Rep 17:11–17. https://doi.org/10.1023/A:1007520314478
4. Gehrig HH, Winter K, Cushman J et al (2000) An improved RNA isolation method for succulent plant species rich in polyphenols and polysaccharides. Plant Mol Biol Rep 18:369–376. https://doi.org/10.1007/BF02825065
5. Wang X, Tian W, Li Y (2008) Development of an efficient protocol of RNA isolation from recalcitrant tree tissues. Mol Biotechnol 38:57–64. https://doi.org/10.1007/s12033-007-0073-6
6. Dos Reis Falcão V, Pedroso Tonon A, Cabral Oliveira M et al (2008) RNA isolation method for polysaccharide rich algae: agar producing *Gracilaria tenuistipitata* (Rhodophyta). J Appl Phycol 20:9–12. https://doi.org/10.1007/s10811-007-9174-7
7. Sim M-C, Ho C-L, Phang S-M (2013) A simple and effective method for RNA isolation and cDNA library construction from the brown seaweed *Sargassum polycystum* (Fucales, Phaeophyceae). J Appl Phycol 25:1277–1285. https://doi.org/10.1007/s10811-013-9980-z
8. Greco M, Sáez CA, Brown MT et al (2014) A simple and effective method for high quality co-extraction of genomic DNA and total RNA from low biomass *Ectocarpus siliculosus*, the model brown alga. PLoS One 9:e96470. https://doi.org/10.1371/journal.pone.0096470
9. Apt KE, Grossman AR (1993) Characterization and transcript analysis of the major phycobiliprotein subunit genes from *Aglaothamnion neglectum* (Rhodophyta). Plant Mol Biol 21:27–38. https://doi.org/10.1007/BF00039615
10. Apt KE, Clendennen SK, Powers DA et al (1995) The gene family encoding the fucoxanthin chlorophyll proteins from the brown alga *Macrocystis pyrifera*. Mol Gen Genet 246:455–464. https://doi.org/10.1007/BF00290449
11. Chan CX, Ho C-L, Othman RY et al (2002) Total RNA extraction for the red seaweed *Gracilaria changii* (Gracilariales, Rhodophyta). Asia-Pacific Conference in Marine Science and Technology, Kuala Lumpur, Malaysia
12. Heinrich S, Valentin K, Frickenhaus S et al (2012) Transcriptomic analysis of acclimation to temperature and light stress in *Saccharina latissima* (Phaeophyceae). PLoS One 7:e44342. https://doi.org/10.1371/journal.pone.0044342
13. Pearson G, Lago-Leston A, Valente M et al (2006) Simple and rapid RNA extraction from freeze-dried tissue of brown algae and seagrasses. Eur J Phycol 41:97–104. https://doi.org/10.1080/09670260500505011
14. Jensen T, Saleh L, Bents D et al (2023) Optimised protocols for RNA extraction from a broad taxonomic range of algae. J Appl Phycol 35:1743–1753. https://doi.org/10.1007/s10811-023-02980-7
15. Fang G, Hammar S, Grumet R (1992) A quick and inexpensive method for removing polysaccharides from plant genomic DNA. BioTechniques 13:52–54, 56
16. Chang S, Puryear J, Cairney J (1993) A simple and efficient method for isolating RNA from pine trees. Plant Mol Biol Rep 11:113–116. https://doi.org/10.1007/BF02670468
17. Porebski S, Bailey LG, Baum BR (1997) Modification of a CTAB DNA extraction protocol for plants containing high polysaccharide and polyphenol components. Plant Mol

Biol Rep 15:8–15. https://doi.org/10.1007/BF02772108
18. Koonjul PK, Brandt WF, Farrant JM et al (1999) Inclusion of polyvinylpyrrolidone in the polymerase chain reaction reverses the inhibitory effects of polyphenolic contamination of RNA. Nucleic Acids Res 27:915–916. https://doi.org/10.1093/nar/27.3.915
19. Wang T, Zhang N, Du L (2005) Isolation of RNA of high quality and yield from *Ginkgo biloba* leaves. Biotechnol Lett 27:629–633. https://doi.org/10.1007/s10529-005-3629-1
20. Lee JTY, Cheung KMC, Leung VYL (2014) Correction for concentration overestimation of nucleic acids with phenol. Anal Biochem 465:179–186. https://doi.org/10.1016/j.ab.2014.08.006
21. Wilfinger WW, Mackey K, Chomczynski P (1997) Effect of pH and ionic strength on the spectrophotometric assessment of nucleic acid purity. BioTechniques 22:474–476, 478–481. https://doi.org/10.2144/97223st01
22. Suresh B, Gassmann M (2016) Assessing integrity of plant RNA with the Agilent 2100 Bioanalyzer System. Agilent Technologies, Inc., USA. https://www.agilent.com/cs/library/applications/5990-8850EN.pdf. Accessed 16 Jan 2025

Check for updates

Chapter 25

A Simple Staining Method to Study Cell Structures of Microalgae

Tobias Lipsewers

Abstract

Using the fluorescence brightener calcofluor white M2R allows the visualization of species-specific cell wall structures of thecate dinoflagellates consisting of cellulose. This method is relatively easy to use and benefits the taxonomy of new as well as the separation of known species. As the dye strongly binds to several molecules, it also visualizes structures of, e.g., other microalgae, plants, and fungi. Calcofluor white M2R can be used on live specimens and samples preserved with a pH-neutral fixative. This chapter aims to provide a detailed protocol for staining environmental water samples and cell cultures of pelagic microalgae including critical factors to consider, highlighting advantages compared to other methods, and it shows practical examples regarding phytoplankton from the Baltic Sea.

Keywords Epifluorescence microscopy, Fluorescence brightener, Calcofluor, Taxonomy, Species identification

1 Introduction

Originally, the fluorescence brightener calcofluor white M2R (referred to henceforth as calcofluor) was developed by Hughes and McCully in 1975 to study higher plants [1]. In 1985, Fritz and Triemer developed a protocol to study thecate dinoflagellates in natural water samples and monoalgal cell cultures using this dye [2]. Later, Dr. Anke Kremp adjusted the method by Fritz and Triemer (1985) for analysing Baltic Sea dinoflagellates. In 2023, this adjusted protocol was published as part of a study on phytoplankton spring bloom dynamics in agreement with Dr. Kremp [3].

Calcofluor strongly binds to callose, carboxylated polysaccharides, and cellulose [1]. The thecae of armored dinoflagellates consist of cellulose and species-specific compositional and structural differences in their cell walls (plate patterns) can be visualized for taxonomy and identification [2]. The analysis of calcofluor-stained samples involves inverted epifluorescence microscopy. When samples

Kristian Spilling (ed.), *Biofuels from Algae*, Methods in Molecular Biology, vol. 2996,
https://doi.org/10.1007/978-1-0716-5031-8_25,

intended for being analyzed using calcofluor are stored, pH-neutral fixatives such as neutral Lugol's solution must be used.

This technique can be used for living specimens as well [1, 2], which is particularly relevant when establishing new cell lines of organisms isolated from environmental samples, which might contain species that can hardly be separated without staining the samples. Also, the development of cultured cells can be followed by staining live samples (e.g., studying different life-cycle stages). Figure 1 shows an example from the Baltic Sea comparing three species of cold-water dinoflagellates (*Gymnodinium corollarium*, *Biecheleria baltica*, and *Apocalathium malmogiense*) and highlights the differences of the cell's appearance with and without using calcofluor. *G. corollarium* is an athecate species and thus not stained by calcofluor at all, which is significant for the separation of these three species (Fig. 1, panel B).

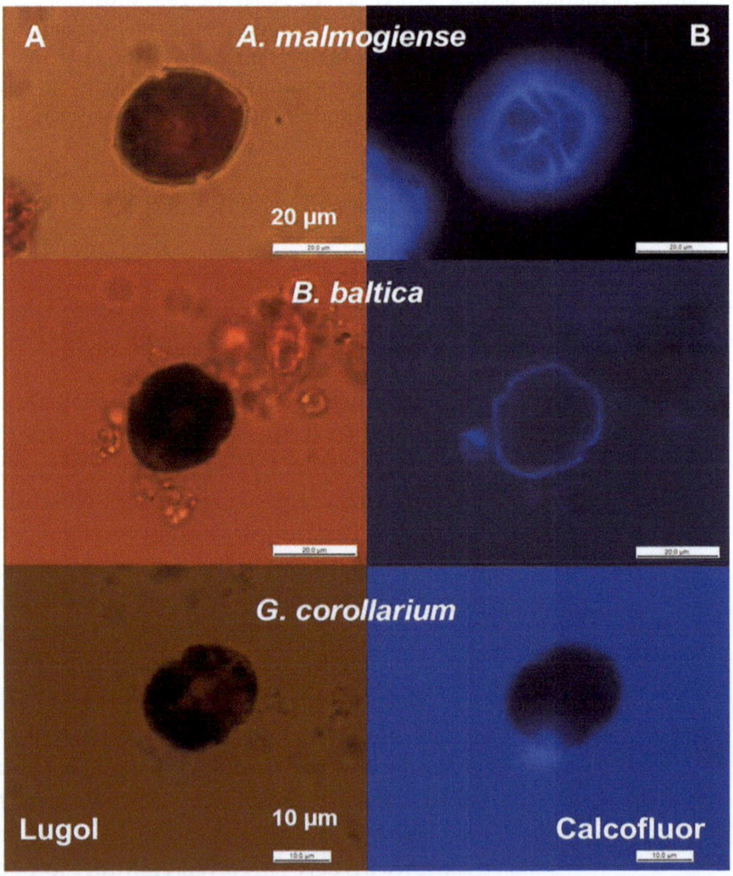

Fig. 1 Comparing the appearances of three dinoflagellates in samples preserved with neutral Lugol's solution (panel A) to the calcofluor-stain of the identical cells (panel B). Note the different scale bars for *Gymnodinium corollarium* (10 μm) and the other two species (20 μm). This figure was reproduced from Lipsewers 2023 [3] with the permission of the Boreal Environment Research Publishing Board

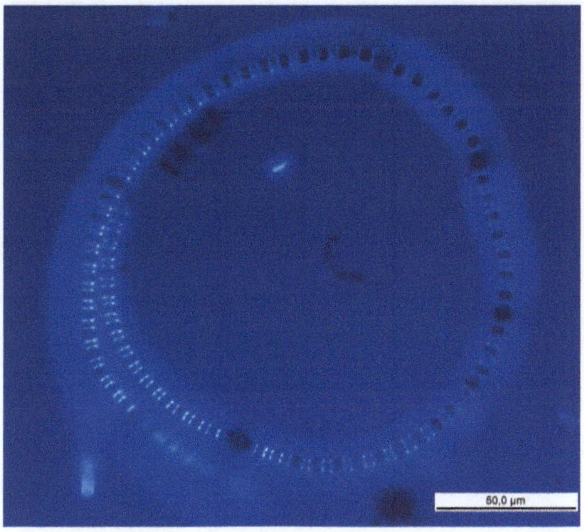

Fig. 2 A chain of auxospores formed by the diatom *Achnanthes taeniata*. The size of the actual auxospores is approximately 5 μm and calcofluor stained structures/molecules in between them

Besides the thecal plates of armored dinoflagellates, calcofluor also stains cell structures of other microalgae. For example, Fig. 2 shows a chain of auxospores formed by the diatom *Achnanthes taeniata*. The use of calcofluor visualized parts between the actual spores, apparently containing cellulose, carboxylated polysaccharides, and/or callose.

This dye can also be used to study fungal parasites of freshwater phytoplankton [4]. This could be a very valuable tool for monitoring potential parasite infections of cultured microalgae to avoid the loss of laboratory cultures and reduce/avoid production losses for commercial algae producers.

Compared to certain molecular tools, this method has multiple advantages: it is easy to learn for a trained phytoplankton microscopist; it can be used on live samples; it allows the separation of different life cycle stages; it results in qualitative and quantitative results; and it is less costly (given the availability of a suitable microscope system). Depending on the previous experience of the microscopist, an initial training using cultured cells and field samples is recommended. After that, several samples can be analyzed daily. In contrast, the advantages of, e.g., quantitative polymerase chain reaction (qPCR) are a lower detection limit and a higher sample throughput. However, this is only relevant when identifying cells and determining their abundances in environmental samples, not for the taxonomy of specific organisms.

2 Materials

The following description of the materials and methods is based on the protocol published in Lipsewers 2023 [3] and the author's experience with the method. More details from other published protocols involving calcafluor were added for this chapter. Be careful to avoid any contamination (*see* **Note 1**).

1. Calcofluor white M2R powder (e.g., Fluorescence Brightener 28, Sigma-Aldrich).
2. Ultrapure water.
3. Appropriate scale.
4. Spoon spatulas.
5. Measuring cylinder.
6. Storage container (ideally a brown glass bottle or one that is wrapped in aluminum foil).
7. Combined plate chambers (phytoplankton counting chambers) and sedimentation cylinders for different sample volumes (manufacturer: e.g., HYDRO-BIOS). As you might be familiar with using the equipment manufactured by other companies, which can be used as well, details relevant for the handling of the HYDRO-BIOS equipment (including PDF files of a data sheet and the manual) and the corresponding accessories can be found on the website [5].
8. Pasteur pipettes (e.g., 3 ml) to be used to add the calcofluor working solution to the sample after carefully removing the sedimentation cylinder from the counting chamber. Pipettes with sterile tips are required to take samples from cell cultures and sub-samples from environmental samples.
9. Containers for sub-samples (e.g., for diluting samples with a high cell density) and/or mixing different cell cultures (see below), such as centrifugation tubes (e.g., 15 ml).
10. Inverted epifluorescence microscope (e.g., Leica DMI 3000 B).
11. Suitable camera (e.g., Leica DFC 490). Calcofluor absorbs ultraviolet (UV)-radiation (340–400 nm range) and re-emits visible blue light [2]. Thus, the microscope equipment and settings must be selected accordingly.

3 Methods

Avoid excess light when working with calcofluor (*see* **Note 2**).

3.1 Sample Preparation, Staining, and Analysis

1. Prepare working solution of calcofluor powder and ultrapure water to a final concentration of 1 mg ml^{-1} w/v. Make sure the

powder is completely dissolved. The working solution can be stored in a refrigerator [2].

2. Acclimate samples and the calcofluor working solution to room temperature before mixing and further preparation.
3. Mixing the sample(s) gently to avoid damaging the cells (inverting them multiple times is better than shaking) but sufficiently before filling the sedimentation cylinder/sample slot of the counting chamber (round opening of the base plate) or taking sub-samples is important.
4. Cells are settled according to Utermöhl [6] and sedimentation volumes of 10, 25, 50, and 100 ml could be used (*see* **Notes 3 and 4**).
5. Five drops (Pasteur pipette, evenly distributed) of a well-mixed (clear) calcofluor working solution (1 mg ml^{-1} w/v, *see* **Note 5**) are added to the concentrated sample after sedimentation (24–72 h, depending on the sedimentation volume). The addition of the dye is done without immersing the pipette tip into the sample to prevent stirring up cells from the bottom of the counting chamber before adding the cover glass.
6. Samples are incubated for at least 5 min in the dark at room temperature before microscopic analysis at, for instance, 40-fold magnification. Micrographs can be taken at different magnifications (*see* **Notes 6 and 7**).

4 Notes

1. Working with particle-free (e.g., void of fibers from paper towels) equipment is essential to ensure a uniform signal created by the dye during microscopy. Sometimes calcofluor overstains the cell structures and closing down the UV field stop can reduce the fluorescence [2]. Ultrapure water should be taken from the source (fresh) just before using it.
2. Working with calcofluor (preparing the working solution, sample preparation, and analysis) should be done under as dark as possible conditions (e.g., dimmed light), since it is sensitive to light.
3. Samples of monoalgal cultures can be used for testing the calcofluor working solution and for validating the identification of species in field samples. Mixing different cultures allows a direct comparison of different species.
4. For dense laboratory cultures, 10 ml might already be too much. Alternatively, the sample slot, fitting 2.973 ml (HYDRO-BIOS) of the counting chamber (see Sect. 2.2 above) can be filled using, e.g., a small measuring cylinder or a pipette.

In this case it should be sufficient to let the cells settle for 1–2 h before adding calcofluor. In general, the mentioned sedimentation times (see below also) refer to samples preserved with Lugol's solution. The sample inside the sample slot should be slightly concave to avoid the inclusion of air bubbles when applying the cover glass, which is more critical for quantitative studies. For samples that have been settled using sedimentation cylinders, this is usually achieved by adding the calcofluor.

5. This protocol was adapted to the brackish water system of the Baltic Sea having a steep salinity gradient. The exact ideal protocol might differ between, e.g., abiotic conditions of sampling sites and the species/structures to be investigated. The method as such is applicable to a variety of samples and serves different purposes. It is advisable to consider different previously published approaches to find the optimal solution for your plan [e.g., 2–4].
6. The comparison of certain species when stained and unstained (Fig. 1) clearly underscores the advantages of calcofluor and validates its applicability in identifying these species. In cases of doubt, characteristic patterns and features of the thecal plates can be studied further (e.g., by studying micrographs with aid of a well-suited image editor) and/or experts may be consulted. The people working in phytoplankton monitoring programs might be a good address for further discussions. In addition, the corresponding author of this chapter can be contacted. This requires the highest possible quality of the micrographs.
7. As this chapter focuses on the staining technique as such, please have a look at Lipsewers 2023 for an approach to obtain quantitative results for environmental samples [3].

Acknowledgments

Special thanks to Dr. Anke Kremp (Leibniz-Institute for Baltic Sea Research, Warnemünde, Germany) for recommending and teaching me the calcofluor method, helping with species identification, and for encouraging me to identify phytoplankton communities to species-level whenever possible and to Dr. Kristian Spilling (Marine and freshwater solutions, Finnish Environment Institute, SYKE, Helsinki) for inviting me to contribute to this book, suggesting the topic, and his helpful comments prior to submission of this chapter. The Baltic Sea examples (Figs. 1 and 2) originate from the C-FluxII project led by Dr. Spilling from 2013 to 2017. During the majority of the practical work behind Figs. 1 and 2, funding was available from the Academy of Finland (decision numbers 259164 and 263862) for C-fluxII and personal grants were awarded to the corresponding author of this chapter by the Onni Talas Foundation and the University of Helsinki. The project utilized SYKE's marine research infrastructure as a part of the national FINMARI consortium.

References

1. Hughes J, McCully ME (1975) The use of an optical brightener in the study of plant structure. Stain Technol 50(5):319–329. https://doi.org/10.3109/10520297509117082
2. Fritz L, Triemer RE (1985) A rapid simple technique utilizing calcofluor white M2R for the visualization of dinoflagellate thecal plates. J Phycol 21(4):662–664. https://doi.org/10.1111/j.0022-3646.1985.00662.x
3. Lipsewers T (2023) Resolving species-specific abundances of three cold-water dinoflagellates using a simple staining technique. Boreal Environ Res 28:195–206. https://www.borenv.net/BER/archive/pdfs/ber28/ber28-195-206.pdf
4. HYDRO-BIOS Apparatebau GmbH. https://www.hydrobios.de/en/product?product_id=39. Accessed 08 December 2024
5. Rasconi S, Jobard M, Jouve L et al (2009) Use of calcofluor white for detection, identification, and quantification of phytoplanktonic fungal parasites. Appl Environ Microbiol 75(8):2545–2553. https://doi.org/10.1128/AEM.02211-086
6. Utermöhl H (1958) Zur Vervollkommnung der quantitativen Phytoplankton-Methodik: Mit 1 Tabelle und 15 Abbildungen im Text und auf 1 Tafel. Mitt Int Ver Theor Angew Limnol 9:1–38

Chapter 26

Life Cycle Assessment of Micro and Macro Algae Products

Kiia Silvennoinen, Sara Saukkonen, Jean-Baptiste Thomas, Léa Braud and Tanja Myllyviita

Abstract

Life cycle assessment (LCA) is a comprehensive methodological framework for evaluating environmental impacts of products and processes throughout their entire life cycle. It can be used for many applications, such as identifying environmental impact hotspots of a product system, solution, and product comparison or marketing. The LCA process is divided into four phases: (1) goal and scope definition, (2) inventory analysis, (3) impact assessment, and (4) interpretation. The goal and scope definition answers to why and how the LCA will be conducted. The collected inventory forms the basis for the following impact assessment. In the end, the results are interpreted, and conclusions are made.

In the evolving algae sector that moves from pilot-scale trials to complex biorefinery concepts, the role of LCAs is becoming ever more important as a key decision-support tool. However, for reliable, comparable, and useful LCAs, more common practices are needed. This chapter describes how to conduct an LCA for both micro- and macroalgae products while providing examples and describing common environmental impact hotspots.

Keywords Life cycle assessment, Environmental impact, Sustainability, Microalgae, Macroalgae

1 Introduction

Over the past decade, both micro- and macroalgae have emerged as promising feedstocks within Europe's growing blue bioeconomy, backed by significant investments from public research grants and private venture capital. As the sector evolves from small-scale, pilot cultivation trials to more complex biorefinery concepts, life cycle assessments (LCAs) have become a key decision-support tool for optimizing production processes, minimizing environmental impacts, and guiding sustainable scale-up. Published LCAs cover diverse cultivation strategies—ranging from sea-based long-line rigs and rafts for macroalgae to various land-based microalgae systems such as raceway ponds, photobioreactors, and hybrid approaches. Early studies often relied on lab-scale or pilot-scale data to explore biofuel potential and single-product systems [1–3]. More recently, however, the focus has

Kristian Spilling (ed.), *Biofuels from Algae*, Methods in Molecular Biology, vol. 2996,
https://doi.org/10.1007/978-1-0716-5031-8_26,

shifted toward assessing a wider range of algae-derived products, particularly high-value compounds such as pigments and fatty acids [4], multi-output biorefineries [5], and integrating algae into circular value chains [6, 7]. Comparative studies have been performed for, e.g., the choice of microalgae species and cultivation systems, growth conditions, downstream processes, scale, and geographical location. These developments underscore the vital importance of LCA in guiding algae-based industries toward environmentally and economically sustainable outcomes.

LCA is a comprehensive methodological framework for evaluating the environmental impacts of products and processes throughout their entire life cycle, from raw material extraction (cradle) to end-of-life disposal (grave) [8]. Originating in the 1960s and 1970s as an effort to quantify resource use and pollutant discharges across interconnected industries, LCA evolved into a standardized and globally recognized practice [9]. Today, LCA is widely employed by policymakers, researchers, and industry to compare technological pathways, inform design optimizations, and shape sustainable development strategies [10]. LCA is also a key part of the Green Claims Directive, which may require companies to prove their green statements through LCA. Key impact categories include global warming potential, acidification, eutrophication, and resource depletion, though certain context-specific or short-term impacts—such as localized ecological disturbances or social repercussions—often lie beyond the traditional scope of LCA [11]. Despite these limitations, LCA's holistic approach to quantifying environmental burdens makes it an indispensable tool for identifying improvement opportunities in fast-evolving sectors, including algae-based biofuel and bioproduct systems.

2 Life Cycle Assessment for Algae Products

Life cycle assessment (LCA) is a method for quantifying the environmental impacts of a product system throughout its lifetime. The method avoids problem-shifting between life cycle phases, regions, or environmental impacts by taking into account the full life cycle.

Use areas of LCAs are, e.g., identifying possibilities for reducing the environmental impacts of a product system, scenario or product comparison, or marketing. There are two main types of LCAs: attributional LCA (ALCA), which assesses the environmental burden of a product as part of the overall global environmental burdens, and consequential LCA (CLCA), which gives an estimate of how the global environmental burdens are affected by the production and use of a product.

The LCA method has been standardized in ISO 14040 [8] and 14044 [12]. ISO 14040 sets the principles and framework, while

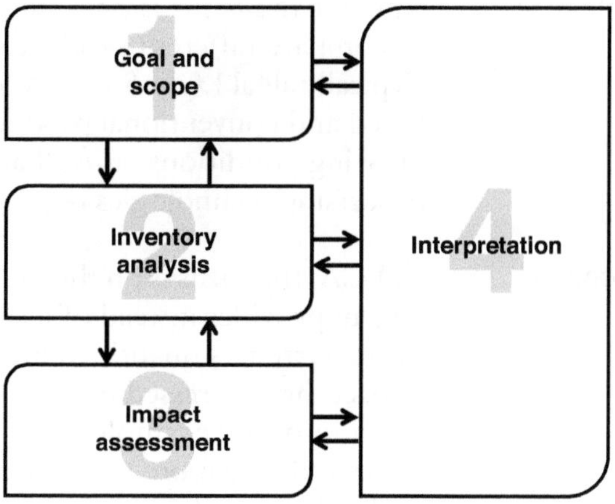

Fig. 1 The four phases of LCA according to ISO 14040

ISO 14044 adds requirements and guidelines for conducting an LCA. In its simplicity, ISO 14040 divides LCA into four phases: (1) goal and scope, (2) inventory analysis, (3) impact assessment, and (4) interpretation (Fig. 1).

In the goal and scope definition phase, the why and the how of the work are defined, i.e., what is the question the LCA tries to answer, what is included, and on the other hand, what is excluded from the LCA and how the work is carried out. While the entire product system is described in the goal and scope phase, the life cycle inventory phase (LCI) includes collecting and quantifying input and output material and energy flows for each process step in the product system. Derived from the inventory data collected, the calculation of the environmental impacts happens in the life cycle impact assessment (LCIA) phase. In the interpretation phase, the results are analyzed, and conclusions are made.

The nature of an LCA is commonly iterative, and it is typical that findings in the later phases require going back to the previous phases, for example, to gather further data or to adjust the scope. The following subsections describe carrying out the four phases of LCA in more detail and give examples of conducting an LCA for both micro- and macroalgae production.

2.1 Goal and Scope Definition

All LCA studies begin with goal and scope definition. The decisions made in this phase guide the work in the following phases and provide the ground on which to build upon. Important steps in this phase include formulating the research question, defining the goal, functional unit, system boundaries, as well as many other relevant modeling aspects.

The goal of the study sets the purpose and focus of the assessment. It forms the basis for the whole study, and all the following methodological decisions reflect onto the goal. The goal determines whether the focus is on a single product and its impacts or if it is to compare different product systems and scenarios to each other. Typically algal LCAs focus on comparison, such as comparing algae-based and conventional products, different cultivation systems and growing conditions, as well as different conversion pathways and processing technologies [4].

2.1.1 Functional Unit

In LCA, the focus is on the function (value or use) that the product system provides instead of just the units of products. The function of the system is quantified into a functional unit (FU), and all the impacts are expressed per this unit. The FU can be based on a system input, for example "one ton of freshly harvested kelp to be preserved," or more typically on the output, for example "one ton of dried seaweed."

Commonly, algae LCAs use energy-based, mostly in algal biofuel studies, and product-based FUs [4]. Moreover, area-, distance-, and time-related FUs have been used. The FU should be clearly defined and measurable, as well as relevant to the goal of the LCA, for example, if the goal is to compare different product systems, the choice of functional unit must facilitate this comparison. It is not always straightforward what is the best choice for the functional unit. The following sections discuss the choice of functional unit separately for macro- and microalgae studies.

(A) Functional Unit in Macroalgae Studies

The choice of FUs in macroalgae LCAs has evolved in parallel with shifting research priorities and technological advancements in seaweed utilization. Early LCAs, starting in 2012, predominantly focused on the use of macroalgae as a feedstock for biofuels and energy products, often employing FUs aligned with fuel output, such as per unit of energy produced (e.g., MJ of biofuel). However, from around 2016 onwards, LCA studies increasingly considered seaweed's potential beyond biofuels, incorporating assessments of novel biorefinery-derived ingredients, food, animal feed, and biomaterials, with some studies focusing on specific sub-processes such as cultivation design and associated area-based functional units, or sub-systems of value chains such as drying, freezing, or alginate extraction [13]. To capture the multi-output nature of integrated seaweed biorefineries, some studies use multiple functional units, while others opt for area-based [5] or intermediate product functional units; for example, fresh-weight biomass can be the functional unit for the production of dried, frozen, and ensiled kelp [14]. Some examples of functional units used macroalgae LCA studies are listed below:

1. "a 1-km trip with a gas-powered car" [3].
2. "cultivation and processing of one tonne of dry seaweed biomass (Laminaria digitata) produced in Denmark for biofuels production" [2].

3. "per MJ_{ex} S. latissima" [15].
4. "1 ha of sea under cultivation" [5].
5. "1 kg protein" [16, 17].
6. "1 ton fresh-weight" [14, 18].
7. "1 kg of seaweed-based bioplastic film" [19].

(B) Functional Unit in Microalgae Studies

Similar to seaweed LCAs, the FUs used in microalgae LCAs have evolved in response to shifting research priorities. Before 2015, biofuels were the primary focus, whereas more recent studies have emphasized high-value compounds such as omega-3 fatty acids and pigments (e.g., astaxanthin, phycocyanin, fucoxanthin) [4]. In addition, recent LCAs have increasingly incorporated FUs that reflect the bioremediation potential of microalgae, such as the use of wastewater as a culture medium.

LCAs of microalgal biofuel production have used energy-related FUs to facilitate the comparison with fossil-based fuels and other biofuels, such as "1 megajoule of energy produced by combustion in a passenger car" [20] and "1 megajoule of microalgae biofuel" [21]. Other studies used product-related FUs such as "1 kilogram of biodiesel produced" [22] to compare different microalgal cultivation systems and conversion pathways.

To date, most LCA studies use a product-related FU that quantifies the amount of biomass produced, such as "1 kilogram of (dried) microalgal biomass." This approach allows for the comparison of microalgae species [23], cultivation systems [24, 25], and growing conditions [26, 27].

The use of input-related FUs is also relevant for microalgal biorefineries, particularly when comparing different conversion pathways for the same biomass [25–28]. Depending on the goal and scope of LCA studies, area-based FUs, as well as FUs encompassing all microalgal biorefinery products, are also relevant. Using a combination of biorefinery outputs allows for the comparison of different configurations of the same biorefinery [29]. However, such FUs hinder the assessment of the influence of changes in biomass composition on the environmental impacts of microalgal biorefineries, for instance due to seasonal variations [29].

Finally, the amount of product of interest is used in a large number of LCA studies, such as "1 kilogram of phycocyanin" [30], "1 kilogram of carotene" [31], "1 kilogram of astaxanthin" [32], "1 kilogram of total fatty acids" [33], and "1 kilogram of nitrogen used as biofertilizer" [34]. Such FUs allow for the comparison of different microalgae species, cultivation systems, and production pathways for a specific product.

Regarding the bioremediation potential of microalgae, FUs such as the volume of treated water have been used (e.g., 1 cubic meter of treated water) [35–37] to compare the potential of microalgae to depollute water using different techniques.

2.1.2 Defining the Product System

The cornerstone of life cycle assessment (LCA) is the concept of the product system, which encompasses all interconnected processes involved in producing, delivering, using, and disposing of a particular product. These processes are typically illustrated by a system (flow) diagram (Fig. 2), which visually represents the network of material and energy flows. Defining the system boundaries constitutes a critical step in LCA because it delimits which processes and impacts will be accounted for.

A commonly used naming convention for system boundaries employs the terms "cradle," "gate," and "grave." In an algae-based production context, *cradle* often denotes the point at which initial raw materials (e.g., nutrients, water, and carbon dioxide sources) are obtained. The *gate* is an intermediate boundary, typically representing the point at which the algae-based product (e.g., dried biomass, extracted oil, or refined fuel) leaves the production facility. Finally, *grave* extends the analysis to the end-of-life stages, incorporating product use and disposal or recycling. When the system boundary encompasses the entire life cycle ("cradle-to-grave"), the analysis

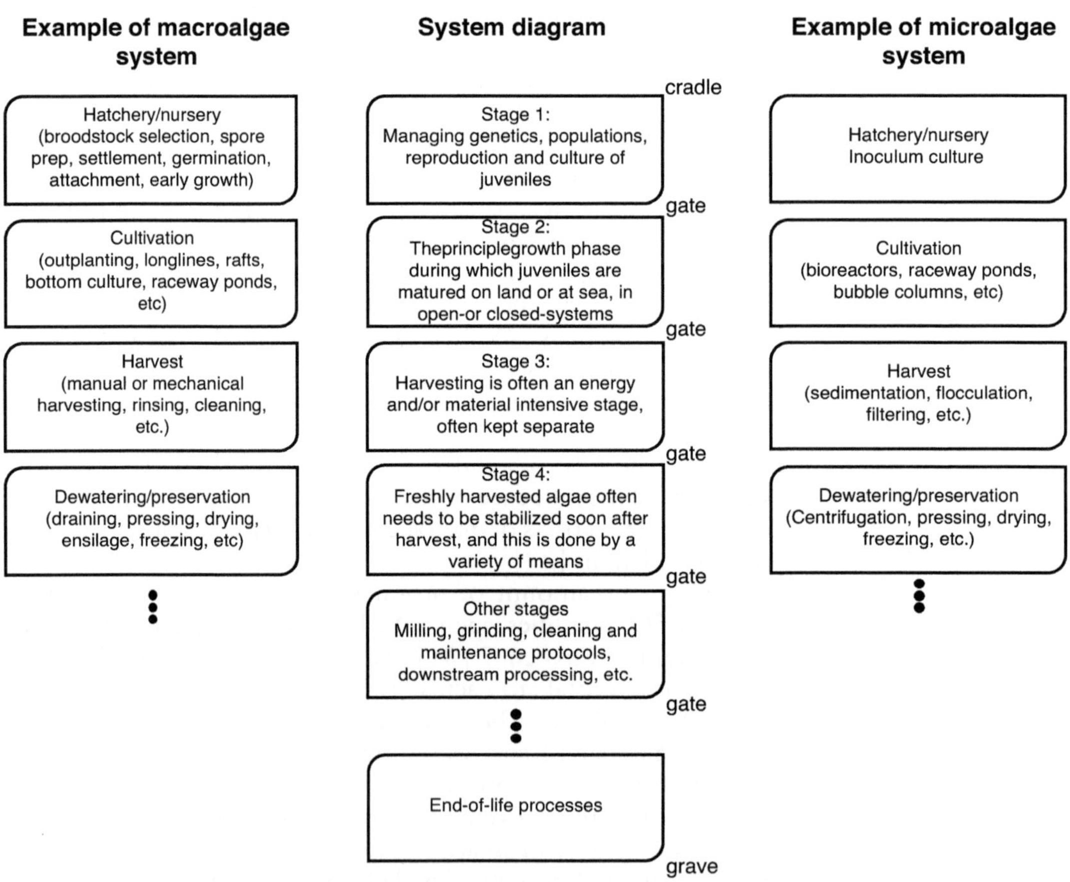

Fig. 2 Example of macro- and microalgae system diagrams

captures all impacts from raw material extraction to the ultimate disposal of the product. In contrast, the more common "cradle-to-gate" approach, used frequently in algae LCAs for food or biofuel applications [4], stops at the factory gate or similar exit point, leaving out the impacts associated with product distribution, use, and end-of-life.

The system can be defined to either include or exclude infrastructure, depending on the goal and the expected impact of the decision. In algae product LCAs, infrastructure has been shown as a possible impact hotspot, and therefore excluding it from the LCA could underestimate the actual environmental burden.

The choice of system boundary is not merely a technical decision; it can substantially influence the LCA results. For example, including downstream processes, such as product distribution or final disposal, may highlight additional hotspots (e.g., transportation emissions or waste management) or reveal trade-offs between the production phase and end-of-life treatment. Therefore, a well-defined product system and carefully justified system boundaries are the key to ensuring the robustness, reproducibility, and clarity of any LCA—particularly in the rapidly evolving field of algae-based products.

In many cases, product value chains do not produce only one product but rather there are multiple products created. A key part in system definition step is deciding how to consider these co- and by-products, i.e., the allocation of burdens. This issue is discussed in the subsection below.

2.1.3 Allocation

The construction of product systems in algae-based LCAs necessitates careful consideration of how to deal with co-products and by-products generated during biomass processing. For example, algae cultivated for biofuel production may concurrently yield high-value compounds such as proteins or pigments. In such cases, the environmental burdens associated with the system must be allocated among all co-generated products. The selection of an appropriate approach for addressing multifunctionality is established during the goal and scope definition phase and has been widely debated in the literature.

ISO 14044 introduced a hierarchical framework to address multifunctionality, beginning with process *subdivision*, where a multifunctional process is divided into multiple unit processes that each produce a single output. If subdivision is infeasible, *system expansion* can be applied, where the system boundaries are expanded to include all the functions associated with the products. If neither of these methods can be applied to avoid multifunctionality, allocation (or partitioning) approaches are used to solve it by distributing inputs and outputs based on underlying physical relationships, such as biochemical composition, biomass proportion, or energy content. In cases where physical relationships cannot be established, allocation is performed based on other criteria, such as economic value. These approaches are commonly referred to as *mass, energy, and economic allocation*.

The definition of system expansion in ISO standards has been criticized for its ambiguity, leading to potential misinterpretations that were later clarified in the ILCD guidance. More recently, a revised classification has been proposed [38], distinguishing four distinct approaches: subdivision, system expansion (or enlargement), substitution, and allocation (or partitioning).

Most algae-based LCAs have employed substitution and partitioning (allocation) methods to handle multifunctionality [4]. However, allocation may introduce challenges, particularly when high-value compounds represent a minor fraction of the total biomass (e.g., fucoxanthin, phycocyanin). The application of mass allocation in such cases may result in disproportionately low environmental impacts assigned to these products, although they are considered main products in the case of product-related FUs [39]. In addition, economic allocation is sensitive to market price fluctuations, necessitating careful consideration. Given the significant influence of allocation choices on LCA results, sensitivity analyses are recommended to assess their impact [39].

Substitution has been widely applied in algae-based LCAs, where products (or functions) outside the FU are substituted. For instance, in an algae system producing pigments, feed, and biofertilizer, if the FU is defined as "1 kg of pigment," the environmental burden of feed and biofertilizer would be substituted. Conversely, in a system using an area-based FU (e.g., "1 ha of facility"), all co-products would be substituted (Fig. 3). Despite its potential advantages, system enlargement, where the FU encompasses all products, has been applied in only a limited number of studies [29]. While this approach allows for a comprehensive assessment, it complicates comparisons with benchmark systems and poses challenges when evaluating algae systems with diverse product outputs. Due to the complexity of these methodologies, the choices made by LCA practitioners must be transparently documented to ensure that researchers and stakeholders can accurately interpret and compare results.

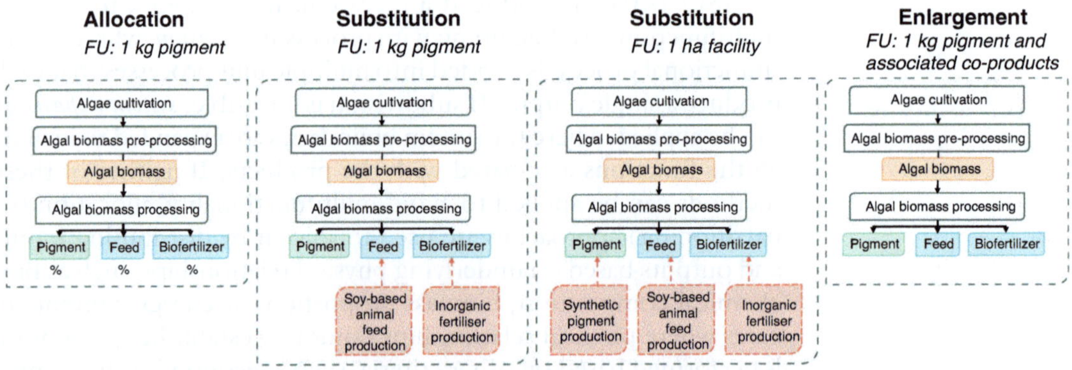

Fig. 3 The use of an allocation, substitution, and enlargement approach to solve or avoid multifunctionality in algae system. Adapted from [4] with permission

2.1.4 Selecting Impact Categories

An LCA study can include varying sets of assessed environmental burdens. Although the sustainability of algae and other products is sometimes evaluated only through one impact category (often carbon footprint), an LCA study should include multiple studied burdens. Key impact categories often studied for micro- and macroalgae products include climate change, acidification, eutrophication, and resource depletion. For microalgae LCAs, especially including the use of acrylic glass in tubular or bubble column reactors, also ozone depletion may be a key impact category to look into. Selecting only one impact category (often climate change impact) may lead to not detecting important environmental aspects and hotspots as trade-offs between different impact categories may exist.

The different impacts are divided into midpoint and endpoint impact categories. While midpoint environmental impacts describe singular environmental burdens and may include varying categories such as climate change, acidification and eutrophication, endpoint impacts aggregate these midpoint impacts typically into three higher areas of protection: human health, biodiversity and resource scarcity.

LCAs are in most conducted considering the environmental effect at midpoint levels. Although assessing damage to endpoint indicators appears a more comprehensible approach to assess the actual impact of a product or process, conducting an LCA for endpoint indicators includes substantial uncertainties as the connections between certain emission sources and the damage to human health and ecosystems are currently poorly understood.

2.2 Inventory Analysis

After defining the goal and scope, it is time to start collecting data for the impact assessment. This phase is called the inventory analysis (LCI). At this phase, the LCA practitioner collects all the relevant inputs and outputs for each of the process steps (Fig. 4). Inputs comprise material, energy, and water consumption, while outputs consist of the produced products and possible side products, as well

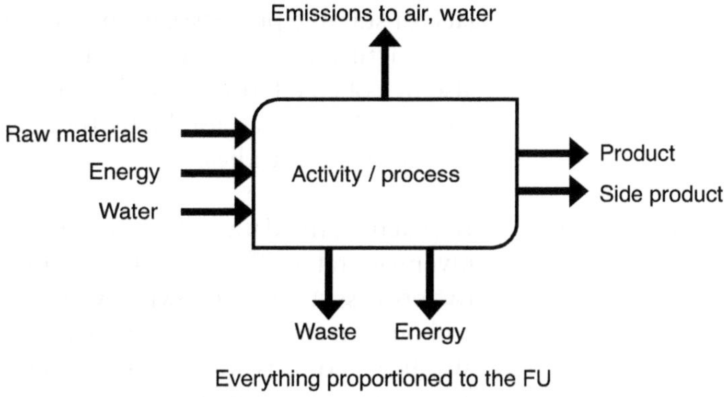

Fig. 4 Inventory collection comprises all energy and mass flows in and out of the system

as generated emissions, waste, and energy. LCI is one of the most time-consuming steps of an LCA study. As input and output flows are the basis for the calculation and the subsequent results, careful and accurate data collection is of utmost importance.

2.2.1 Foreground System and Primary Data

When collecting inventory data, it is crucial to distinguish between *foreground* and *background* systems. The foreground system comprises processes that are typically under the direct control or influence of the practitioner carrying out the LCA, and for which site-specific or company-specific data can often be collected. This is also called primary data. For an algae production facility, foreground processes include cultivation (in open ponds or photobioreactors), harvesting, dewatering, extraction, and any on-site energy or resource usage. A good practice is to aim to collect as much of primary data as possible, as this can enhance the relevance and accuracy of the assessment—at the very least, the major and most relevant flows should be primary data.

In the case of algae product LCAs, collecting energy consumption data for different process phases and equipment is one of the main challenges often faced when collecting primary data. However, understanding the process energy consumption is important for trustworthy results as well as for identifying opportunities to optimize processes. Data sensitivity may also become an issue with primary data collection and result reporting, when a company may not want to share their personal data publicly. Masking and aggregating data points is a possibility, but it adds complexity to the result analysis and reporting.

2.2.2 Background System

The background system consists of processes that support the foreground system but are generally outside the direct control of the practitioner (e.g., electricity generation, production of fertilizers and chemicals, transportation infrastructure, and machinery manufacturing). These background processes often rely on secondary data from databases such as the ecoinvent LCA database or from literature sources.

In terms of secondary data for algae processes, there is often a lack of suitable processes in the main databases. This often leads to the practitioner having to resort to proxies. As background data can also be collected from published literature, publishing the data used for algae LCAs may also help other practitioners in their data collection for future studies.

2.2.3 Collecting Data

In practice, the data collection is usually done and reported with an inventory table. The table lists all the inputs and outputs for each of the process steps as its own row. In the case of infrastructure inputs and outputs, information on the infrastructure expected lifetime can also be listed. It is also beneficial to include information on the data accuracy or uncertainty. Unit management throughout the collection process and reporting is paramount, both for the practitioner themselves as well as future readers of the LCA.

A relevant issue to keep in mind throughout the inventory analysis is the data quality. In many cases, there can be large differences in the data quality between different processes in the system—some data can be accurate and measured on-site, while others can be just a rough estimation. Furthermore, data can be accounted for using different FUs, and thus combining data from different sources requires meticulous calibration. Keeping note of whether a collected data point is an estimation, or an actual measurement helps later when performing an uncertainty analysis. Some LCA guidelines also require reporting of the data quality.

When collecting the LCI, flows with low expected (e.g., 5% or lower) environmental significance to the system on singular processing step can be excluded from the study. This is called a cut-off, and the initial limit and criteria for it can be based on either mass, energy, or environmental relevance.

Most algae LCAs rely on case-specific data, which limits their generalizability and comparability across the sector. Variations in system boundaries, methodological choices, and operational factors—such as differing maintenance protocols—further complicate cross-study assessments. This challenge is particularly relevant for algae systems at low technological readiness levels, where processes are still evolving, and for which the systems are often subject to LCAs. Harmonizing LCA inventories through standardized data collection and reporting helps overcome some of these limitations [4, 13], enabling more robust comparisons, reducing uncertainties, and improving the broader applicability of findings. This is essential for supporting industry scaling, policy development, and informed decision-making.

2.3 Life Cycle Impact Assessment

2.3.1 From Life Cycle Inventory to Environmental Impacts

In the phase of life cycle impact assessment (LCIA), environmental impact values for different impact categories are derived from inputs and outputs of the LCI through different steps. First, relevant impact categories to be calculated are chosen. Figure 5 gives an overview of all the steps within the impact assessment phase.

The inputs and outputs are classified into impact categories, i.e., which inputs and outputs contribute to which impact category. For example, greenhouse gas emissions contribute to the impact category of climate change. After classification, each of these inputs and outputs is multiplied by a characterization factor that describes how much a quantity of the input or output contributes to an impact score. In this step, the inputs and outputs of different units contributing to the same impacts are converted into the same unit. Greenhouse gas emissions such as methane are transformed into carbon dioxide equivalents by multiplying the emissions with a characterization factor reflecting the harmfulness of the emission. For instance, as methane has a higher global warming potential than carbon dioxide, it is multiplied by a characterization factor of 29.8. Finally,

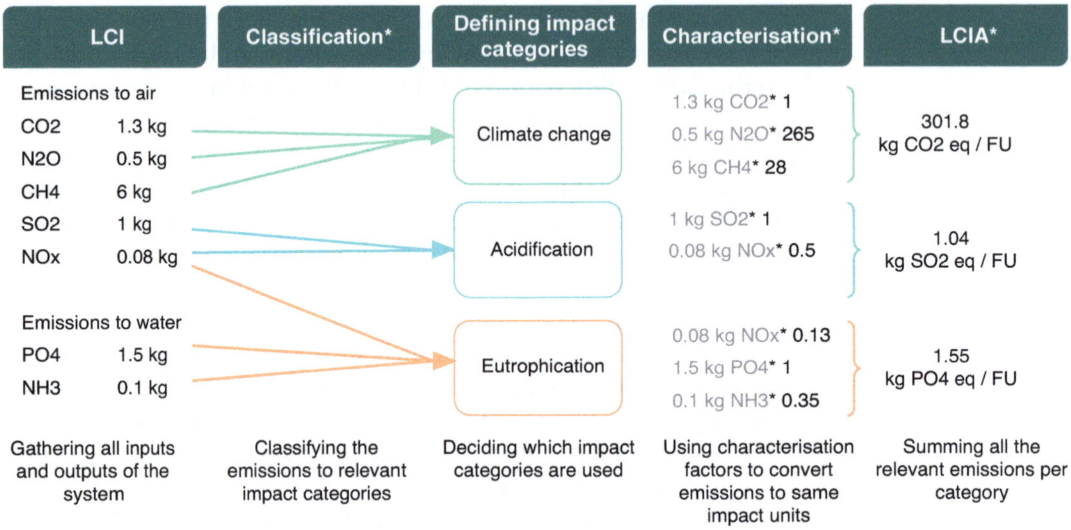

Fig. 5 Conceptual diagram of steps in life cycle impact assessment

the characterized scores are summed into midpoint environmental impact results within each impact category. These impact results for each impact category have their own units (such as carbon dioxide or phosphorus equivalents) and they cannot be added or compared with each other, but they should be looked at separately.

The steps of an LCIA are usually performed within a modeling software, according to a chosen LCIA method. These methods already contain characterization factors for a list of possible impact categories that can be calculated with the method in question. The most common LCIA methods used in the case of algae products, in the past, have been CML and ReCiPe. However, at least in the EU, the environmental footprint (EF) method has been gaining more attention.

In LCA modeling software, the product system is built, it is connected to the background data, and results are calculated. LCA modeling software helps dealing with large systems and datasets. Background datasets are used within the modeling software although it is possible to conduct an LCA with, e.g., MS Excel. There are several free software, such as openLCA and Brightway, and commercial software including SimaPro and GaBi a modeler can choose from.

2.3.2 Normalization and Weighting

If the results from different impact categories need to be compared, or put to a context, normalization and weighting can be performed. In normalization, the result of each impact category is compared to the corresponding impact of a larger unit, such as a geographical area. This can mean either comparison with the total impact caused annually by one citizen in Europe or in the world, for example. The results can further be turned into a single-score result by multiplying each impact with a weighting factor for that certain impact category. Weighting the LCIA results provides the possibility to compare scenarios or

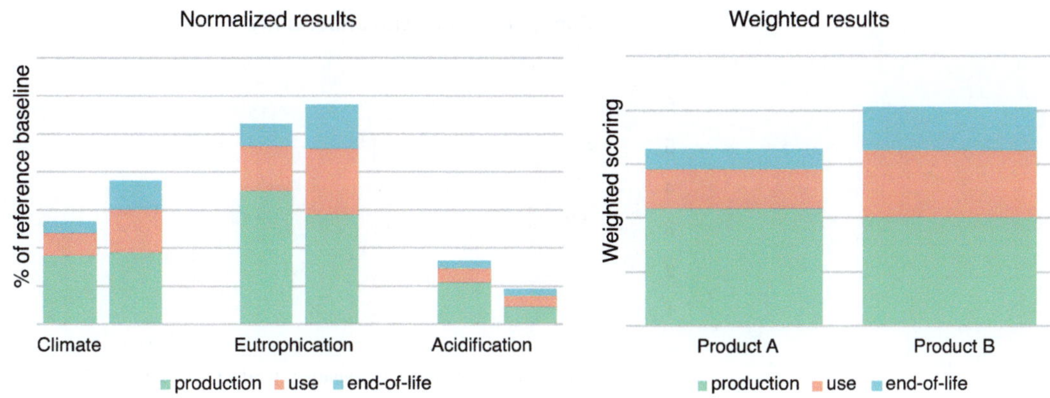

Results table

		Climate kg CO2 eq/FU	Eutrophication kg PO4 eq/FU	Acidification kg SO2 eq/FU
Product A	production	181.08	0.92724	0.624
	use	60.36	0.30908	0.208
	end-of-life	30.18	0.15454	0.104
Product B	production	189.5	0.762	0.2625
	use	113.7	0.4572	0.1575
	end-of-life	75.8	0.3048	0.105

Fig. 6 Normalization and weighting can be used to present result comparison

products with each other using only a single result for each of them. An example of presenting normalized and weighted results is given in Fig. 6. The weights on different impact categories are typically defined by a panel of experts and/or stakeholders. Even though normalization and weighing can be useful for simplifying result comparison, their application should be carefully considered, as they introduce additional uncertainty and blur the original impact category results.

2.4 Interpretation and Reporting

Interpretation is the procedure during which the results are evaluated. The interpretation should reflect the goal of the study, what were the key questions that needed to be answered? Based on the results, it is possible to see what the environmental impacts of a product or a service are, and what are the areas that could still be improved. For instance, is it possible to replace energy source to less emission-intensive alternative or to increase energy efficiency? Also replacing raw materials with a more environmentally friendly alternative can decrease the environmental burden of a product. Conducting an LCA is a highly iterative process, and in many cases, it is noted during interpretation of the results that some modifications are needed on, e.g., input parameters.

It is also crucial to assess the sensitivity and uncertainty of the results, even though these steps are often left out. In sensitivity analysis, selected data and methodological choices are tested for their

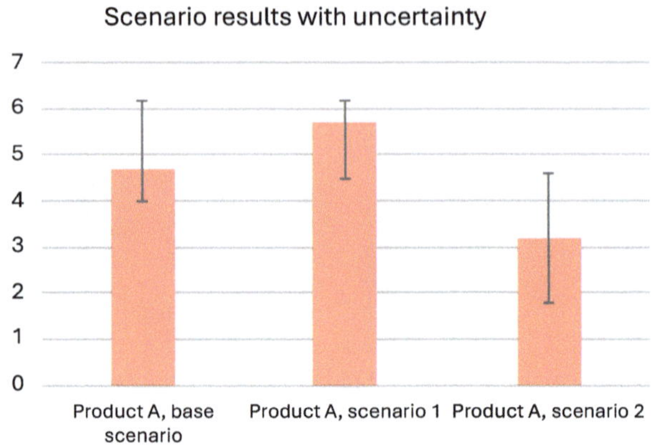

Fig. 7 Results can be presented together with uncertainty

influence on the results. A sensitivity analysis in LCA is a fundamental step as it the findings may lead to returning to the LCI phase for data reiteration. Typically, LCA practitioners have been assessing the sensitivity of the input parameters by generating alternative scenarios. In uncertainty analysis, the effect of data uncertainties on the reliability of the end results is modeled. The uncertainty of the input parameters can be assessed by using e.g. the pedigree matrix method or minimum and maximum values together with probability distributions. The uncertainty of the results can then be assessed using Monte Carlo simulations. The uncertainty can be presented together with the initial results in the results figures (see example Fig. 7).

If an LCA is conducted for internal purposes only, critical review is not mandatory. On the contrary, if an LCA is used for external communication, a critical review becomes obligatory. The aim of a critical review is to ensure consistency of an LCA. The conductors on an LCA should provide a clear documentation of the applied methodology, data sources, and limitations of the study. The critical review is compiled by LCA experts with a sufficient knowledge on LCA and independent from the organizations involved in the LCA and the products. In cases where results are not used in comparative assessments, a single LCA expert is considered sufficient. However, if comparative assessments are being made, e.g., stating that a product is superior to an alternative product, a critical review should be conducted by a panel of at least three experts.

2.4.1 Other Guidelines for Conducting an LCA

As the ISO LCA standards leave many choices to the user, many other additional guidelines have been created to further refine the rules of LCA for certain purposes, like allowing for better comparison between different LCA studies and products. These additional guidelines comprise, e.g., European Commission's Product Environmental Footprint (PEF), Environmental Product Declaration (EPD) guidance, and greenhouse gas (GHG) protocol product standard.

PEF is a detailed guidance meant for harmonizing LCA studies within product groups to make them comparable through product category rules (PEFCRs). The PEFCR for algae products has not yet been created. EPD has initially been established for communicating the environmental impacts of products offering certain functions. Also, the EPD guidance offers different product category rules (PCRs). The GHG protocol product standard is a guidance based on a guidance for calculating environmental impacts on an organizational level, modified for the assessment of a singular product.

3 Environmental Impacts of Algae Production

3.1 Micro Algae Production

In the case of microalgae products, key environmental hotspots include electricity consumption and mineral fertilizer use, which significantly impact climate change, eutrophication, and acidification. As a result, cultivation is a major contributor to the overall impacts of microalgal value chains. Energy consumption during cultivation accounts for 40–75% of the total environmental impacts, primarily due to temperature regulation, air supply, and agitation of the culture medium [25]. For example, heat pumps can contribute up to 44% of the climate impacts in open raceway ponds (ORPs) [24], while artificial lighting also plays a significant role [40]. In fermenters, the duration of the fermentation process is a key factor influencing environmental impacts [41].

Mineral fertilizers, such as sodium nitrate, sodium bicarbonate, and urea, contribute significantly to environmental impacts due to their energy-intensive production processes and nutrient runoff in wastewater [4, 24]. In heterotrophic cultivation, organic carbon sources also represent a major hotspot [42]. The environmental impact of these carbon sources varies, with, for instance, sorghum stem being more sustainable than corn glucose [41]. Downstream processing, including harvesting and drying, is a major contributor to environmental impacts due to its high energy consumption [43, 44]. Using residual heat, such as that from a co-located anaerobic digestion plant, and solar drying is more sustainable than energy-intensive methods like oven, freeze, or spray-drying [29, 43]. In terms of infrastructure materials, in PBRs, acrylic glass is a significant contributor to ozone depletion [25], while in ORPs, materials such as concrete, PVC, EPDM rubber, and steel can account for up to 36% of the total climate impacts [24].

Microalgal cultivation systems vary in environmental impacts based on design and operational parameters. Closed systems, such as vertical tubular PBRs and bubble columns, offer advantages like reduced contamination risks but are energy-intensive, particularly for mixing and lighting, contributing significantly to their environmental impacts [25]. Heterotrophic cultivation can lower environmental impacts through optimized fermentation and the use of non-food carbon sources [41]. Key parameters for environmental impact mitigation can be identified by coupling microalgal growth models with LCA, with factors such as production period, location, and system configuration influencing environmental impacts. Higher biomass productivity and year-round production help reduce climate impacts [24, 44]. Technological choices in harvesting, dewatering, and drying play a significant role in determining environmental impacts. Conversion pathways also affect sustainability, with, for instance, hydrothermal liquefaction showing lower impacts compared to gasification in aviation fuel production [45].

The bioremediation potential of microalgae can reduce environmental impacts in cultivation by using wastewater and CO_2 from flue gas, decreasing the need for mineral fertilizers and synthetic CO_2 while treating wastewater and lowering energy use at treatment plants. However, transportation and energy for waste stream injection can increase environmental impacts. Alternative organic nutrient sources, including domestic wastewater, brewery effluents, compost, manure, and liquid digestate from anaerobic digestion, also reduce reliance on mineral fertilizers [4]. Waste CO_2 from industries like coal-fired power plants, distilleries, paper factories, and biomass boilers serves as a key inorganic carbon source for microalgal cultivation [4].

3.2 Macro Algae Production

Common impact hotspots across seaweed value chains at sea typically include cultivation infrastructure, harvesting and processing steps, and transport. In near-shore rope systems typical for kelp, such as *Saccharina latissima*, infrastructure can contribute 15–30% of the total climate impact due to embodied energy in ropes, buoys, chains and anchoring systems [13]. These can increase to an even larger share in higher-energy offshore conditions, where cultivation infrastructure requires more robust materials to withstand the higher currents and wave action. Once mature, seaweed is typically harvested using diesel-driven vessels; this step alone can therefore represent a significant impact hotspot, particularly if repeated trips and distant grow-out sites are involved [13].

For land-based cultivation, energy demands for pumping, mixing, and temperature control in raceway ponds or tanks can represent over 50% of total GHG emissions, overshadowing downstream burdens in some cases [46]. This is also reflected in the literature for some microalgae production systems, where temperature control in particular can be a significant impact hotspot [47]. Hatchery or seaweed nurseries have also been shown to be an impact hotspot, primarily through energy use, though the extent of impact depends on the species, specific protocols, conditions (e.g., temperature), and maturation timeframes.

After harvesting, processing can be particularly energy-intensive, especially for products requiring drying, blanching, or freezing—common practices for seaweed destined for food applications or as a raw material in high-value biochemicals [14]. In many cases, seaweed has a naturally high water content, so dewatering, removing iodine with blanching, or preserving biomass quality through freezing can represent some of the largest operational energy demands, at over 50% of the total energy demand [14]. Finally, although seaweed farming typically requires fewer fertilizers and pesticides than most terrestrial systems, some large-scale or land-based operations may use nutrient inputs or other additives, which have upstream production burdens. Taken together, these hotspots underscore the importance of system-wide optimizations, from infrastructure selection and site planning to refined processing techniques and improved logistics management, in order to lower the overall environmental impacts of seaweed value chains.

Beyond the core hotspots, several broader considerations can significantly influence the overall environmental performance of seaweed value chains. First, the choice of system boundary—whether it encompasses only cultivation and primary processing or extends from cradle to grave—can shift the apparent significance of specific impacts and highlight additional stages, such as downstream product distribution and end-of-life disposal. Second, species selection and cultivation methods play a pivotal role: certain brown seaweeds, for instance, may thrive in cooler offshore conditions, while red seaweeds might be better suited for nearshore or warmer regions, each with its own resource requirements and potential yields. Third, strategies for managing co-products, such as high-value proteins or hydrocolloids, can substantially improve the net environmental balance by distributing impacts across multiple product streams. Finally, adopting integrated or complementary approaches—ranging from pairing seaweed farms with existing infrastructure to leveraging anaerobic digestion of leftover biomass—can help optimize resource use and reduce waste. When these considerations are addressed in tandem with careful site planning and process optimization, they can lead to more sustainable and economically viable seaweed value chains.

4 Discussion and Conclusion

As the algae sector evolves from small-scale trials to larger-scale processes producing more complex products, LCA is a key decision-making tool to support environmentally sustainable choices and process optimization. LCA allows for quantifying environmental impacts of products and processes throughout their value chains, avoiding problem-shifting between life cycle phases. Reasons for utilizing LCA for algae products may vary from recognizing environmental hotspots of processes and value chains to comparing options for raw materials, cultivation strategies, process and value chain options, or even entire products.

4.1 Current Trends and Evolvement of Algae LCAs

Recently, the use of LCA in the algae sector has expanded from studying simple product systems to assessing more complex processes and algae-derived products, circular value chains and broader environmental benefits. As the use of LCA increases, it is vital to recognize that the method can be applied in many ways, and thus both the assessment and result interpretation must be done with care to avoid the possible pitfalls. As discussed previously, the choices of the LCA practitioner may have an impact on the obtained results and their comparability. Over the past decades, diverse approaches, varying impact categories, and different methods have been used in algae LCAs which makes the comparison of results difficult or even impossible. When numerous LCA studies are used to inform decision-makers and actions, different approaches and assumptions can, in the worst case, lead to misinterpretation of the results if careful attention is not given to methodological choices.

Harmonizing the ways of utilizing the LCA method with different choices and assumptions as well as the inventory collection gives ground to make more robust and comparable results of different algae production processes. Lately, there have been efforts to make clear of the different methodological choices used in algae product LCAs and harmonize the calculation of inventory data in published studies [1, 7]. Review of previous studies highlights both the similarities and the dissimilarities and gives understanding of the common practices among the LCA practitioners in the field.

There are also several ongoing EU-funded algae projects that work together to boost the harmonization of algae product LCAs on an EU level.

Besides harmonized ways of utilizing LCA, networking and sharing information between LCA practitioners are becoming more and more important. There is a need to move forward from a black-box way of conducting LCAs, where assumptions and background choices and data are not openly shared. More active and open sharing of data would increase the reliability and accuracy of the LCA results as well as make conduction of LCAs easier.

The algae sector is always evolving, and new products and processes are piloted. LCA provides support to assess the products' possible environmental hotspots already when designing the production systems and is a useful tool already when piloting products. Still, as the goal is to provide a product to markets on a larger scale, there is a need for LCAs in larger scale production systems than lab- and pilot-scales to assess impacts of a larger-scale production. This can be achieved with so-called prospective LCAs which try to forecast future systems based on current smaller-scale systems. However, there is only so much prospective LCAs can scale up with current information before the results become unrealistic and unreliable.

4.2 Limitations in LCA

Although LCA is a useful tool for assessing sustainability of whole value chains from many viewpoints, it holds its limitations that include for example the freedom of choices, dependence on accessible as well as accurate data, and the disability to assess certain type of impacts.

As discussed before, even though there is a common ISO standard, the standard leaves room for decisions and interpretation for the LCA practitioner, and the standard itself does not guarantee that two LCAs made for the same product are identical. Notably, there is no single right way to conduct an LCA, and therefore, the different choices have to be taken into account when comparing different LCA studies.

Credibility of LCA is tightly connected to the use of reliable datasets. However, data scarcity and reliability is often a challenge. This is the case especially for new emerging technologies. On the other hand, it has been considered important to conduct an LCA at an early stage to get insights on environmental impacts of a product or a service.

Although LCA is a moderately efficient tool to assess environmental impacts of a product or a service, social aspects are seldom included. Methods to assess social sustainability, however, have been actively developed, but so far there is no consensus on how the social impacts should be assessed. Although halting biodiversity loss is considered an urgent target, biodiversity is seldom included into an LCA. Furthermore, biodiversity effects in LCA are typically assessed considering by considering only land use, and biodiversity effects on water ecosystems are not included. Moreover, LCA is not suitable to assess the impacts on local scale, such as local habitats or species population. A holistic sustainability assessment requires additional methods to assess these impacts.

4.3 Stakeholder Engagement

Engaging stakeholders early and consistently in an LCA—particularly during primary data collection—can greatly enhance both the quality and perceived credibility of the assessment. Stakeholder involvement in LCA can be thought of as existing on a continuum or "ladder" of participation. At the lower rungs, stakeholders may simply be informed about the scope, objectives, and methods of the assessment; this level of engagement is common for more straightforward or "non-wicked" problems where consensus on basic definitions and boundaries is relatively easy to achieve. As one ascends the ladder, participation moves through consultation—where stakeholders are asked for input on specific issues—into collaboration and co-production, in which they actively shape the LCA approach and data collection methods. A recent study [48] focusing on stakeholder involvement in seaweed LCAs emphasizes that even if the production system itself is well-structured (e.g., with standardized cultivation techniques), targeted consultation can improve data accuracy, uncover context-specific constraints (like local regulations or community preferences), and ultimately yield more relevant conclusions. In contrast, for more complex or contested systems—where multiple objectives, trade-offs, or regulatory uncertainties exist—comprehensive engagement becomes essential, helping reconcile competing viewpoints and ensuring the study incorporates a broad range of socio-environmental considerations.

Findings on effective environmental assessments [49] echo this principle: inviting stakeholders to participate not merely as data providers but as co-creators of the assessment process increases transparency, fosters buy-in, and can lead to more robust decision-making. In all cases, the degree of participation should align with the complexity of the problem at hand. For seaweed LCAs, this often means at least some level of meaningful engagement—whether through questionnaires or workshops—to ensure that key local insights (e.g., regarding harvest timing, transport logistics, or community impacts) are captured. By appropriately matching the depth of stakeholder involvement to the nature of the question being asked, LCAs gain both scientific rigor and practical legitimacy, thus making their results more actionable for policymakers, industry practitioners, and community members alike.

4.4 Future Developments and Concluding Remarks

Both the algae sector and the LCA methodology are continuously developing, which leads to new needs and improvements in the algae LCAs. In the future, new legislation and standardization are also expected to guide the use and development of algae LCAs, especially in the EU. At the time of writing this chapter, the EU's green claims directive under preparation is expected to require production of well-prepared, verified LCA studies before making environmental claims, increasing the importance and amount of LCA studies for products. Also new LCA standards and guidelines have been introduced over the years. Currently, there is no specific standard or guideline for algae products. However, the development of PEF and its related product category rules are ongoing in the EU and gaining more importance. If a set of category rules is introduced for algae products, it will make it easier for everyone to follow the same guidelines and increase the comparability of the studies. Moreover, with the lack of any all-guiding rule, the networking and discussions between the algae LCA practitioners have a key role in developing and harmonizing the algae LCA practices with a common understanding.

Besides changes through new legislation and guidance, new impact categories are being developed for those environmental impacts that are not yet covered by the categories in the common LCA methods. With the pressing issue of global biodiversity loss, many research projects are studying the ways in which biodiversity impacts can be quantified. The impacts on marine biodiversity are especially important for algae product systems where algae are cultivated in marine environments. Another rising issue is the quantification of microplastic pollution and its effects on the environment.

As a conclusion, with the algae sector constantly evolving, the role of LCA is ever more important as comprehensive and holistic sustainability assessments of products become essential. LCAs can and should be used in product development from the very start to support sustainable product design as well as guiding sustainable system scale-up. However, for reliable, comparable, and useful LCAs, common practices are needed. These can be achieved through open communication and cooperation between researchers, industry, as well as creating and using common guidelines through standardization.

Acknowledgments

This study was supported by the AlgaeProBANOS project, co-funded by the European Union through the Horizon Europe program (project number 101112943).

References

1. Aitken D, Bulboa C, Godoy-Faundez A et al (2014) Life cycle assessment of macroalgae cultivation and processing for biofuel production. J Cleaner Prod 75:45–56. https://doi.org/10.1016/j.jclepro.2014.03.080
2. Alvarado-Morales M, Boldrin A, Karakashev DB et al (2013) Life cycle assessment of biofuel production from brown seaweed in Nordic conditions. Bioresour Technol 129:92–99. https://doi.org/10.1016/j.biortech.2012.11.029
3. Langlois J, Sassi J-F, Jard G et al (2012) Life cycle assessment of biomethane from offshore-cultivated seaweed. Biofuels Bioprod Biorefin 6:387–404. https://doi.org/10.1002/bbb.1330
4. Braud L, McDonnell K, Murphy F (2023) Environmental life cycle assessment of algae systems: critical review of modelling approaches. Renew Sustain Energy Rev 179:113218. https://doi.org/10.1016/j.rser.2023.113218
5. Seghetta M, Hou X, Bastianoni S et al (2016) Life cycle assessment of macroalgal biorefinery for the production of ethanol, proteins and fertilizers – a step towards a regenerative bioeconomy. J Cleaner Prod 137:1158–1169. https://doi.org/10.1016/j.jclepro.2016.07.195
6. Seghetta M, Tørring D, Bruhn A, Thomsen M (2016) Bioextraction potential of seaweed in Denmark — an instrument for circular nutrient management. Sci Total Environ 563–564:513–529. https://doi.org/10.1016/j.scitotenv.2016.04.010
7. Thomas J-BE, Sinha R, Strand Å et al (2022) Marine biomass for a circular blue-green bioeconomy? A life cycle perspective on closing nitrogen and phosphorus land-marine loops. J Ind Ecol 26:2136–2153. https://doi.org/10.1111/jiec.13177
8. ISO 14040 (2006) Environmental management. Life cycle assessment. Principles and framework. International Organization for Standardization, Geneve
9. Guinée JB, Heijungs R, Huppes G et al (2011) Life cycle assessment: past, present, and future. Environ Sci Technol 45:90–96. https://doi.org/10.1021/es101316v
10. Hauschild MZ, Rosenbaum RK, Olsen SI, (2018) Life cycle assessment: theory and practice. Springer, Cham
11. Finnveden G, Hauschild MZ, Ekvall T et al (2009) Recent developments in Life cycle assessment. J Environ Manage 91:1–21. https://doi.org/10.1016/j.jenvman.2009.06.018
12. ISO 14044 (2006) Environmental management. Life cycle assessment. Requirements and guidelines. International Organization for Standardization, Geneve
13. Thomas J-BE, Ahlgren E, Hornborg S, Ziegler F (2024) Life cycle environmental impacts of kelp aquaculture through harmonized recalculation of inventory data. J Cleaner Prod 450:141987. https://doi.org/10.1016/j.jclepro.2024.141987
14. Thomas J-BE, Sodré Ribeiro M, Potting J et al (2021) A comparative environmental life cycle assessment of hatchery, cultivation, and preservation of the kelp *Saccharina latissima*. ICES J Mar Sci 78:451–467. https://doi.org/10.1093/icesjms/fsaa112

15. Taelman SE, Champenois J, Edwards MD et al (2015) Comparative environmental life cycle assessment of two seaweed cultivation systems in North West Europe with a focus on quantifying sea surface occupation. Algal Res 11:173–183. https://doi.org/10.1016/j.algal.2015.06.018
16. van Oirschot R, Thomas J-BE, Gröndahl F et al (2017) Explorative environmental life cycle assessment for system design of seaweed cultivation and drying. Algal Res 27:43–54. https://doi.org/10.1016/j.algal.2017.07.025
17. Koesling M, Kvadsheim NP, Halfdanarson J et al (2021) Environmental impacts of protein-production from farmed seaweed: comparison of possible scenarios in Norway. J Cleaner Prod 307:127301. https://doi.org/10.1016/j.jclepro.2021.127301
18. Li J, Bergman K, Thomas J-BE et al (2023) Life cycle assessment of a large commercial kelp farm in Shandong, China. Sci Total Environ 903:166861. https://doi.org/10.1016/j.scitotenv.2023.166861
19. Ayala M, Goosen N, Michalak L et al (2024) Prospective LCA of brown seaweed-based bioplastic: upscaling from pilot to industrial scale. Sustain Prod Consump 52:416–426. https://doi.org/10.1016/j.spc.2024.11.020
20. Passell H, Dhaliwal H, Reno M et al (2013) Algae biodiesel life cycle assessment using current commercial data. J Environ Manage 129:103–111. https://doi.org/10.1016/j.jenvman.2013.06.055
21. Zhang Y, Kendall A (2019) Effects of system design and co-product treatment strategies on the life cycle performance of biofuels from microalgae. J Cleaner Prod 230:536–546. https://doi.org/10.1016/j.jclepro.2019.05.137
22. Adesanya VO, Cadena E, Scott SA, Smith AG (2014) Life cycle assessment on microalgal biodiesel production using a hybrid cultivation system. Bioresour Technol 163:343–355. https://doi.org/10.1016/j.biortech.2014.04.051
23. Foteinis S, Antoniadis-Gavriil A, Tsoutsos T (2018) Life cycle assessment of algae-to-biodiesel shallow pond production systems in the Mediterranean: influence of species, pond type, by(co)-product valorisation and electricity mix. Biofuels Bioprod Biorefin 12:542–558. https://doi.org/10.1002/bbb.1871
24. Duran Quintero C, Ventura A, Lépine O, Pruvost J (2021) Eco-design of spirulina solar cultivation: key aspects to reduce environmental impacts using life cycle assessment. J Cleaner Prod 299:126741. https://doi.org/10.1016/j.jclepro.2021.126741
25. Pechsiri JS, Thomas J-BE, Bahraoui NE et al (2023) Comparative life cycle assessment of conventional and novel microalgae production systems and environmental impact mitigation in urban-industrial symbiosis. Sci Total Environ 854:158445. https://doi.org/10.1016/j.scitotenv.2022.158445
26. Porcelli R, Dotto F, Pezzolesi L et al (2020) Comparative life cycle assessment of microalgae cultivation for non-energy purposes using different carbon dioxide sources. Sci Total Environ 721:137714. https://doi.org/10.1016/j.scitotenv.2020.137714
27. Yadav G, Dubey BK, Sen R (2020) A comparative life cycle assessment of microalgae production by CO2 sequestration from flue gas in outdoor raceway ponds under batch and semi-continuous regime. J Cleaner Prod 258:120703. https://doi.org/10.1016/j.jclepro.2020.120703
28. Ahlgren S, Björklund A, Ekman A et al (2015) Review of methodological choices in LCA of biorefinery systems – key issues and recommendations. Biofuels Bioprod Biorefin 9:606–619. https://doi.org/10.1002/bbb.1563
29. Braud L, McDonnell K, Murphy F (2025) Beyond phycocyanin: environmental life cycle assessment of a European pilot scale *Spirulina* biorefinery. Bioresource Technol Rep 29:102027. https://doi.org/10.1016/j.biteb.2025.102027
30. Papadaki S, Kyriakopoulou K, Tzovenis I, Krokida M (2017) Environmental impact of phycocyanin recovery from *Spirulina platensis* cyanobacterium. Innov Food Sci Emerg Technol 44:217–223. https://doi.org/10.1016/j.ifset.2017.02.014
31. Espada JJ, Pérez-Antolín D, Vicente G et al (2020) Environmental and techno-economic evaluation of β-carotene production from *Dunaliella salina*. A biorefinery approach. Biofuels Bioprod Biorefin 14:43–54. https://doi.org/10.1002/bbb.2012
32. Onorato C, Rösch C (2020) Comparative life cycle assessment of astaxanthin production with *Haematococcus pluvialis* in different photobioreactor technologies. Algal Res 50:102005. https://doi.org/10.1016/j.algal.2020.102005
33. Gaber K, Rösch C, Biondi N (2022) Life cycle assessment of Total Fatty Acid (TFA) production from microalgae *Nannochloropsis oceanica* at different sites and under different sustainability scenarios. Bioenerg Res 15:1595–1615. https://doi.org/10.1007/s12155-021-10279-z
34. de Souza MHB, Calijuri ML, Assemany PP et al (2019) Soil application of microalgae for nitrogen recovery: a life-cycle approach. J Cleaner Prod 211:342–349. https://doi.org/10.1016/j.jclepro.2018.11.097

35. Maga D (2017) Life cycle assessment of biomethane produced from microalgae grown in municipal waste water. Biomass Conv Bioref 7:1–10. https://doi.org/10.1007/s13399-016-0208-8
36. Arashiro LT, Montero N, Ferrer I et al (2018) Life cycle assessment of high rate algal ponds for wastewater treatment and resource recovery. Sci Total Environ 622–623:1118–1130. https://doi.org/10.1016/j.scitotenv.2017.12.051
37. Arashiro LT, Josa I, Ferrer I et al (2022) Life cycle assessment of microalgae systems for wastewater treatment and bioproducts recovery: natural pigments, biofertilizer and biogas. Sci Total Environ 847:157615. https://doi.org/10.1016/j.scitotenv.2022.157615
38. Heijungs R, Allacker K, Benetto E et al (2021) System expansion and substitution in LCA: a lost opportunity of ISO 14044 amendment 2. Front Sustain 2. https://doi.org/10.3389/frsus.2021.692055
39. Sills DL, Van Doren LG, Beal C, Raynor E (2020) The effect of functional unit and co-product handling methods on life cycle assessment of an algal biorefinery. Algal Res 46:101770. https://doi.org/10.1016/j.algal.2019.101770
40. Gurreri L, Calanni Rindina M, Luciano A et al (2024) Microalgae production in an industrial-scale photobioreactors plant: a comprehensive life cycle assessment. Sustain Chem Pharm 39:101598. https://doi.org/10.1016/j.scp.2024.101598
41. Lu Y, Mu D, Xue Z et al (2021) Life cycle assessment of industrial production of microalgal oil from heterotrophic fermentation. Algal Res 58:102404. https://doi.org/10.1016/j.algal.2021.102404
42. Davis D, Morão A, Johnson JK, Shen L (2021) Life cycle assessment of heterotrophic algae omega-3. Algal Res 60:102494. https://doi.org/10.1016/j.algal.2021.102494
43. de Souza MP, Colares GS, Wiesel PG et al (2024) Life cycle assessment of different scenarios for pigment production from an algal turf scrubber (ATS) system. Biomass Conv Bioref 14:28163–28174. https://doi.org/10.1007/s13399-022-03530-w
44. Zarra T, Senatore V, Zorpas AA et al (2024) Advanced membrane photobioreactors in algal CO2 biofixation and valuable biomass production: integrative life cycle assessment and sustainability analysis. Sustain Chem Pharm 40:101658. https://doi.org/10.1016/j.scp.2024.101658
45. Marangon BB, Castro J de S, Assemany PP et al (2024) Wastewater-grown microalgae biomass as a source of sustainable aviation fuel: life cycle assessment comparing hydrothermal routes. J Environ Manage 360:121164. https://doi.org/10.1016/j.jenvman.2024.121164
46. Eikenbusch JK (2024) Life cycle assessment of a land-based seaweed cultivation system: Oregon Coast case study on the base ingredient for "seaweed bacon." Master's Thesis, KTH
47. Pérez-López P, de Vree JH, Feijoo G et al (2017) Comparative life cycle assessment of real pilot reactors for microalgae cultivation in different seasons. Appl Energy 205:1151–1164. https://doi.org/10.1016/j.apenergy.2017.08.102
48. Potting J, Thomas J-BE, Gröndahl F (2022) Stakeholder participation in sustainability assessment of non-wicked problems: the case of a future seaweed industry in Sweden. Ambio 51:901–913. https://doi.org/10.1007/s13280-021-01609-8
49. Eckley N (2001) Designing effective assessments: the role of participation, science and governance, and focus. Environmental Issue Report No. 26. Environment Agency. Copenhagen, Denmark

Check for updates

Chapter 27

Life Cycle Assessment (LCA) of Algal Fuels

Homa Hosseinzadeh-Bandbafha, Mohammadali Kiehbadroudinezhad, Meisam Tabatabaei, Mortaza Aghbashlo, Alawi Sulaiman and Abbas Ghassemi

Abstract

In today's world, life cycle assessment (LCA) stands as a revolutionary tool for environmental policymakers and business leaders striving for sustainable production and consumption. This dynamic, standardized approach meticulously quantifies resource use, pollutant emissions, and their associated environmental and health impacts throughout the entire life cycle of goods and services. When it comes to algal fuels, LCA is indispensable for unlocking their true environmental potential. This chapter dives deep into the technical intricacies of applying LCA to algal fuels, offering expert insights on major international standards, i.e., ISO 14040 and 14044. Designed for LCA practitioners and technical experts in both the public and private sectors, these guidelines offer essential tools not only to assess but also to optimize the sustainability of algal biofuel production—driving innovation toward a greener future.

Keywords Algal fuels, Life cycle assessment, ISO standards, Sustainable development

1 Introduction

The increasing global energy demand, along with the adverse impacts of petroleum-based fuels on ecosystems and human health, has driven substantial interest in renewable energy alternatives like biofuels [1]. These fuels are often regarded as "carbon neutral," since the amount of carbon dioxide (CO_2) emitted during combustion is roughly equivalent to what the plants absorb during their growth phase [2]. This equilibrium positions biofuels as a viable substitute for fossil fuels, presenting an opportunity to reduce greenhouse gas emissions and support global sustainability efforts.

Biofuels, produced from diverse feedstocks, are generally categorized into different generations, each presenting unique challenges. First-generation biofuels are often criticized for contributing to the food-versus-fuel debate [3], while second-generation biofuels struggle with high production costs and technological complexity [4].

Kristian Spilling (ed.), *Biofuels from Algae*, Methods in Molecular Biology, vol. 2996,
https://doi.org/10.1007/978-1-0716-5031-8_27,

In contrast, third-generation biofuels, particularly those derived from algae, have garnered attention for their remarkable advantages. Algae exhibit exceptional productivity [5], rapid growth rates [6], enhanced CO_2 absorption [7], and high lipid content [8]. Moreover, algae cultivation does not require fertile agricultural land, preventing competition with food crops [9]. Beyond theoretical applications, algal fuels are positioned to significantly contribute to renewable energy targets set by the U.S. Renewable Fuel Standards under the Energy Independence and Security Act of 2007 [10]. Although the algae-based fuel industry remains in its early stages, it holds immense potential to transform liquid transportation fuel production, particularly biodiesel, offering a cleaner and more sustainable solution that reduces reliance on imported oil while enhancing national energy security [11].

However, before advancing to large-scale commercialization of biofuel production, it is essential to evaluate its environmental footprint and comply with established sustainability guidelines to drive responsible technological progress [12]. Throughout their life cycle, products like algal fuels can present environmental challenges—some of which may not be immediately evident. Among the various tools available for sustainability assessment, life cycle assessment (LCA) stands out for its comprehensive approach to evaluating the environmental impacts of products and processes throughout their entire life cycle [13]. By examining every stage, from raw material extraction and production to utilization and disposal, LCA systematically tracks resource use and emissions, providing a complete perspective on environmental consequences [14]. What makes LCA particularly valuable is its ability to prevent burden shifting—ensuring that mitigating impacts in one phase does not unintentionally exacerbate other environmental concerns [15].

LCA also serves as a transformative tool in sustainability, enabling decision-makers to assess the environmental impact of various production methods and identify the most sustainable solutions [16]. Beyond just environmental analysis, LCA offers a comprehensive perspective by incorporating economic and social dimensions, facilitating well-rounded, sustainable decision-making [17]. Its adaptability makes it instrumental in shaping policies, enhancing green marketing strategies, and supporting businesses committed to sustainability leadership. By using LCA, industries and governments can drive innovation, craft effective policies, and accelerate the shift toward a cleaner and more resilient future [18].

Overall, LCA stands as a state-of-the-art sustainability assessment method, delivering in-depth insights into a product's environmental impact across its entire life cycle [19]. Unlike traditional evaluation techniques, LCA takes a holistic and scientifically grounded approach, continuously evolving with technological advancements to remain at the cutting edge of environmental analysis. It systematically tracks emissions across air, water, and land, pinpoints key areas for improvement [20], and ensures that reducing environmental harm in one stage does not inadvertently increase impacts elsewhere. Through precise environmental footprinting, LCA equips industries, policymakers, and NGOs with essential data to foster sustainable innovation [21]. By enabling informed comparisons between products and processes, it ensures that the most environmentally responsible choices are made [22]. With its capacity to shape strategic planning and enhance environmental performance, LCA stands as an essential tool in advancing sustainability, particularly in optimizing the environmental impact of algal fuel production.

In line with that, LCA has become a key tool in evaluating the environmental impact of both small- and large-scale algal fuel production systems, providing valuable insights into their sustainability across various production scales [23–26]. A significant portion of LCA research on algal fuel pathways focuses on similar cultivation and conversion processes—algae are grown in open ponds, harvested, and then undergo lipid extraction and transesterification to produce biodiesel (Fig. 1). However, despite following ISO 14040 and 14044 standards [27, 28], these studies often produce widely differing results due to variations in assumptions, system boundaries, and methodological approaches. To address these discrepancies and improve the consistency of LCA outcomes, it is essential to create clear, standardized protocols for LCA in algal fuel production. A cohesive methodology not only enhances the reliability and usefulness of the results but also helps accelerate the development of sustainable algal fuels, emphasizing their role as a viable alternative to fossil fuels.

Therefore, the primary objective of this chapter is to address and resolve key inconsistencies in the assumptions used in existing research while providing a set of standardized guidelines designed to improve the consistency and accuracy of LCAs for algal biofuel production in the future. This chapter also offers a structured, step-by-step approach for performing a thorough and dependable LCA, outlining the essential data needed at each stage to ensure reliable and actionable results. By promoting a more transparent, coherent, and harmonized methodology, the chapter aims to strengthen the credibility and effectiveness of sustainability evaluations in the realm of algal fuels. Ultimately, it seeks to empower more informed decision-making and accelerate the adoption of cleaner, more sustainable energy solutions.

2 Materials/Tools Required for Conducting an LCA of Algal Fuels

2.1 Software Guidelines

Specialized software currently serves as the primary tool for conducting LCAs, streamlining the complex calculations and data management essential to this methodology. De Wolf et al. [33] highlighted that the successful completion of an LCA study is largely dependent

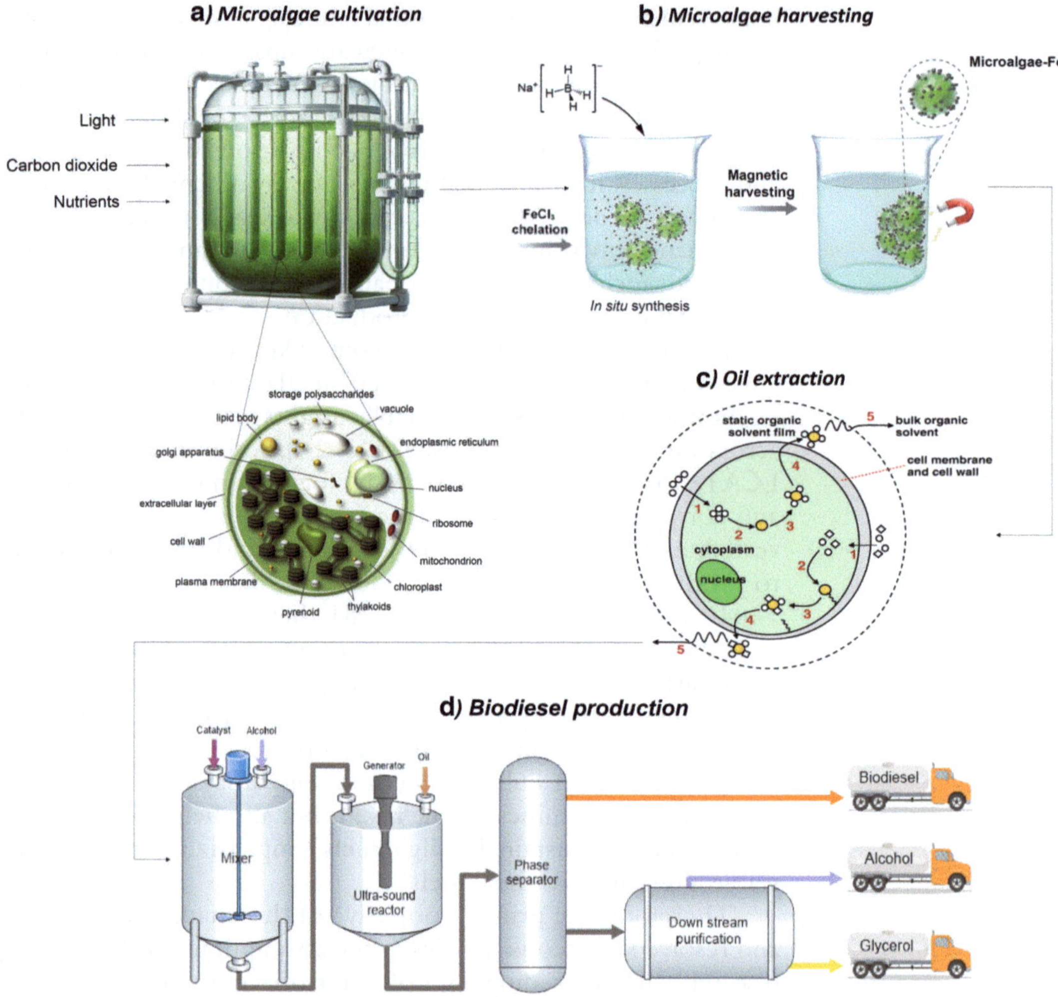

Fig. 1 Biodiesel production chain from algae (**a**) microalgae cultivation, adopted from Bernaerts et al. [29]. With permission from Elsevier. Copyright© 2019; (**b**) microalga harvesting, adopted from Li et al. [30]. With permission from Elsevier. Copyright© 2020; (**c**) oil extraction from microalga, ● lipids, ○ non-polar organic solvent, ◇ polar organic solvent, reprinted from Halim et al. [31]. With permission from Elsevier. Copyright© 2012; (**d**) biodiesel production, reprinted from Tabatabaei et al. [32]. With permission from Elsevier. Copyright© 2019

on the use of such software tools. In the early stages of LCA development, assessments were performed manually, relying on spreadsheets or basic computational techniques to process and analyze data. This method was not only time-consuming but also vulnerable to inaccuracies. The introduction of LCA software has since transformed the process, making it more efficient. As a result, LCA outcomes now depend on the methodologies and impact assessment models embedded in specific software tools, which enable more accurate and streamlined development of LCA studies [34].

The six leading LCA software tools—SimaPro, GaBi, Ecochain Mobius, OneClick LCA, OpenLCA, and Umberto—are widely recognized for conducting LCAs and calculating environmental impacts in accordance with scientific standards. While tools like SimaPro and GaBi are tailored to meet the advanced needs of experienced LCA professionals [35], options such as Ecochain Mobius are designed for users with less LCA experience [36]. They provide a more accessible platform for assessing the environmental impacts of their products. Therefore, the choice of LCA software for evaluating the life cycle of algal fuels depends on the specific goals of the assessment and the expertise of the user or team. Any of the mentioned tools can effectively support the LCA process, provided they are selected according to these criteria:

1. SimaPro software: For conducting reliable LCAs and gaining actionable sustainability insights, powerful and dependable software is essential. SimaPro, developed by PRé, has been a trusted tool for LCA professionals for nearly 35 years, offering advanced features and a comprehensive life cycle inventory database (https://simapro.com/). Widely recognized by industry leaders and academics across more than 80 countries, SimaPro stands as one of the most respected sustainability and LCA software solutions available. The software supports LCA by offering advanced tools to assess the environmental impact of products across their life cycle. Based on robust scientific methods and life cycle thinking (LCT), SimaPro is widely used by LCA experts, sustainability analysts, product developers, and decision-makers. Its integrated databases and consistent methodologies allow users to examine their results at each life cycle stage thoroughly. Designed by experts for professionals, SimaPro effectively transforms raw data into valuable, actionable insights, thereby supporting informed decision-making and driving sustainability efforts. For more detailed information, including the latest features and updates, visit the official SimaPro website at https://simapro.com/.
   a. To perform an LCA for algal fuels, the latest version, SimaPro 10 (available in various editions from educational to professional in 2025), offers a powerful toolset to gather, analyze, and track the environmental performance of algal fuels. It enables the modeling and analysis of complex life cycles in the production and use chains of algal fuels systematically and transparently, adhering to the ISO standards. Whether users are in the data collection phase, analyzing systems, or tracking sustainability metrics, SimaPro 10 (as of 2025) equips them with the professional tools necessary to drive environmental innovation with precision and clarity.

2. GaBi software: GaBi is a well-established and widely used LCA software that combines advanced modeling features with a comprehensive environmental database covering a wide range of industries and regions (https://sphera.com/). With access to over 20 sector-specific databases, GaBi provides reliable, consistent analytics to assess the environmental impacts across a product's entire life cycle, supporting fact-based decision-making. A key strength of GaBi is its flexibility in customizing LCA models. Users can adjust parameters and inputs to tailor analyses to specific industry sectors or regional contexts, ensuring that results are both accurate and relevant. Additionally, GaBi simplifies the presentation of results by generating clear, concise, and customizable reports, which makes it easier for stakeholders to interpret the findings and take informed actions based on the outcomes. GaBi's robust capabilities and intuitive reporting features make it an effective tool for conducting detailed, application-specific LCAs. For more in-depth information, including detailed features and updates, visit the official GaBi website at https://sphera.com/.
   a. To conduct an LCA on algal fuels, GaBi software, available in various editions from educational to professional, offers a powerful platform to assess the environmental, cost, and social aspects of algal fuels throughout their life cycle. With its extensive database and advanced modeling capabilities, GaBi allows users to perform detailed LCAs, delivering valuable insights into the sustainability performance of algal fuels across multiple dimensions.
3. OpenLCA software: OpenLCA is a leading open-source, high-performance software for LCA and sustainability evaluations (https://www.openlca.org/). Known for its speed and reliability, it provides in-depth calculations for both LCA and sustainability assessments, offering valuable insights that help identify key drivers across the life cycle, whether by process, flow, or impact category. OpenLCA also features advanced visualization tools to map and present results with clarity. With top-tier import and export functions, the software facilitates easy sharing of models and collaboration among users. It seamlessly integrates life cycle costing (LCC) and social assessments into the life cycle model, delivering a comprehensive sustainability analysis. OpenLCA's user-friendly interface, available in multiple languages, combined with ongoing improvements in its repository and collaboration features, enhances its accessibility and usability. Regular updates ensure that OpenLCA remains at the forefront of LCA software, continuously evolving to meet the changing needs of its users. For further details, including updated features and functionalities, visit the official OpenLCA website at https://www.openlca.org/.

   a. To analyze the LCA of algal fuels, OpenLCA is a suitable choice if users are looking for a free, open-source platform to model and evaluate algal fuel systems. It provides flexible import and export functionalities, allowing for seamless collaboration and easy sharing of models.

4. Ecochain Mobius software: Ecochain Mobius is a cloud-based platform designed to facilitate detailed LCA of products (https://ecochain.com/). It provides tools for companies to track and minimize their products' environmental impact, offering industry-specific customization for sectors such as food and beverage, manufacturing, and energy. Key features include advanced data management, impact analysis, scenario modeling, and visualization tools. Ecochain Mobius supports benchmarking, supplier evaluation, and decision-making, enabling businesses to drive sustainability, enhance product development, and meet environmental goals. It also helps set and monitor sustainability objectives, improve internal and external communication, and comply with ISO standards. For more detailed features and updates, visit the official Ecochain website at https://ecochain.com/.
   a. Use Ecochain Mobius to assess and improve the environmental impact of algal fuels, particularly if users are new to LCA. The platform's data management and visualization tools cover the entire life cycle, from cultivation to processing, with flexible modules for customized solutions. Track sustainability metrics, benchmark performance, and make data-driven decisions to optimize supplier selection and innovation.
5. OneClick LCA software: OneClick LCA is an automated, cloud-based software designed for performing LCAs and Environmental Product Declarations (EPDs) within the construction sector (https://oneclicklca.com/). It integrates global standards, comprehensive databases, and design software, making it adaptable for a wide range of construction projects, including buildings, infrastructure, renovations, and materials. For more in-depth details, including features and updates, visit the official OneClick LCA website at https://oneclicklca.com/.
   a. To evaluate the environmental impacts of building algae-based fuel plants or algal biorefineries using LCA methods, OneClick LCA software provides essential insights for promoting sustainable infrastructure development in the algae fuel sector.
6. Umberto Software: Umberto is a versatile software tool designed for comprehensive LCA and environmental impact analysis across a product's entire life cycle (https://lca-software.org/umberto/). It offers features such as graphical modeling,

inventory analysis, and the evaluation of key impact categories such as carbon emissions, water consumption, and energy usage. Fully aligned with ISO standards, Umberto supports scenario planning and sensitivity analysis to guide informed decision-making. Its capabilities also include LCC and material flow cost accounting (MFCA), providing a complete perspective on sustainability. With its intuitive interface and seamless database integration, Umberto is well-suited for companies looking to enhance product sustainability, improve eco-efficiency, and boost competitiveness.

a. To assess the environmental and economic impacts of algal fuels, including LCC and MFCA, Umberto is an excellent tool. It provides comprehensive analysis capabilities for evaluating the full scope of sustainability impacts, ensuring a holistic approach to understanding both environmental and economic aspects of algal fuel production.

Table 1 presents an in-depth comparison of various LCA software tools used to assess the environmental impact of algal fuels. It outlines the features, advantages, limitations, and ideal scenarios for each software, providing a clear overview to guide users in selecting the most appropriate tool for their specific needs in algal fuel environmental evaluations.

2.2 Databases

A crucial element for conducting an LCA is the use of a dependable and validated database. Since LCA involves extensive data analysis, having access to accurate and sufficient data is essential for carrying out a comprehensive evaluation [37]. These databases provide extensive environmental data for assessments, especially when specific information is lacking. To address concerns about confidentiality, various global databases have been established to guarantee reliable LCA results [38]. Several databases are available for LCA studies, including EcoInvent, Agri-footprint, GaBi, and Agribalyse [39], each with its own advantages and limitations. When performing an LCA of algal fuels, careful consideration must be given to six key factors, i.e., scope, completeness, transparency, comprehensiveness, updates, and licensing, before selecting the most appropriate database, as emphasized by Martínez-Rocamora et al. [40]:

1. Check the scope of the LCA database by first identifying the region in which algal fuel production is taking place (e.g., Europe, North America, and Asia) and then reviewing the material categories included (e.g., metals, plastics, and chemicals) across the various stages of the algal fuel production process.
2. Verify the comprehensiveness of the database by ensuring that all relevant material families are included while also considering

Table 1
A comprehensive overview of LCA software for algal fuel assessments

Software	Description	Key features	Strengths	Weaknesses	Use case for algal fuels
SimaPro	A professional LCA tool used for collecting, analyzing, and monitoring environmental performance, with a focus on complex and detailed analysis	Advanced modeling, extensive life cycle inventory database, ISO 14040 and 14044 compliance, system transparency, and precision	Comprehensive analysis of complex life cycles, robust scientific foundations, ideal for detailed, high-accuracy results	Expensive, steep learning curve for beginners, requires advanced expertise in LCA to maximize its capabilities	Model and analyze the life cycle of algal fuels, focusing on sustainability and actionable insights. Supports environmental innovation in algal fuel production
GaBi	A widely recognized LCA software that integrates advanced modeling capabilities with sector-specific environmental data	Customizable modeling, sector-specific databases, clear reporting tools, ISO compliance	Extensive sector-specific databases, flexible modeling, and strong reporting features	Expensive, requires significant setup time to tailor models, and can be complex for new users	Evaluate the environmental, cost, and social profiles of algal fuels through detailed life cycle assessments, tailored for specific industry conditions
OpenLCA	Open-source LCA software with fast and reliable calculations, designed to offer flexibility and ease of use	Open-source, life cycle costing integration, social assessments, advanced import/export, multilingual	Free and open-source, highly flexible with multiple integration options, user-friendly interface	Limited support and documentation compared to paid software, can be challenging for users unfamiliar with LCA concepts	Model and assess algal fuel systems using this free tool, with a focus on flexibility and ease of sharing models for collaboration and sustainability assessments

(continued)

Table 1
(continued)

Software	Description	Key features	Strengths	Weaknesses	Use case for algal fuels
Ecochain Mobius	Cloud-based software that helps measure, reduce, and optimize the environmental impact of products, tailored for various industries	Data management, impact assessment, visualization, scenario analysis, ISO 14040 and 14044 compliance	Easy-to-use cloud interface, customizable tools for different industries, transparent sustainability tracking	Cloud-based only may require a stable internet connection, less suitable for large-scale or highly customized models	Evaluate and optimize the environmental performance of algal fuels, leveraging data management, scenario analysis, and tracking sustainability metrics throughout the production cycle
OneClick LCA	Automated cloud-based tool for LCA and EPD in the construction industry, integrating global standards	Cloud-based, automated, integrates with design software, supports various standards	Automated reports and integration with design software simplify LCA for non-experts	Primarily designed for the construction industry, limited customization for non-construction sectors like algal fuels	Assess the environmental impacts of constructing algae-based biofuel plants or biorefineries, supporting sustainable infrastructure development
Umberto	Comprehensive LCA tool for analyzing environmental impacts across the entire product life cycle, supporting various international standards	Graphical modeling, scenario analysis, material flow cost accounting, carbon footprint calculation, ISO compliance	User-friendly graphical modeling, scenario analysis, and strong integration with databases	High cost, requires significant user training, can be overwhelming for beginners, and needs powerful hardware for large datasets	Evaluate both the environmental and economic impacts of algal fuel production, focusing on carbon emissions, water usage, and energy consumption

whether all material variations within each category, specific to the algal fuel production chain, are accounted for.

3. Evaluate the transparency of the database, which is crucial for conducting an LCA of algal fuel production. This step involves assessing the clarity and traceability of the methodological process, including how the material life cycle in algal fuel production is outlined, such as through flow diagrams and life cycle inventories. The traceability of an LCA study is enhanced if the database provides clear references to supporting literature.
4. Assess the comprehensiveness of the database by evaluating the integrity of the information for each material involved in algal fuel production. Notably, beyond flow charts and inventories, a significant amount of additional data is often provided on the factors influencing the analysis. These details are essential for modifying and adapting the data for use in other studies.
5. Verify the update frequency of the database by assessing how recent the data entries are relative to the timeframe of the project under evaluation. Ensuring that the data is up to date helps confirm that the information reflects the latest advancements and trends in the field of algal fuel production.
6. Examine the licensing requirements of the database to determine if a paid subscription is necessary for access. This step is essential for understanding the costs associated with using the database and ensuring that the chosen database aligns with the budget and usage needs of the algal fuel project.

By adhering to this systematic approach, researchers can effectively navigate the large variety of over 40 available LCA databases [40] and select the ones that provide the most relevant, accurate, and comprehensive data for assessing algal fuel production systems, as outlined below:

1. Use EcoInvent databases (accessible directly or integrated into platforms like SimaPro and OpenLCA) when researchers require secondary data on algal fuel production across various global regions, with a particular emphasis on Europe. These databases offer extensive data collection on energy flows, raw material usage, emissions, and environmental impacts specific to algal fuel systems.
2. Opt for the GaBi database when seeking secondary data related to algal fuel production in European nations, especially for data tied to industrial operations. GaBi's extensive database offers detailed insights into the environmental footprint of various production processes. It is particularly beneficial for conducting LCA studies that require specific, high-quality, industry-focused data. This resource is fully integrated into both GaBi

software and OpenLCA, allowing users to access a wealth of reliable data for in-depth, scientifically robust assessments.

3. Utilize the US LCI database when free access to data on the production processes of algal fuels in the United States is needed. This database serves as a centralized resource, offering comprehensive details on the environmental impacts linked to energy consumption and resource usage across various commercial materials, products, and systems.
4. Consider using the OpenLCA database when free, global data on the production of algal fuels is needed. This platform provides an extensive collection of datasets from multiple regions, facilitating comprehensive LCAs for algal fuels. It is ideal for users seeking to conduct global analyses, offering valuable insights into various environmental impacts and helping users evaluate the sustainability of algal fuel production across different locations.
5. Consider the ELCD database (which can be accessed directly or through LCA software such as SimaPro and GaBi) when seeking cost-free data related to algal fuel production within European regions. This database is a comprehensive resource for environmental data, with a strong emphasis on European industries and production processes.
6. Leverage the Environmental Footprint (EF) database to support the LCA of algal fuels while adhering to the established product environmental footprint category rules (PEFCR) and organization environmental footprint sector rules (OEFSR). As a key element of the European Commission's efforts to create a unified market for green products, the EF database offers a reliable and standardized approach for evaluating the environmental impacts of biofuels. It is seamlessly incorporated into platforms like GaBi, OpenLCA, and SimaPro, providing a wealth of high-quality, comprehensive data to conduct precise, transparent, and comparable LCAs that encourage sustainable biofuel production and consumption practices.
7. Consider using specialized databases such as plastics Europe (https://plasticseurope.org/) for data on plastic production impacts or the world steel association (worldsteel) database (https://worldsteel.org/) for detailed information on the environmental impacts of metal production. These databases provide highly specific data related to materials commonly involved in the algal fuel production process. Accessing these resources can enhance the accuracy of an LCA, particularly during phases where plastics, metals, or other materials play a significant role in the fuel production cycle. These specialized sources allow for a more precise and targeted assessment of material-related environmental impacts.

3 Methods to Conduct an LCA of Algal Fuels

When evaluating the environmental sustainability of products from a product-centered perspective, adopting an LCT framework is critical [41]. LCT offers an in-depth approach that goes beyond conventional evaluation methods, looking at all stages involved in the product's creation and use [42]. This approach assesses sustainability impacts not only from an environmental standpoint but also in terms of social and economic factors, helping identify opportunities to reduce adverse effects while preventing issues from being shifted to other stages of the product's life cycle [42, 43]. Methods such as LCA, LCC, and social life cycle assessment (SLCA) are vital tools within LCT, offering robust mechanisms for sustainability evaluation across multiple dimensions [44].

In the evaluation of algal fuels, adopting LCT is crucial for understanding both the environmental and socio-economic impacts throughout its entire life cycle, from cultivation and harvesting to processing and usage. LCA, as a key method within LCT, provides a systematic and quantitative approach to assess the environmental implications of algal fuel production and consumption. By considering the entire life cycle, LCT ensures that algal fuels contribute positively to sustainable energy development, supporting the transition to cleaner, more eco-friendly energy solutions.

To carry out an LCA for algal fuel production, practitioners adhere to two essential international standards: ISO 14040 and ISO 14044 [27, 28]. These standards set out the general principles and specific requirements for performing LCA studies on products and services across different economic activities worldwide. As per the guidelines outlined in ISO [27, 28], four primary steps should be followed when assessing the life cycle of algal fuels:

1. Goal and scope definition: This initial step involves clearly defining the purpose of the LCA study, identifying the system boundaries, and determining the functional unit (FU) of the study. The goal sets the context for the analysis, while the scope defines the extent and limitations of the system being studied.
2. Life cycle inventory (LCI) analysis: In this step, all inputs and outputs associated with the production of algal fuels are quantified. The analysis includes raw materials, energy consumption, emissions, and waste generated across each stage of the product or process life cycle, such as that of algal fuels.
3. Life cycle impact assessment (LCIA): LCIA evaluates the environmental impacts based on the inventory data collected in the previous step. It involves categorizing and quantifying potential impacts at midpoint and endpoint levels.
4. Life cycle interpretation: The final step involves interpreting the results from the LCIA or LCI to make conclusions and

recommendations. This step helps in identifying hotspots, assessing trade-offs, and offering suggestions for improving the sustainability performance of the production system such as algal fuels.

By following these structured steps, practitioners can systematically evaluate the environmental sustainability of algal fuels, ensuring the results align with global standards and provide valuable insights into optimizing fuel production processes (Fig. 2).

1. Clearly outline the goals of the study by addressing the following questions: why is this research being conducted? What specific questions does it aim to answer? Who is the intended audience for the results?
2. Based on the defined goals, develop an initial scope. The scope should explain the boundaries of the study (e.g., the stages of the algal fuel production system that should be analyzed), the key parameters to be evaluated (e.g., carbon footprint, energy consumption, and resource use), the geographical context (e.g., global, regional, and specific countries), and the time frame and data sources to be used in the study.
3. Compile LCI for the foreground and background systems:
   a. Collect data related to the direct processes involved in algal fuel production, such as cultivation, harvesting, processing, and distribution.

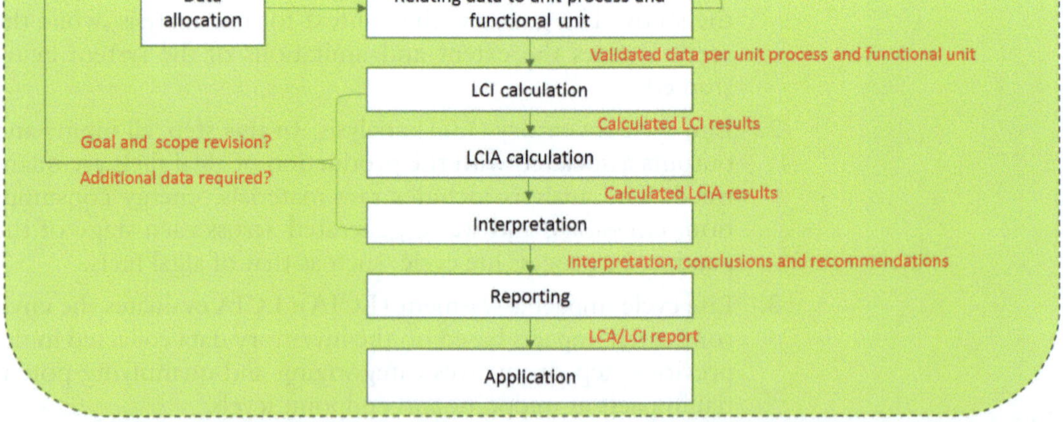

Fig. 2 Details of the iterative approach to LCA, with a focus on inventory data collection and modeling Adopted from ISO [27, 28]

   b. Collect data on indirect processes, such as the production of energy, materials, and services that support the foreground processes.
   c. Model the entire production cycle of algal fuels, capturing both direct and indirect processes while aligning the data with the goal and scope of the study.
4. Using the available data, calculate the environmental impacts of algal fuel production by applying appropriate LCIA methods. Ensure that the methods selected align with the goals and scope of the study.
5. Identify significant issues and perform sensitivity, consistency, and completeness checks:
   a. Analyze the results to identify any significant issues or uncertainties in the data.
   b. Perform sensitivity analysis to determine which variables have the greatest impact on the results. Ensure the data consistency and completeness by checking if all relevant stages of the life cycle have been included and whether the data for each stage is robust and reliable.
6. Based on the findings from Step 5, revise the goal, scope, data, and model if necessary.
7. Iterate to achieve final results:
   a. Expect multiple rounds of revisions and improvements as more data becomes available and the study progresses. Iteration depends on the complexity of the analyzed algal fuel systems, the quality of the data, and the specific questions being investigated. Ensure continuous refinement to improve the study's precision and reliability.
8. From the beginning, document the key aspects of the study, including the goal, scope, data sources, LCIA methods, and initial results from sensitivity checks. Use this documentation as a working report and continuously revise it through the iterations of the study. Maintaining such records ensures transparency and supports the efficient preparation of the final study report or dataset.

By following these structured steps and incorporating iterative refinement, the LCA of algal fuels will be comprehensive, scientifically robust, and aligned with the study's initial goals and scope. This process will enable researchers and stakeholders to make informed decisions about the sustainability of algal fuel production.

4 Goal and Scope Definition

4.1 Goal Definition

The primary objective of conducting an LCA is to evaluate and quantify the environmental impacts that arise throughout the entire life cycle of a product, from its raw material extraction to its final disposal or recycling stage. For algal fuels, the study's goal is to assess the environmental effects tied to their production, which can also extend to their usage and disposal phases, depending on the study's scope [45, 46]. The specific focus of the LCA can vary according to the intended objectives. For example, one common goal might be to assess the environmental impacts during each production phase, such as algae cultivation, processing into fuel, and its end use. By conducting such an analysis, it is possible to identify stages within the life cycle that have a disproportionate environmental impact—so-called hotspots. Once these hotspots are recognized, targeted actions can be taken to reduce their negative effects, thereby improving the sustainability of the entire algal fuel production process.

The objectives of an LCA study may also involve comparing different production methods to determine which one has the least environmental impact [47]. This type of analysis provides valuable data to assess how algal fuels stack up against conventional energy sources or other biofuels in terms of sustainability. By conducting such comparisons, researchers can pinpoint the most eco-friendly and efficient methods for fuel production [48]. In addition, LCA can be an important tool for promoting environmental transparency. It can help companies fulfill regulatory obligations or obtain certifications that prove their products are environmentally sustainable. A report by de Bruijn et al. [49] from 1996 underscored the significant role of LCA in ecolabeling within the EU.

It is essential to recognize that different stakeholders may have varying needs and priorities. For businesses, the emphasis is often on cost-effectiveness and ensuring compliance with relevant regulations. On the other hand, consumers may be more concerned with sustainability practices, ethical sourcing, and the environmental impact of products. To effectively address these differing concerns, the LCA approach must be adaptable and specific to each group's perspective. Tailoring the study ensures that the findings are communicated in a way that is most relevant and engaging for each audience. At the outset, it is vital to define the objectives of the study clearly. Doing so guarantees that the scope, boundaries, and methodologies are aligned with the overall purpose of the analysis [27]. By ensuring the LCA is structured around specific goals, the resulting findings are reliable and provide actionable insights that are meaningful to all relevant parties.

To ensure a thorough and effective goal definition phase in an LCA study of algal fuel production, the documentation should comprehensively cover six critical aspects based on the guidance provided by Bjørn et al. [50] and Albertí et al. [51]. These components are crucial for clarity, transparency, and ensuring that the study serves its intended purpose:

1. Intended application(s) of the results/deliverables.
2. Limitations due to the method, impact, and coverage assumptions.
3. Reasons for study and decision-context.
4. Target audience of the results/deliverables.
5. Comparative studies to be disclosed to the public.
6. Commissioner of the study and other influential actors.

When conducting an LCA study, the decision to use either an attributional or consequential approach is a fundamental one. The attributional method focuses on measuring the environmental effects of the flows of materials, energy, and emissions within a defined system based on current data or average practices [15]. In the case of algae fuel production, attributional LCA is used to assess the environmental impact of resource consumption and emissions based on the present system, offering insights into the environmental performance of current practices. However, this approach has limitations in forecasting how environmental impacts might shift due to potential technological advancements, market dynamics, or changes in policy [52].

Conversely, consequential LCA takes a broader perspective by evaluating how environmental impacts might evolve in response to a particular decision or action [53]. It looks at the effects of specific choices or scenarios, such as shifts in production methods, market conditions, or regulatory changes, and provides insights into the environmental consequences of those choices. Though this method can be more complex and less precise than attributional LCA, it is invaluable for understanding how future actions or policy interventions could influence environmental outcomes [54]. For algal fuels, consequential LCA can evaluate how various production techniques, policy reforms, or market changes could impact overall environmental performance, guiding the most sustainable strategies moving forward.

In an LCA study, the FU, system boundaries, and allocation approach are fundamental elements that guide the overall analysis and ensure the study's relevance and consistency. These elements are defined during the goal and scope phase and are essential to achieving accurate and comparable results. They are analyzed as follows.

4.1.1 Details of Functional Unit and Reference Flow

In an LCA study, determining the FU and/or reference flow(s) is a key step in quantifying and qualitatively assessing the environmental impacts of the system under investigation. The choice of FU should align with the study's objective and be revised if it no longer serves the purpose or represents the system accurately. There is no universal consensus on the choice of FUs in LCA studies of algal fuels, making comparisons across studies more difficult. However, four categories of FUs can be identified in the context of algal fuel production [55]:

1. Input-oriented unit: The FU is defined based on a physical or chemical characteristic of the primary input to the algal fuel system, often referring to algal biomass. This approach is suitable when the focus is on investigating the environmental impacts of converting biomass into fuel. This FU is particularly relevant when the goal is to examine the sustainability and environmental performance of biomass production and conversion. Common examples include:
   a. One ton of algal biomass: Used when the study focuses on the environmental impacts associated with cultivating, processing, and converting biomass to biofuel.
   b. One MJ of energy derived from algal biomass: This FU quantifies the energy contained in the biomass and can be useful for studies analyzing energy efficiency in biofuel production.
2. Output-oriented unit: In this approach, the FU is defined by the physical or chemical characteristics of the primary output of the system, such as the energy produced from algal fuels or the distance that can be traveled using the biofuel. This FU is suitable when the study's objective is to evaluate the environmental impacts of using algal fuels compared to other energy sources or fuels. Examples include:
   a. One ton of algal fuel: Focuses on the final output of the fuel, particularly when comparing algal fuels to conventional fuels in terms of quantity produced.
   b. One MJ of energy from algal fuel: Commonly used when comparing the energy produced by algal fuels to other energy sources, allowing for a more direct comparison of environmental impacts in terms of energy generation.
   c. Kilometers of transportation facilitated by algal fuel: A service-based FU that links the output to practical use, such as evaluating how much transportation can be facilitated per unit of algal fuel.
3. Agricultural land-based unit: This FU is related to the land area required to produce the algal biomass feedstock. It is particularly relevant when the study aims to assess the land-use efficiency or impacts of different algal cultivation methods. The agricultural land-based unit is suitable for comparing land efficiency and resource use for algae cultivation, particularly when different types of biofuels are compared based on land-use impacts.

4. Year-based unit: This FU is especially useful in studies where different outputs are produced over the course of the year, and it helps to assess the overall environmental performance on a temporal scale. The year-based unit is helpful for studies that aim to analyze the long-term environmental impacts of algal fuel production and consumption.

Attention should be drawn to the fact that the choice of FU significantly influences the interpretation of LCA results, as it defines the unit for which environmental impacts are calculated and reported. Therefore, it is recommended to present the LCA results using multiple FUs whenever possible. This approach enhances the comparability of the study's findings with other studies in the literature and provides a more comprehensive understanding of the environmental impacts from different perspectives.

4.1.2 Defining System Boundary

In an LCA study, defining the system boundary is crucial for ensuring that the analysis captures all relevant processes and impacts related to the product's life cycle. By carefully selecting the processes that will be included and establishing cut-off criteria during the goal and scope phase, the study can maintain a focused and meaningful assessment [56].

In conducting an LCA for algal fuels, it is essential to define the system boundary to cover all critical life cycle stages involved in the production process. In algal fuels, this includes algae cultivation, harvesting, conditioning, and conversion into biofuels, as well as subsequent transportation and usage. The initial system boundary can be revised as new information becomes available or further analysis is conducted [57]. Establishing a clear boundary is necessary to facilitate comparisons between different LCA studies of competing systems. The boundary defines which processes and activities are included in the analysis [58], such as algae cultivation, harvesting, drying, transportation, and biofuel production using various methods. For algal fuels such as bioethanol, biodiesel, and, to a lesser extent, algae biomass and oil, system boundaries can typically be categorized into three main perimeters [55]:

1. Well to Gate (WtG): Encompasses algae cultivation, harvesting, conditioning, and conversion of algae into fuels;
2. Well to Tank (WtT) or Well to Pump (WtP): Includes the stages from algae cultivation through to biofuel conversion, as well as the transportation of the biofuels to storage or fueling stations;
3. Well to Wheel (WtW): Covers the full life cycle, from algae cultivation to fuel conversion, transportation to fueling stations, and the final use in engines or power generation equipment.

The selection of the system boundary is crucial and directly linked to the specific objectives of the LCA study. When using the

WtT approach, it serves as an intermediary boundary, incorporating numerous variables but not considering the final usage of the product. In terms of unit processes, the system boundary represents the dividing line between the modeled process and the broader technosphere. This boundary includes all inputs (such as algae, raw materials, and energy) that flow into the biofuel production step, as well as all outputs (fuels, by-products, and emissions) leaving it. Furthermore, all direct flows crossing from or to the ecosphere—whether entering or exiting the system—are to be accounted for in the inventory. Essentially, any material or energy exchange across this boundary should be included in the LCA.

In an LCA study of algal fuels, it is beneficial to present the system boundary using a semi-schematic diagram that clearly outlines the stages of the life cycle, indicating which parts are initially included and excluded (see Fig. 3 [59]). This preliminary diagram should be flexible and subject to adjustments if changes to the goal and scope arise throughout the study. It is also advisable to develop a technical flow chart for the foreground system, which will display the key steps involved in algal fuel production. This chart can be refined as data collection progresses to ensure it remains accurate and comprehensive.

4.1.3 Defining Cut-off

In LCA studies, the system boundaries define the life cycle stages of the analyzed system by specifying which unit processes are included or excluded [60]. The cut-off rule serves to exclude material or energy flows that do not significantly impact the system, thereby simplifying the modeling process and easing the workload for practitioners. Determining which processes to include within the system boundaries requires the application of well-defined and justifiable criteria, such as mass, energy consumption, or environmental relevance. However, the application of these criteria can vary, and consistency is not always maintained across different studies [60].

In an LCA of algal fuels, establish quantitative cut-off criteria unless already defined during the goal-setting step. These criteria should be revisited and refined if the study involves comparisons. If the desired level of completeness cannot be achieved due to data limitations or resource constraints, adjustments may be necessary. In some cases, this could mean that the overall objective of the study is no longer achievable, necessitating a revision of the study's goals.

For algal fuel production, following the ILCD handbook [61], a 5% cut-off should be applied to each relevant impact category. However, as per the handbook's recommendation, any missing flows should still be identified and recorded in the inventory, though their quantities need not be specified. These flows will be categorized as either "missing relevant" or "missing irrelevant" based on their significance.

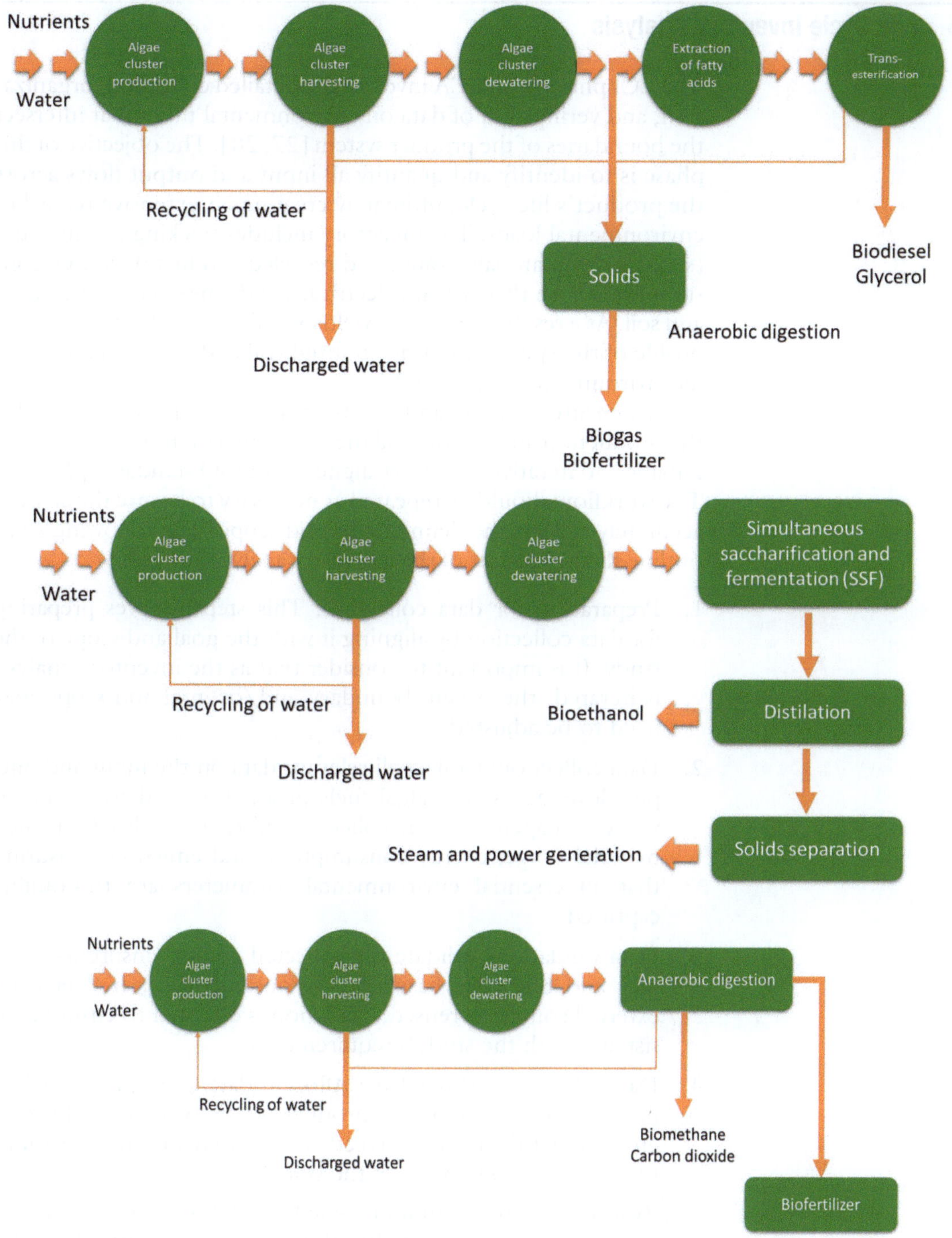

Fig. 3 Boundaries of the system including different types of production of algal fuels. Adopted from Singh and Olsen [59] With permission from Elsevier. Copyright© 2010

5 Life Cycle Inventory Analysis

The LCI phase in an LCA involves the detailed collection, organization, and verification of data on environmental flows that intersect the boundaries of the product system [27, 28]. The objective of this phase is to identify and quantify all input and output flows across the product's life cycle, ultimately creating an extensive record of environmental loads. The inventory includes tracking resource use (such as raw materials, water, and recycled products), energy consumption (both thermal and electrical), and emissions to air, water, and soil. As a result, the process yields a well-rounded environmental profile of the system, offering an insightful and realistic account of its environmental impact [62].

To ensure a robust and comprehensive inventory analysis for the production and use of algal fuels, it is important to follow a systematic and iterative workflow aligned with ISO standards [27, 28]. The workflow should be repeated as necessary to ensure the analysis accurately reflects the defined goal and scope. The following steps outline the key phases of the inventory analysis process:

1. Preparation for data collection: This step involves preparing for data collection by aligning it with the goal and scope of the study. It is important to consider that as the inventory analysis is iterated, the system boundary and/or goal and scope may need to be adjusted.
2. Data collection: Collect all relevant data on the input and output flows across the algal fuels production and use chain or waste management. The collected information should include material usage, energy consumption, and emissions, ensuring that all essential environmental parameters are thoroughly captured.
3. Data validation: Validate the collected data to ensure its accuracy and relevance based on various methods. If data from external sources is reused, validation is essential to ensure consistency with the study's requirements.
4. Data allocation (if needed): Allocate data appropriately when processes involve multiple products or by-products. This step ensures that the data is correctly distributed among the various functions and outcomes of the system.
5. Relating data to the unit process: Link the collected data to the specific unit processes within the system boundary. This helps to trace the flow of materials, energy, and emissions through the different stages of the algal fuel production and use chain.
6. Relating data to the FU: Relate the data to the FU, which is the reference unit of analysis that allows the comparison of

different systems or options. This ensures that the data can be consistently applied to different scenarios or alternatives within the study.

7. Data aggregation: Aggregate the collected and validated data to create a comprehensive environmental profile. This involves compiling the data from all unit processes and linking them to the overall system to evaluate the full environmental impact of algal fuels.

5.1 Data Collection

Data collection in LCI is a crucial phase in LCA as it provides the foundational data necessary for evaluating the environmental impacts of a product or process. This process involves gathering both foreground and background data to ensure a comprehensive and accurate evaluation. Foreground data is obtained directly from experimental measurements or direct data collected from the production process, such as material input and output, energy use, and emissions [63]. Meanwhile, background data refers to the data that represents the environmental impacts associated with upstream processes, such as the production, transportation, and distribution of fuels, chemicals, electricity, and other materials that interact with the foreground production system [64]. This data is essential for capturing indirect environmental impacts that arise from the supply chain. A common challenge in LCI is the scarcity of local data for upstream supply chains, particularly for emerging technologies and novel production methods [65].

In such cases, reputable databases like EcoInvent, US LCI, and others become vital resources for filling in these gaps and providing a broader context. These databases help to mitigate the lack of direct data by offering generalized data for various materials, processes, and regions. However, even with the extensive development of LCI databases, significant data gaps still exist, especially for newer technologies and less-studied processes. This limitation constrains the ability to perform fully comprehensive LCA studies and highlights the need for ongoing database expansion and refinement [62]. Continual improvement in data availability and accuracy is key to enhancing the robustness of LCA and enabling better decision-making for sustainable development.

In an LCA of algal fuels, it is crucial to gather foreground data during experimental or production processes at both pilot and industrial scales to ensure the data's accuracy and reliability. For indirect data, background data can be sourced from computer databases, expert input, scientific studies, or records from relevant industrial and agricultural sectors. In cases where precise measurements, such as emissions during algae cultivation, are difficult to obtain, estimations may be made based on extrapolated results from lab-scale studies. The inventory data should be compiled from three key categories, as identified by Carneiro et al. [55]:

1. Information extracted from previously published resources such as books, research papers, technical reports, manuals, and databases related to the life cycle of algal fuel production and use;
2. Data gathered from controlled laboratory experiments examining the various phases of algal fuel production and utilization;
3. Real-world data collected from industry operations, algae farming, and harvesting processes or through discussions with industry experts involved in these activities.

Make sure that the models account for all key capital assets involved in the algae cultivation and harvesting processes, such as [67]:

1. Photobioreactors and bioreactors
2. Raceway systems
3. Harvesting equipment, such as centrifuges
4. Disposable materials used in the facility, considering a facility lifetime of 20 years.

Take the following factors into account during the production of algal biomass:

1. Define the amount of algae grown in open ponds for each operational day throughout the year.
2. Calculate the total nutrients required for algae growth.
3. Analyze the concentrations of proteins, lipids, and carbohydrates, as well as the overall yield of the biomass.
4. Evaluate the direct emissions associated with algae cultivation and harvesting using multiple methods, such as guidelines, laboratory techniques, and literature reviews. In the case of carbon capture by algae, if direct CO_2 consumption measurements are not feasible, literature data relevant to the specific algae species and technology used can offer a suitable alternative.
5. List the machinery involved, specify their energy sources, and estimate their emissions using various approaches, often based on database information.

During the production of algal fuels, consider the following aspects:

1. Determine the sources of heat and energy used and estimate their emissions using different methods, commonly based on database analysis.
2. Identify the machinery utilized, along with the energy supplies required for their operation, and calculate their emissions using a variety of methods, typically relying on database information.

3. Identify the co-products generated at each phase of production.
4. Estimate the direct emissions from the algal fuel production process, particularly those arising from wastewater in biodiesel production and other pollutants that may affect the environment, using various calculation techniques.

During the combustion phase of algal fuels, consider the following:

1. Identify the type of engine or energy generation system being used and assess the emissions through various methods, commonly based on database calculations.
2. Define the operating duration of the engine or system.
3. Estimate the amount of fuel required throughout the operational period.
4. Assess the emissions from the exhaust gases.

Notably, in the majority of LCA studies, the distribution phase of algal fuels is generally excluded, as it is common to all fuel types, both fossil-based and bio-derived [67]. This means that transportation-related emissions do not offer specific insights for algal fuels. In accordance with the Renewable Energy Directive, it is assumed that the algae refinery is within 10 km of the fueling station, thereby allowing transportation emissions to be excluded from the analysis. Under these assumptions, such emissions contribute to less than 1% of the total impact across all categories [67].

5.2 Data Allocation

In systems that generate multiple products, distributing environmental burdens presents a key challenge. Allocation becomes necessary when emissions and resource consumption need to be assigned to individual products within a shared production system [68]. This step is both critical and widely debated in LCA due to its substantial impact on results, which can shape economic strategies and policy decisions [69]. When allocation cannot be avoided, ISO recommends distributing impacts among co-products based on chemical or physical properties like mass or energy content [66]. If a chemical or physical basis is not applicable, economic value can serve as an alternative criterion for allocation [70]. The choice of allocation approach is crucial, as different methods can yield varying interpretations and influence comparisons [68]. To manage this complexity, system expansion is often proposed as an alternative to direct allocation in multiproduct systems. As outlined in ISO standards, this approach involves broadening the system boundaries to incorporate additional functions of co-products. This technique facilitates more comprehensive assessments and is particularly useful for evaluating the utilization of residual flows, making it a valuable strategy for handling multifunctional systems effectively [71].

In an LCA analysis of algal fuels with multiple outputs, allocate environmental impacts among all co-products derived from multi-output processes. The allocation follows established methodologies, including the following [55, 72]:

1. Null allocation: This method assigns the entire environmental burden to the primary product, leaving co-products without any associated impacts or energy demands. While simple and direct, it overlooks the actual environmental contributions of co-products. Therefore, this approach is generally discouraged, as it does not accurately reflect the environmental footprint of algal fuels [55].

2. Cut-off allocation: This method excludes the environmental burdens associated with certain co-products based on specific criteria, such as a predetermined threshold or when their contribution is deemed negligible. It involves excluding certain flows or impacts from the system to focus on the main product, often to simplify the analysis or avoid overly complex calculations.
3. Economic or physical-chemical allocation: Environmental impacts are apportioned among co-products based on either their economic worth or their physical and chemical attributes, such as mass or energy content. This method accounts for each product's relative significance, using measurable characteristics or market value to distribute environmental burdens. While mass or energy-based allocation tends to remain stable over time, economic allocation is influenced by fluctuating market conditions, introducing greater uncertainty [55]. Nevertheless, in cases where co-products hold minimal economic value, economic allocation may offer more reasonable outcomes [73].
4. System expansion, displacement, or substitution method: This method accounts for co-products by assigning them a compensatory environmental burden, often referred to as the "credit" or "avoided burden" approach. It expands the system boundaries by incorporating the alternative function of co-products and then subtracting their impacts, ensuring that only the primary product remains in focus. This technique is beneficial as it removes the need for allocation, avoids artificial fragmentation of unit processes, and offers a broader sustainability perspective [74]. However, its effectiveness depends on the availability of high-quality LCI data. If precise data on co-products are lacking or if multiple alternatives exist for substitution, applying this approach can become complex and less reliable [75].

It is worth emphasizing that the choice of allocation method in LCA is critical, as different approaches can lead to significantly varying results. In the case of algal fuels, where both storable and non-storable co-products are generated, selecting an allocation method becomes even more complex. No single approach has been universally recognized as the most appropriate, highlighting the need for careful consideration of system boundaries, co-product functions, and sustainability objectives [67].

6 Life Cycle Impact Assessment

LCIA represents the third step of an LCA, where raw inventory data is processed to derive relevant environmental impact categories [76]. This step is fundamental in assessing, quantifying, and interpreting the environmental consequences of producing and utilizing algal fuels. It aids in making informed sustainability choices and pinpointing areas for improvement. LCIA evaluates the magnitude of potential environmental impacts by applying characterization factors to emissions and resource consumption, providing a clearer depiction of their effects [77]. As outlined by ISO [27, 28], LCIA includes mandatory components—classification and characterization—as well as optional elements such as normalization and weighting. In the context of algal fuel production, careful execution of these steps is essential for ensuring accurate impact assessments, which are further examined in the subsequent sections.

6.1 Classification and Characterization

The LCI is a vast dataset, typically encompassing over 1000 elementary flows, each carrying various environmental significance [78]. In LCIA, these flows serve as the foundation for uncovering the true ecological footprint of a product or process. Based on ISO [27, 28], the LCIA begins with "Classification," where each elementary flow is meticulously assigned to one or more relevant impact categories. Following this, the "Characterization" phase comes into play—where inventory results are systematically multiplied by scientifically derived impact factors from the chosen LCIA method. Once characterized, these impacts—referred to as midpoints—are aggregated and grouped into broader environmental themes, painting a comprehensive picture of sustainability performance. This structured approach aligns seamlessly with ISO [27, 28], ensuring scientific credibility and robust decision-making for a greener future.

Various environmental impact quantification methods are available to derive comprehensive environmental impact values, enabling a more holistic sustainability assessment. Among these, CML, Eco-indicator 99, ReCiPe, IMPACT 2002+, and IMPACT World+ are widely recognized methodologies integrated into LCA software. These tools streamline the process, offering advanced impact assessment techniques within a structured framework. In addition to comprehensive methods, single-issue indicators such as carbon footprint and water footprint are also available in LCA software, allowing for targeted environmental evaluations. Historically, single-issue assessments were conducted manually using emission factors for various substances, but this approach was highly time-consuming and prone to inconsistencies. With modern LCA tools, these calculations have become more efficient, accurate, and scalable, empowering researchers and industries to make informed, data-driven sustainability decisions.

Various methods are available to quantify environmental impacts, enabling holistic sustainability assessments for algal fuels. Some of the most important are as follows:

1. Use Eco-indicator 99 to assess resource use, ecosystem quality, and human health impacts of algal fuels, applying three perspectives (Egalitarian, Hierarchist, and Individualist) to interpret results based on different value choices. Ideal for cradle-to-gate assessments of land-based algal fuel systems. However, it lacks a comprehensive assessment of marine environments (e.g., marine eutrophication), making it less suitable for ocean-based algae cultivation.
2. Use ReCiPe on a global scale, applying three perspectives (Egalitarian, Hierarchist, and Individualist) for a comprehensive evaluation of algal fuels, addressing 18 midpoint categories and three endpoint categories (human health, ecosystems, and resources). It effectively quantifies climate change effects, water depletion, and metal resource scarcity, making it ideal for large-scale sustainability assessments of algal fuels. With the broadest set of midpoint impact categories, ReCiPe enables scenario comparisons for different cultivation techniques and refining processes.
3. Use IMPACT 2002+ for a comprehensive LCIA of algal fuels, integrating both midpoint and damage categories to evaluate environmental impacts. It utilizes 14 midpoint categories, which are then translated into four main damage categories: human health, ecosystem quality, climate change, and resource depletion. Although this method is designed with a global scope, its primary characterization factors are based on data from Western Europe. For use in other regions, spatial differentiation can be applied, making it particularly applicable within Europe. However, with appropriate regional data, it can still be effectively used in global assessments of algal fuel sustainability.
4. Use IMPACT World+ for high-resolution global assessments of algal fuels. By differentiating short and long-term climate change impacts, it provides insights into long-lasting environmental trade-offs. This method is highly applicable to both localized and worldwide algal fuel supply chains, offering improved resolution on ecosystem service impacts.

5. Use CML for midpoint-focused life cycle impact assessment of algal fuels, emphasizing acidification, eutrophication, global warming, and abiotic resource depletion. Best suited for regulatory compliance and benchmarking sustainability metrics, it provides straightforward impact quantification. However, its exclusion of endpoint-level damage categories limits its applicability in holistic sustainability evaluations.
6. Use TRACI, developed by the US EPA, for North America-specific analysis of algal fuels, aligning with US environmental policies and regional impact factors. Its region-specific emission factors make it ideal for evaluating algal biofuel production within the US energy sector. However, its geographical limitation reduces its applicability for global assessments. Also, its exclusion of endpoint-level damage categories limits its applicability in holistic sustainability evaluations.
7. Use the EF method for EU-standardized, midpoint-oriented assessments of algal fuels, covering 16 impact categories. Developed by the European Commission, EF ensures consistency across product environmental footprint and organizational footprint studies, making it highly applicable for regulatory compliance and environmental benchmarking of algae-based fuels within European markets. However, its geographical limitation reduces its applicability for global assessments.
8. Use the EPD method for standardized, third-party verified environmental assessments of algal fuels, ensuring transparency and comparability. EPD quantifies the life cycle impact of algal fuel production across multiple impact categories, such as global warming potential, eutrophication, and energy demand. However, EPD does not provide a direct weighting of environmental impacts, requiring complementary LCIA methods for deeper analysis.
9. Use single-issue methods such as carbon footprint and water footprint for targeted assessments of algal biofuel sustainability, focusing on greenhouse gas emissions and water consumption. While these methods provide valuable insights, they lack broader environmental trade-offs, requiring integration with comprehensive LCIA models for a complete sustainability analysis.

6.2 Normalization

As an optional step, the results from the LCIA can be refined through normalization, where the inventory data is compared to a reference, such as the total emissions of a specific country or the average individual. This approach transforms the results into dimensionless, normalized values [79]. The outcomes can be displayed either at the midpoint level or at the endpoint level [80]. Presenting the normalized data across various impact categories allows for a clearer understanding of each category's relative contribution to the overall environmental impact of the system.

Normalization of LCIA results for algal fuels can serve two main purposes [81]:

1. Normalization for result interpretation: As an optional step outlined by ISO, normalization aids in interpreting the results of the environmental assessment of algal fuels.
2. Normalization for cut-off quantification: To assess the completeness of the study or the cut-off criteria, the results from various impact categories can be normalized by comparing them to a common reference, known as the normalization basis.

The normalized results for algal fuels should not be summed across different impact categories, as this would incorrectly assume that each category holds equal significance [61]. This would only be appropriate if an explicit and clearly communicated weighting is applied to each category as part of the analysis [61].

6.3 Weighting

In the final step of LCIA, the normalized results for algal fuels can be refined through weighting, a process that transforms them into weighted results for both midpoint and endpoint categories. This step involves multiplying each impact category's normalized results by a designated weighting factor, generating the "weighted LCIA results" [55]. Ultimately, this approach consolidates all environmental impacts into a single, comprehensive score [82], offering a streamlined yet powerful assessment of overall sustainability. However, weighting is inherently subjective, as it relies on value-based decisions that can significantly shape the prioritization of environmental concerns [58]. To maintain credibility and balance, it is crucial to ensure transparency in the selection and justification of weighting factors, preventing bias and reinforcing the integrity of the analysis.

Apply weighting in LCIA of algal fuels for two main purposes:

1. Weighting for enhanced interpretation: As an optional step, weighting can be used to provide additional value to the normalized or non-normalized results from different impact categories in the context of algal fuels [61]. This approach supports the deeper interpretation of the environmental performance of the fuel production process.
2. Weighting for cut-off analysis: Weighting is also useful for analyzing the completeness of the study, helping to assess and quantify the significance of different impact categories by assigning weights across various indicators [81].

In general, both normalization and weighting play an essential role in refining the overall analysis of life cycle impacts. These methods ensure a more organized and clear comparison of the environmental effects of algal fuels. Researchers can rely on assumptions or leverage advanced LCA software like SimaPro for more effective standardization and quantification of results. These tools enable a more nuanced understanding of the relative importance of environmental impacts and support comprehensive decision-making.

7 Interpretation

Interpretation marks the final step of the LCA process, where the data derived from LCIA is critically examined to facilitate well-informed decisions [27, 28]. This step is not just about concluding; it is about uncovering opportunities for improvement and aligning findings with the original goals. It also involves benchmarking against industry standards or sustainability thresholds to highlight areas where algal fuel production could be optimized. In evaluating algal fuels, the following key considerations should be kept in mind:

1. Pinpoint the critical factors within the algal fuel production or usage cycle—these could include key processes, influential parameters, assumptions, and essential flows [83].
2. Assess how each of these factors impacts the overall outcomes of the LCA for algal fuel production or use [84].
3. Leverage these insights to draw informed conclusions and provide recommendations for optimizing the algal fuel production chain. When comparing multiple algal fuel production systems, additional considerations must be incorporated into the interpretation to ensure comprehensive analysis.
4. Interpret the implications of any identified data gaps, uncertainties, or limitations in the study. Recommendations should address the need for more accurate or refined data in key areas to ensure more reliable future assessments. Addressing these gaps can also support decision-makers in prioritizing areas for further research and development.
5. If applicable, integrate socio-economic factors (e.g., cost, market feasibility, and policy considerations) into the interpretation, especially when the results will inform policy or regulatory decisions. This broader perspective helps in assessing the practical and economic viability of algal fuels alongside their environmental impacts.

The interpretation step in the LCA of algal fuels serves three main purposes:

1. Interpretation in iterative LCA of algal fuels: The interpretation step plays a pivotal role in the LCA of algal fuels, especially in comparative studies. It serves as the bridge that connects the LCI data and LCIA results to actionable insights and

recommendations. During this phase, the results from the previous stages of the LCA are scrutinized to answer the questions posed in the goal-setting phase. The interpretation considers the broader implications of the data in alignment with the study's objectives. Additionally, it critically evaluates the accuracy, completeness, and precision of the data and assumptions used, allowing for well-founded recommendations that can guide improvements in algal fuel production, utilization, and sustainability [81].

2. Evaluating LCIA and LCI results: When both LCI and LCIA are part of the study, the interpretation phase becomes crucial for deriving meaningful conclusions or recommendations that are closely aligned with the study's goals. This phase focuses on verifying the appropriateness of the FU and system boundaries, ensuring they are consistent with the study's objective and scope [85]. Equally important is the assessment of the data quality, which underpins the credibility of the entire analysis. The interpretation process should effectively communicate the results in a clear, understandable format, enabling users to assess the robustness of the findings. Moreover, it should highlight potential limitations or uncertainties, helping to contextualize the results and provide more reliable guidance for decision-makers in the algal fuels sector.
3. Timing of conclusions and recommendations: Conclusions and recommendations should only be drawn at the conclusion of the study after a comprehensive evaluation of all LCI and LCIA results. This approach ensures that decisions are based on a full understanding of the environmental impacts of algal fuels. By postponing recommendations until this final stage, the study can provide robust, data-driven conclusions that are both reliable and actionable for stakeholders. This step ensures that the insights offered are well-supported by a thorough analysis, making them more impactful for future decision-making and policy formation in the algal fuel industry.

7.1 Uncertainty Test

An uncertainty test is crucial in an LCA to assess the reliability and robustness of the results, especially when data quality or assumptions are uncertain [86]. Conduct an uncertainty test in the context of algal fuels to assess data variability, identify key parameter uncertainties, quantify confidence in results, prioritize research efforts, and inform decision-making:

1. Evaluate the potential variations in environmental impacts due to uncertainty in input data (e.g., emissions, energy use, or resource consumption): Assess how variations in the data used for the LCI affect the environmental impact results. Using probability distributions for these inputs, one can simulate

a range of possible outcomes and observe the variation in impacts. This step provides an understanding of how uncertain inputs can alter the overall environmental profile.

2. Determine the range of possible outcomes based on variations in key parameters (e.g., algae yield, energy efficiency, or transportation distances): Identify key parameters such as algae yield, the energy efficiency of algal fuel conversion processes, and transportation distances, all of which may exhibit uncertainty due to variability in environmental conditions or technological performance. Using Monte Carlo simulations or similar techniques, you can estimate the possible range of outcomes that reflect the uncertainty in these parameters.
3. Identify which inputs have the greatest influence on uncertainty, helping prioritize areas for more accurate or detailed data collection. Perform a sensitivity analysis alongside the uncertainty test to identify which inputs have the most significant effect on the variability of the results. This process helps to pinpoint critical data gaps or parameters with the greatest impact on uncertainty. For instance, uncertainty in energy use during algae cultivation may have a larger influence on the environmental results than the uncertainty in transportation. By identifying these key parameters, researchers can prioritize efforts to collect more accurate or specific data, improving the overall reliability of the LCA.
4. Assess the level of confidence in the results and whether the conclusions drawn from the LCA are robust enough to support decision-making. Once the range of uncertainty has been analyzed, evaluate whether the results are sufficiently reliable for decision-making purposes. If the uncertainty is high, the conclusions drawn from the LCA may be less reliable.

7.2 Sensitivity Analysis

Sensitivity analysis is a vital tool in LCA studies, especially for assessing the robustness of results in the context of algal fuel production [87]. It identifies the most influential parameters and helps in understanding the variability in outcomes due to changes in input data. Here is how sensitivity analysis is applied in the context of LCA for algal fuels:

1. Test the assumptions and data used for algal fuels LCA: Evaluate the accuracy, completeness, and assumptions underlying the LCI and LCIA. Sensitivity analysis can help highlight which assumptions about inputs (e.g., cultivation methods, energy sources, or feedstock) influence the results the most. By varying these assumptions systematically, researchers can better understand the robustness of the conclusions.
2. Identify the key parameters affecting the embodied energy of an element or algal fuel: Embodied energy refers to the total

energy required to produce, process, and transport the materials used in algal fuel production. Sensitivity analysis can identify which stages or inputs have the most significant impact on the embodied energy of the fuel, allowing researchers to focus on optimizing these areas for energy efficiency.

3. Test the extent to which parameters are important to the life cycle of an algal fuel (e.g., maintenance and replacement rates): In many algal fuel production systems, equipment maintenance and replacement cycles are key factors influencing the overall environmental performance. Sensitivity analysis can test how variations in these rates affect the environmental impact. By identifying the most sensitive parameters, researchers can focus on reducing maintenance-related emissions or optimizing equipment longevity.
4. Determine which materials require accurate data regarding energy spent on transportation: Transportation energy can be a significant contributor to the overall environmental impact of algal fuel production. Sensitivity analysis can pinpoint which materials or processes (e.g., feedstock transportation, distribution of biofuels) are most sensitive to transportation-related energy use.
5. Determine the importance of life cycle embodied energy compared to operational energy for an algal fuel: The embodied energy (energy required for production, including materials and processes) is contrasted with operational energy (energy used during the fuel's production or consumption stages). Sensitivity analysis can help assess the relative importance of these two types of energy in the overall environmental footprint of algal fuels. Understanding which phase of the life cycle dominates in terms of energy consumption allows for better-targeted efforts in optimizing the environmental performance of algal fuels.

Through these targeted tests, sensitivity analysis ensures that the LCA of algal fuels can provide meaningful insights and guide decisions for improvements, resulting in a more sustainable production process.

8 Case Study

The following is a step-by-step LCA of algal biodiesel production and consumption conducted according to the ISO 14040/14044 standards. This example outlines the four key LCA steps.

8.1 First Step: Goal and Scope Definition

1. Define the goal of the study. For example, assessment of the environmental impacts of a pond-to-wheel LCA of algal

biodiesel using real-world commercial facility data to identify environmental opportunities and limitations.

2. Define the study approach clearly. Determine whether the purpose aligns with an attributional LCA, focusing on describing the environmental impacts of a system, or a consequential LCA, aimed at understanding the broader impacts of changes within the system. For example, in this case, the aim can be achieved through an attributional LCA approach.
3. Define the FU. For example, the production of 1 MJ of energy from algal biodiesel combustion in the engine of a passenger car.
4. Clearly define and draw which processes are included in the LCA, as presented in Fig. 4.
5. Identify and allocate co-products generated in different stages of algal biodiesel production based on the study's objectives. For instance, an energy-based allocation method assigns 42% to crude algae oil, 28% to hydrocarbons, and 30% to algae residue, reflecting their respective contributions to the total energy output [45].

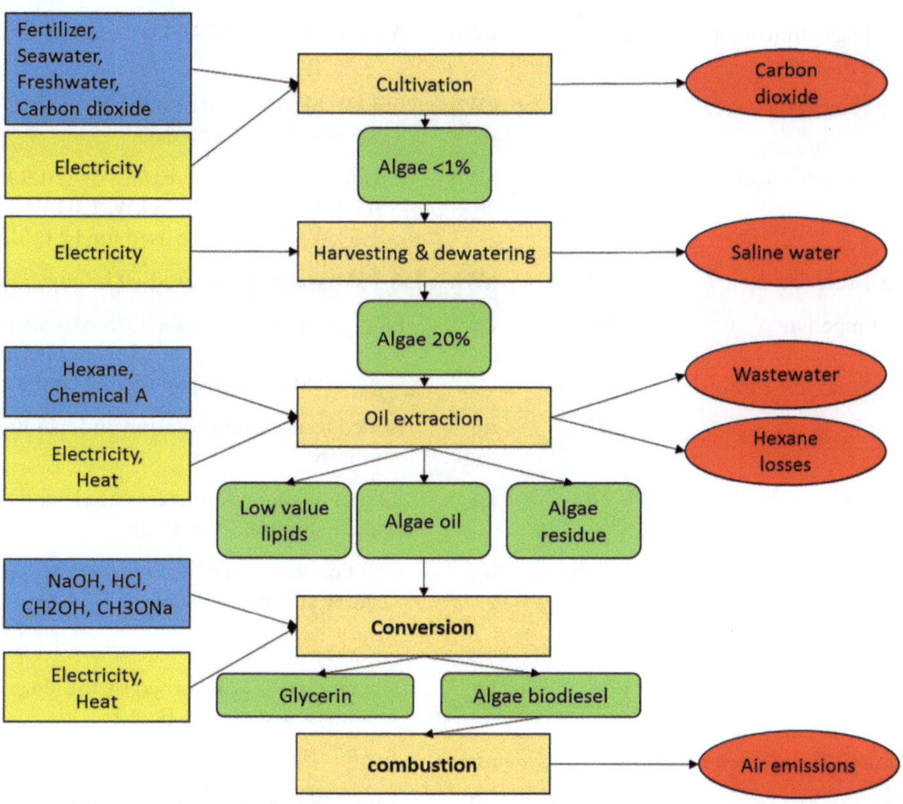

Fig. 4 System boundaries depicting the unit processes considered (adopted from Passell et al. [45]. With permission from Elsevier. Copyright© 2013

8.2 Second Step: Data Collection

1. Collect data tailored to the study's nature, whether in a laboratory, pilot, or industrial-scale setting. For laboratory data, gather information from multiple replicates to ensure reliability. For larger-scale studies, collect data over various periods to capture variability and trends. An example dataset for algal biodiesel production is presented in Table 2.
2. Gather data that cannot be measured directly by conducting thorough literature reviews or utilizing equations to ensure the inclusion of accurate and relevant information.
3. Organize the data in a Microsoft Excel spreadsheet. Assess statistical parameters for the data and exclude outliers from the calculations, if possible.

Table 2
LCI data for algal biodiesel production/combustion

Process/inputs	Value	Unit	Comment
Cultivation: inputs (kg^{-1} dry biomass)			
Electricity (algae inoculant prep)	6.77	kWh	Florescent light (0.846 kW) used 24 h/day
Electricity (algae inoculant prep)	14.9	kWh	Air conditioner (0.746 kW) used 24 h/day
Electricity (flue gas blower)	12	kWh	A 3 kW blower used to bubble the CO_2 into the ponds for 12 h/day
Electricity (paddlewheels)	30	kWh	8 paddle wheels of increasing size (0.5 kW, 0.5 kW, 0.75 kW, 0.75 kW, 1.0 kW, 1.0 kW, 1.5 kW, 1.5 kW) are used for 12 h/day
Electricity (water pump)	3.33	kWh	A 1 kW water pump used 10 h/day
Flue gas pumped in	−181	kg	Flue gas at 180 m^3/h for 12 h/day with a CO_2 concentration of 13–15% This input is modeled as an environmental credit
Freshwater	1.67	m^3	Added to offset evaporation and maintain salinity
Nitrogen fertilizer	0.26	kg	52 g of nitrogen fertilizer is used per kg of algae slurry at 20% solids
Phosphorus fertilizer	0.045	kg	9 g of phosphate fertilizer is used per kg algae slurry at 20% solids
Cultivation: outputs (kg^{-1} dry biomass)			
CO_2 emitted	179	kg	The ratio of CO_2 absorbed by the algae to dry biomass assumed to be 2:1
Harvesting & dewatering: inputs (kg^{-1} dry biomass)			
Electricity	2	kWh	A 0.5 kW harvesting pump is used for 12 h/day

Table 2
continued

Process/inputs	Value	Unit	Comment
Electricity	16	kWh	A 4 kW centrifuge is used for 12 h/day to produce algae slurry
Oil extraction: inputs (kg⁻¹ algae oil)			
Chemical A	0.08	kg	Unidentified chemical. It is modeled by using a generic EcoInvent process for organic chemicals
Electricity	0.21	kWh	Electricity used per kg oil
Electricity (belt filter press)	2.0	kWh	Energy input for belt filter press to dewater the biomass. Energy needed for this process is assumed to be comparable to the feed dryer energy
Heat (feed dryer)	6417	BTU	Energy input for drying the biomass
Heat (oil separation)	1805.57	BTU	Energy input for processing of the oil to separate the oil and other lipids
Heat (pretreatment and extraction)	4933	BTU	Energy input for the pretreatment and extraction of the oil
Heat (solvent recovery)	6278	BTU	Energy input for recovery of hexane
Hexane	0.33	kg	Solvent extraction method is used to extract the oil. Hexane is used as the solvent
Oil extraction: outputs (kg⁻¹ algae oil)			
Algae oil	1	kg	Primary output of the process. Allocated based on the high heat value (HHV) of 16200 BTU/lb
Algae residue (oilcake)	1.87	kg	Co-product of algae oil production. Allocated based on the HHV of 6107 BTU/lb (21)
Hexane losses	0.0038	kg	A small amount of hexane is lost during the oil extraction as fugitive emissions
Low-value lipids	0.67	kg	Co-product of algae oil production. Allocated based on the HHV of 16000 BTU/lb (assumed to be comparable to biodiesel)
Wastewater	17.35	l	The slurry contains 80% water and 20% solids. The water is sent to a treatment facility. Out of the 20% solids, about 3% remain in the water and are also sent to wastewater treatment. Chemical A is removed as wastewater also

Conversion to biodiesel: Transesterification from GREET 1_2011.
Biodiesel combustion: Combustion in a CIDI vehicle from GREET 1_2011.
Adopted from Passell et al. [45]. With permission from Elsevier. Copyright© 2013.

8.3 Third Step: Define the Project and Import the Data into SimaPro Software

This section provides an overview of the key steps in setting up a project, selecting a database, importing data, choosing an LCA method, and extracting results for the LCA of algal biodiesel using SimaPro. Detailed installation and licensing instructions can be found in the official SimaPro guidelines at https://simapro.com/. This analysis utilizes the faculty version of SimaPro.

1. Open SimaPro and select the appropriate database, typically labeled "Professional" (Fig. 5). Then, either create a new project or open an existing one (Fig. 6).
2. Choose the necessary databases from the library tab, such as the EcoInvent database, available in the "Inventory" menu on the left panel (Fig. 7).
3. Navigate to "Processes" under the "Inventory" menu (Fig. 8) to access different process categories. Here, either select an existing category or create a new one that aligns with the study objectives. To create a category, right-click on the page, then define key processes such as cultivation, harvesting, dewatering, oil extraction, biodiesel conversion, and combustion (Fig. 9).
4. Enter input and output data based on collected information. For input data requiring database integration, right-click the relevant row (e.g., materials, fuel, and electricity) and use the search function in the left pane to locate and select the appropriate database entry. If necessary, apply allocation methods for co-products such as oil extraction (Fig. 10).

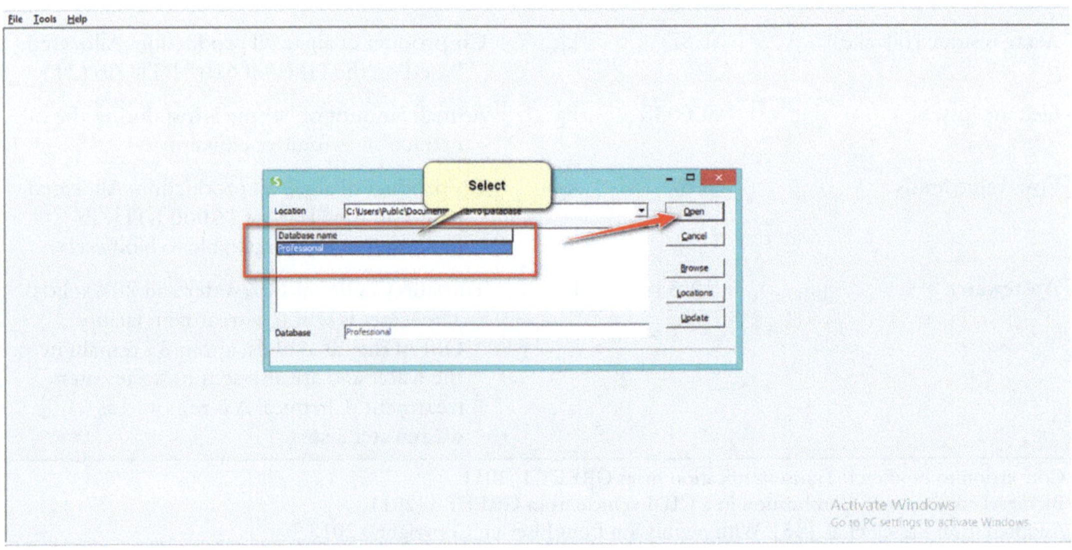

Fig. 5 Selection of "Professional" database

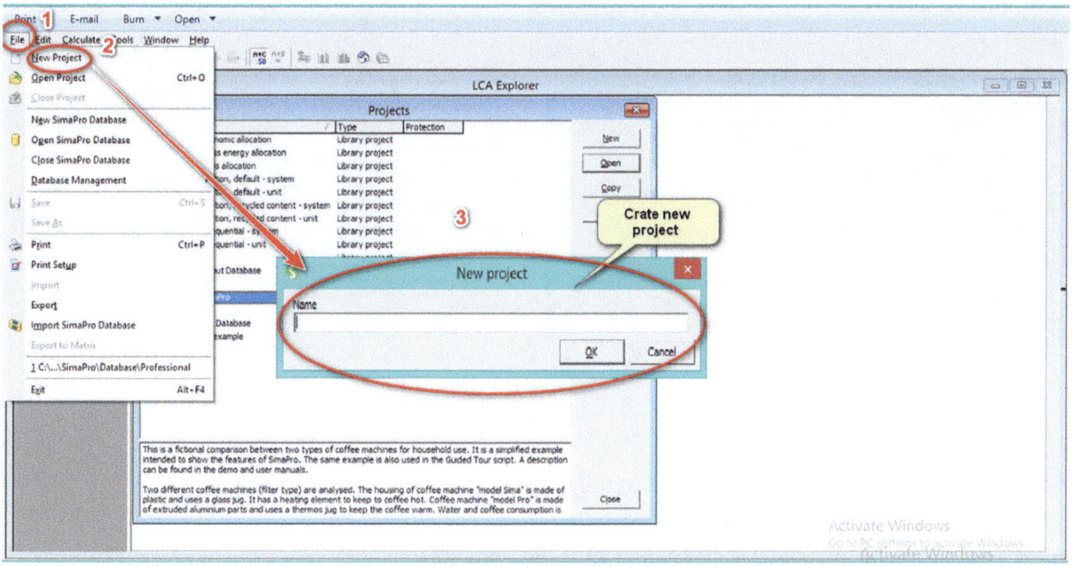

Fig. 6 Creation of a new project: LCA of algal biodiesel production/combustion in SimaPro

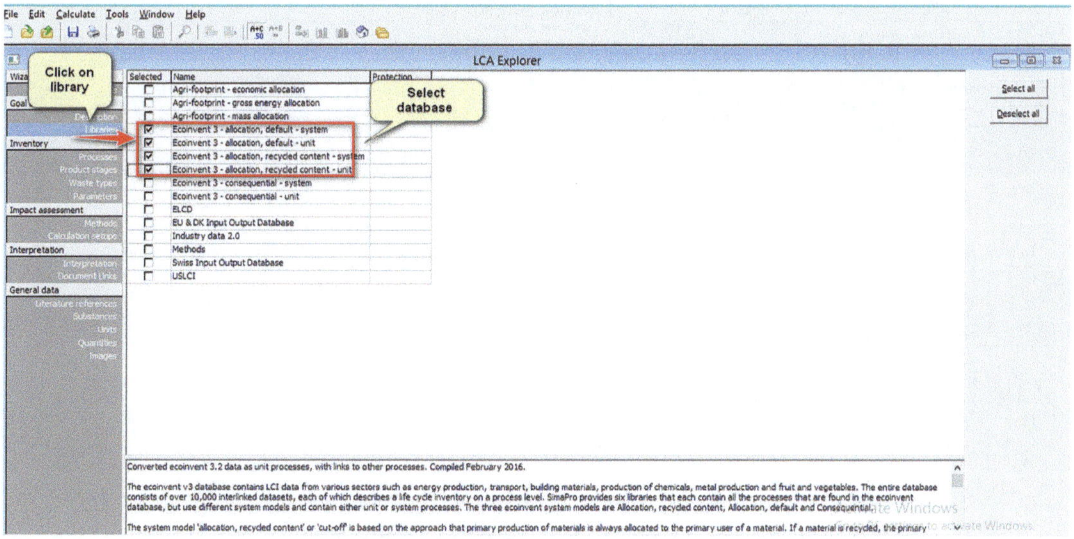

Fig. 7 Selection of EcoInvent database from the library

5. Once data entry is complete, click the "Analyze" button at the top of the page (Fig. 10) to proceed to LCIA method selection and FU definition, ensuring alignment with the study goal (Fig. 11).
6. Compute the environmental impact or damage by selecting the "Calculate" button (Fig. 11).

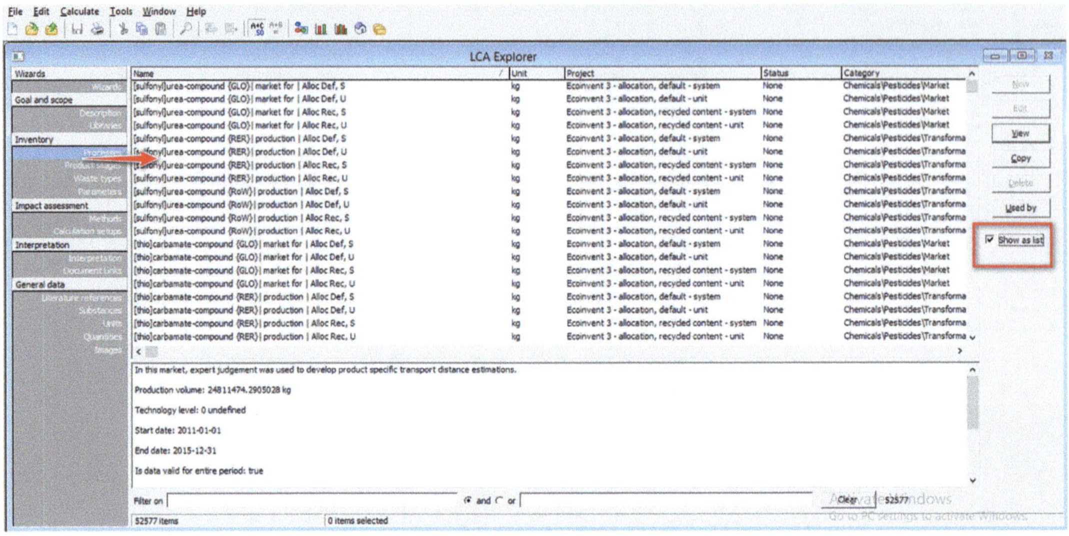

Fig. 8 Default view of processes in libraries when starting SimaPro

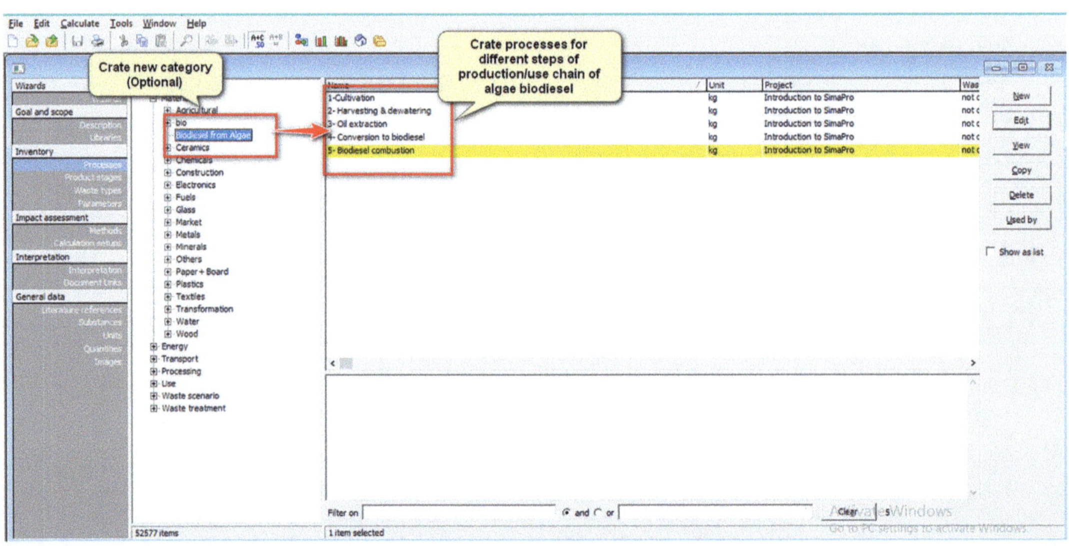

Fig. 9 Defined processes of algal biodiesel production/combustion in SimaPro

7. Refer to the SimaPro software guide to extract results and generate classification tables, characterization charts, damage assessments, normalization, and weighting visualizations (Fig. 12).

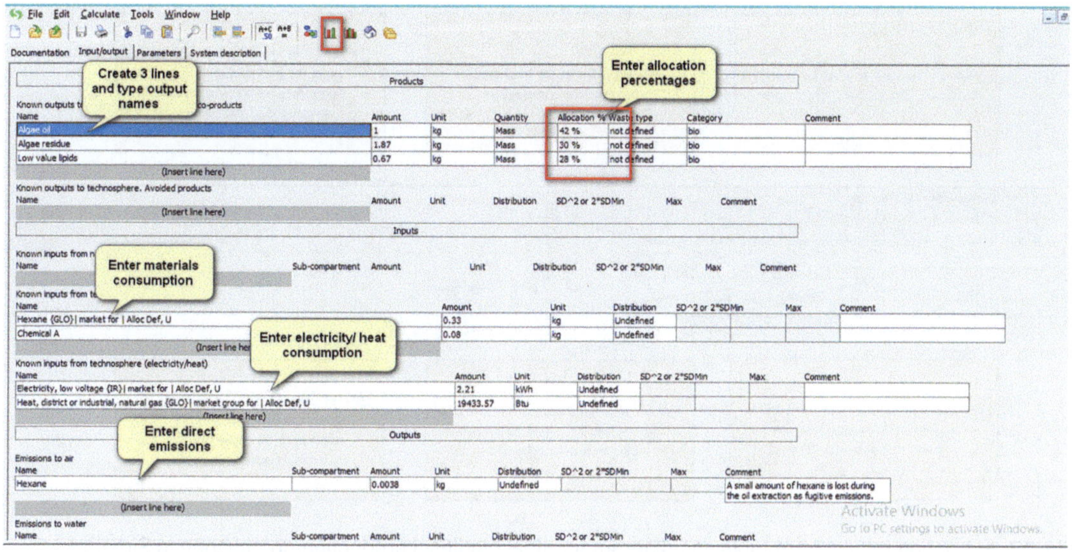

Fig. 10 Inputs and outputs in sheet for algae oil production in SimaPro

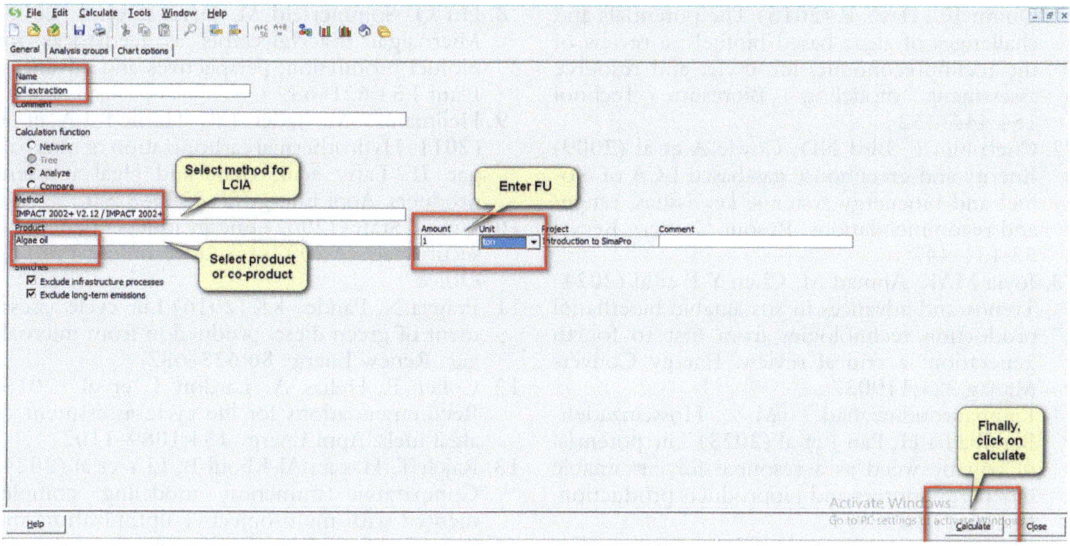

Fig. 11 Choosing the LCIA method and FU for algal biodiesel production/combustion in SimaPro

8. Repeat these steps for additional stages of algal biodiesel production and combustion. Conduct a sensitivity analysis using standard methodologies and perform a Monte Carlo analysis if supported by the selected SimaPro version.

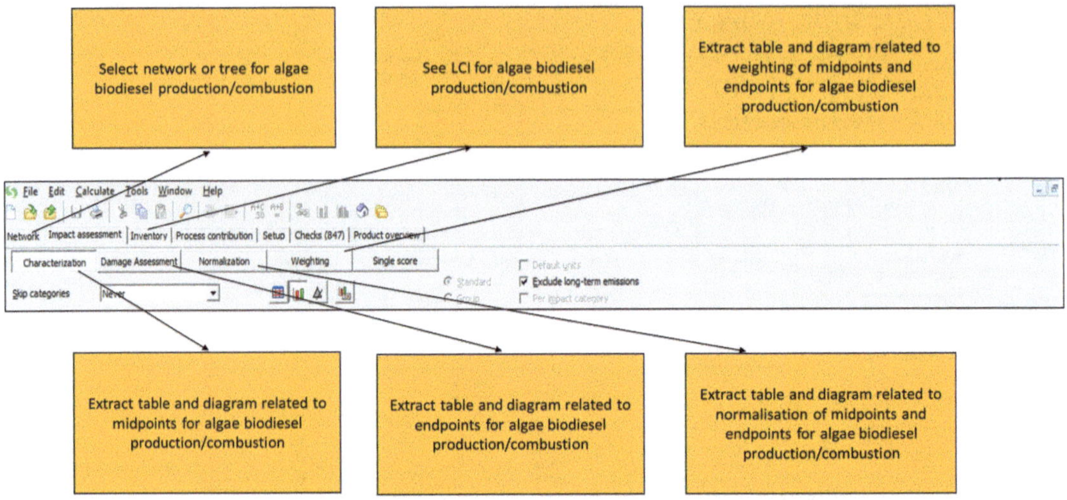

Fig. 12 Extraction results from table and diagram for algal biodiesel production/combustion in SimaPro

References

1. Quinn JC, Davis R (2015) The potentials and challenges of algae based biofuels: a review of the techno-economic, life cycle, and resource assessment modeling. Bioresour Technol 184:444–452
2. Cherubini F, Bird ND, Cowie A et al (2009) Energy-and greenhouse gas-based LCA of biofuel and bioenergy systems: key issues, ranges and recommendations. Resour Conserv Recycl 53:434–447
3. Joyia MAK, Ahmad M, Chen Y-F et al (2024) Trends and advances in sustainable bioethanol production technologies from first to fourth generation: a critical review. Energy Convers Manag 321:119037
4. Kiehbadroudinezhad M, Hosseinzadeh-Bandbafha H, Pan J et al (2023) The potential of aquatic weed as a resource for sustainable bioenergy sources and bioproducts production. Energy 278:127871
5. Chhandama MVL, Satyan KB, Changmai B et al (2021) Microalgae as a feedstock for the production of biodiesel: a review. Bioresour Technol Reports 15:100771
6. Yang J, Li X, Hu H et al (2011) Growth and lipid accumulation properties of a freshwater microalga, Chlorella ellipsoidea YJ1, in domestic secondary effluents. Appl Energy 88:3295–3299
7. Jorquera O, Kiperstok A, Sales EA et al (2010) Comparative energy life-cycle analyses of microalgal biomass production in open ponds and photobioreactors. Bioresour Technol 101:1406–1413
8. Hu Q, Sommerfeld M, Jarvis E et al (2008) Microalgal triacylglycerols as feedstocks for biofuel production: perspectives and advances. Plant J 54:621–639
9. Heilmann SM, Jader LR, Harned LA et al (2011) Hydrothermal carbonization of microalgae II. Fatty acid, char, and algal nutrient products. Appl Energy 88:3286–3290
10. United States (2007) Energy independence and security act of 2007. US Government Printing Office
11. Pragya N, Pandey KK (2016) Life cycle assessment of green diesel production from microalgae. Renew Energy 86:623–632
12. Collet P, Hélias A, Lardon L et al (2015) Recommendations for life cycle assessment of algal fuels. Appl Energy 154:1089–1102
13. Rajeh T, Hassan Al-Kbodi B, Li Y et al (2024) Comparative numerical modeling complemented with multi-objective optimization and dynamic life cycle assessment of coaxial ground heat exchangers with oval-shaped and typical circular-shaped configurations. Appl Therm Eng 244:122673
14. Larsen VG, Tollin N, Sattrup PA et al (2022) What are the challenges in assessing circular economy for the built environment? A literature review on integrating LCA, LCC and S-LCA in life cycle sustainability assessment, LCSA. J Build Eng. 50:104203
15. Finnveden G, Hauschild MZ, Ekvall T et al (2009) Recent developments in life cycle assessment. J Environ Manage 91:1–21

16. Joseph SJP, Subramani N (2024) Advancing sustainability: the role of epiplastic microalgae and their enzymatic pathways in freshwater plastic biodegradation. J Water Process Eng 67:106134
17. Samani P (2023) Synergies and gaps between circularity assessment and Life Cycle Assessment (LCA). Sci Total Environ 903:166611
18. Razman KK, Hanafiah MM, Mohammad AW (2022) An overview of LCA applied to various membrane technologies: progress, challenges, and harmonization. Environ Technol Innov 27:102803
19. Torabi F, Ahmadi P (2020) Battery technologies. Simulation of battery systems: fundamentals and applications. Academic, pp 1–54
20. Park J-H, Seo K-K (2006) A knowledge-based approximate life cycle assessment system for evaluating environmental impacts of product design alternatives in a collaborative design environment. Adv Eng Informatics 20:147–154
21. Barbhuiya S, Das BB (2023) Life cycle assessment of construction materials: methodologies, applications and future directions for sustainable decision-making. Case Stud Constr Mater e02326
22. Tulbure I, Palamariu M (2021) Environmental impacts mitigation odds by soft-computing methods. Climate and land use impacts on natural and artificial systems. Elsevier, pp 287–306
23. Roostaei J, Zhang Y (2017) Spatially explicit life cycle assessment: opportunities and challenges of wastewater-based algal biofuels in the United States. Algal Res 24:395–402
24. Kiehbadroudinezhad M, Hosseinzadeh-Bandbafha H, Karimi K et al (2023) Production of chemicals and utilities in-house improves the environmental sustainability of phytoplankton-based biorefinery. Sci Total Environ 899:165751
25. Wei C, Xu Y, Li Y et al (2024) Life-cycle assessment of microalgae liquid biofuel production in biofilm cultivation system via conversion technologies of transesterification, hydrothermal liquefaction and pyrolysis. J Clean Prod 436:140559
26. Mu D, Xin C, Zhou W (2020) Life cycle assessment and techno-economic analysis of algal biofuel production. Microalgae Cultiv Biofuels Prod 2020:281–292
27. ISO (2006) 14044 International standard. Environmental management–life cycle assessment–requirements and guidelines. International Organisation for Standardization, Geneva
28. ISO (2006) 14040 International standard. Environmental management–life cycle assessment–principles and framework. International Organisation for Standardization, Geneva.
29. Bernaerts TMM, Gheysen L, Foubert I et al (2019) The potential of microalgae and their biopolymers as structuring ingredients in food: a review. Biotechnol Adv 37:107419
30. Li X, Liu B, Lao Y et al (2021) Efficient magnetic harvesting of microalgae enabled by surface-initiated formation of iron nanoparticles. Chem Eng J 408:127252
31. Halim R, Danquah MK, Webley PA (2012) Extraction of oil from microalgae for biodiesel production: a review. Biotechnol Adv 30:709–732
32. Tabatabaei M, Aghbashlo M, Dehhaghi M et al (2019) Reactor technologies for biodiesel production and processing: a review. Prog Energy Combust Sci 74:239–303
33. De Wolf C, Cordella M, Dodd N et al (2023) Whole life cycle environmental impact assessment of buildings: developing software tool and database support for the EU framework Level(s). Resour Conserv Recycl 188:106642
34. Lopes Silva DA, Nunes AO, Piekarski CM et al (2019) Why using different Life Cycle Assessment software tools can generate different results for the same product system? A cause–effect analysis of the problem. Sustain Prod Consum 20:304–315
35. Speck R, Selke S, Auras R, Fitzsimmons J (2016) Life cycle assessment software: selection can impact results. J Ind Ecol 20:18–28
36. Koščáková M, Korba P, Koščák P et al (2024) Comparative analysis of life cycle assessment tools for aviation. In: New trends in aviation development (NTAD). IEEE, pp 96–101
37. Takano A, Winter S, Hughes M, Linkosalmi L (2014) Comparison of life cycle assessment databases: a case study on building assessment. Build Environ 79:20–30
38. Frischknecht R (2006) Notions on the design and use of an ideal regional or global LCA database. Int J Life Cycle Assess 11:40–48
39. Cooreman-Algoed M, Boone L, Uitterhaegen E et al (2023) Environmental life cycle assessment of nutraceuticals: a case study on methylcobalamin in different packaging types. Sci Total Environ 893:164780
40. Martínez-Rocamora A, Solís-Guzmán J, Marrero M (2016) LCA databases focused on construction materials: a review. Renew Sustain Energy Rev 58:565–573
41. Alvarenga RAF, Dewulf J, Guinée J et al (2019) Towards product-oriented sustainability in the (primary) metal supply sector. Resour Conserv Recycl 145:40–48
42. Arodudu O, Helming K, Wiggering H, Voinov A (2017) Towards a more holistic sustainability assessment framework for agro-bioenergy systems-a review. Environ Impact Assess Rev 62:61–75

43. Jacob-Lopes E, Zepka LQ, Deprá MC (2021) Assistant's tools toward life cycle assessment. Sustain Metrics Indic Environ Impact 2021:77–90
44. De Laurentiis V, Caldeira C, Sala S, Tonini D (2024) Life cycle thinking for the assessment of waste and circular economy policy: status and perspectives from the EU example. Waste Manag 179:205–215
45. Passell H, Dhaliwal H, Reno M et al (2013) Algae biodiesel life cycle assessment using current commercial data. J Environ Manage 129:103–111
46. Ahmed S, Li T, Zhou XY et al (2025) Quantifying the environmental footprints of biofuels for sustainable passenger ship operations. Renew Sustain Energy Rev 207:114919
47. Bradley T, Rajaeifar MA, Kenny A et al (2023) Life cycle assessment of microalgae-derived biodiesel. Int J Life Cycle Assess 28:590–609
48. Tsang M, Fox Lent C, Wallace S et al (2015) Life cycle impacts of soybean and algae biodiesel: case study of US marine vessels. Biofuels Bioprod Biorefining 9:567–580
49. deBruijn, H, vanDuin, R, Huijbregts, MAJ et al, (2002.) Goal and scope definition. Springer, Dordrecht, pp. 455–476
50. Bjørn A, Owsianiak M, Molin C, Laurent A (2018) Main characteristics of LCA. In: Hauschild MZ, Rosenbaum RK, Olsen SI eds. Life cycle assessment. Springer, Cham, pp 9–16
51. Albertí J, Brodhag C, Fullana-i-Palmer P (2019) First steps in life cycle assessments of cities with a sustainability perspective: a proposal for goal, function, functional unit, and reference flow. Sci Total Environ 646:1516–1527
52. Yang Y (2016) Two sides of the same coin: consequential life cycle assessment based on the attributional framework. J Clean Prod 127:274–281
53. Shirzad M, Kazemi Shariat Panahi HB, Dashti B et al. (2019) A comprehensive review on electricity generation and GHG emission reduction potentials through anaerobic digestion of agricultural and livestock/slaughterhouse wastes in Iran. Renew Sustain Energy Rev 111:571–594
54. Buonocore E, Mellino S, De Angelis G et al (2018) Life cycle assessment indicators of urban wastewater and sewage sludge treatment. Ecol Indic 94:13–23
55. Carneiro MLNM PradelleF, Braga SL et al (2017) Potential of biofuels from algae: comparison with fossil fuels, ethanol and biodiesel in Europe and Brazil through life cycle assessment (LCA). Renew Sustain Energy Rev 73:632–653
56. Bhatnagar A, Härri A, Levänen J, Niinimäki K (2024) Exploring the role of social life cycle assessment in transition to circular economy: A systematic review. Resour Conserv Recycl 207:107702
57. Wolf MA, Pant R, Chomkhamsri K et al (2012) The International Reference Life Cycle Data System (ILCD) handbook-JRC reference reports
58. Jose S, Archanaa S (2017) Environmental and economic sustainability of algal lipid extractions: an essential approach for the commercialization of algal biofuels. In: Algal biofuels. Springer, Cham, pp 281–313
59. Singh A, Olsen SI (2011) A critical review of biochemical conversion, sustainability and life cycle assessment of algal biofuels. Appl Energy 88:3548–3555
60. Avadí A, Fréon P (2013) Life cycle assessment of fisheries: a review for fisheries scientists and managers. Fish Res 143:21–38
61. European Commission (2010) ILCD handbook-general guide for life cycle assessment-detailed guidance. European Commission, Joint Research Centre. Publications Office of the European Union, Luxembourg
62. Ferrari AM, Volpi L, Settembre-Blundo D, García-Muiña FE (2021) Dynamic life cycle assessment (LCA) integrating life cycle inventory (LCI) and Enterprise resource planning (ERP) in an industry 4.0 environment. J Clean Prod 286:125314
63. Alengebawy A, Ghimire N, Abdelkhalek ST, Samer M (2024) Conversion of bioenergy materials to secondary fuels. Encyclopedia of renewable energy, sustainability and the environment. Elsevier, pp 825–838
64. Lombardi L, Mendecka B, Carnevale E (2018) Comparative life cycle assessment of alternative strategies for energy recovery from used cooking oil. J Environ Manag 216:235–245
65. Landi D, Consolini A, Germani M, Favi C (2019) Comparative life cycle assessment of electric and gas ovens in the Italian context: an environmental and technical evaluation. J Clean Prod 221:189–201
66. Saad M, Zhang Y, Jia J, Tian J (2024) Decision tree-based approach to extrapolate life cycle inventory data of manufacturing processes. J Environ Manage 360:121152
67. Bradley T, Maga D, Antón S (2015) Unified approach to life cycle assessment between three unique algae biofuel facilities. Appl Energy 154:1052–1061
68. Michiels F, Hubo L, Geeraerd A (2021) Why mass allocation with representative allocation factor is preferential in LCA when using residual livestock products as organic fertilizers. J Environ Manag 297:113337
69. da Silva JAM, Santos JJCS, Carvalho M, de Oliveira S (2017) On the thermoeconomic and LCA methods for waste and fuel allocation in multiproduct systems. Energy 127:775–785

70. Ardente F, Cellura M (2012) Economic allocation in life cycle assessment: the state of the art and discussion of examples. J Ind Ecol 16:387–398
71. Ma Y, Preveniou A, Kladis A, Pettersen JB (2022) Circular economy and life cycle assessment of alumina production: simulation-based comparison of Pedersen and Bayer processes. J Clean Prod 366:132807
72. Dominguez Aldama D, Grassauer F, Zhu Y et al (2023) Allocation methods in life cycle assessments (LCAs) of agri-food co-products and food waste valorization systems: systematic review and recommendations. J Clean Prod 421:138488
73. Kendall A, Yuan J (2013) Comparing life cycle assessments of different biofuel options. Curr Opin Chem Biol 17:439–443
74. Collet P, Spinelli D, Lardon L et al (2014) Life-cycle assessment of microalgal-based biofuels. In: Biofuels from algae. Elsevier, pp 287–312
75. Börjesson P, Tufvesson LM (2011) Agricultural crop-based biofuels-resource efficiency and environmental performance including direct land use changes. J Clean Prod 19:108–120
76. Villagrán E, Romero-Perdomo F, Numa-Vergel S et al (2023) Life cycle assessment in protected agriculture: where are we now, and where should we go next? Horticulturae 10:15
77. Li X, Qian Y, Xie M et al (2024) An endpoint model for life cycle impact assessment in China and preliminary normalization values: a case study of vehicles. J Clean Prod 434:140326
78. Hosseinzadeh-Bandbafha H, Tabatabaei M, Aghbashlo M et al (2020) Data supporting consolidating emission indices of a diesel engine powered by carbon nanoparticle-doped diesel/biodiesel emulsion fuels using life cycle assessment framework. Data Br 2020:105428
79. Fieschi M, Pretato U (2018) Role of compostable tableware in food service and waste management. A life cycle assessment study. Waste Manag 73:14–25
80. Roigé Montornés N (2014) Structural and environmental optimization of DWTDN trenches.
81. Hauschild MZ, Bjørn A (2018) LCA Cookbook. In: Life cycle assessment. Springer, Cham, pp 963–1048
82. Johnsen FM, Løkke S (2013) Review of criteria for evaluating LCA weighting methods. Int J Life Cycle Assess 18:840–849
83. Reinhardt R, Pautzke F, Schröter M, Wiemers M (2017) A case study of sustainable manufacturing strategy: comparative LCA of wheel hub engine for solar car application. In: 2017 International conference on Research and Education in Mechatronics (REM). IEEE, pp 1–6
84. Davis J, De Menna F, Unger N et al (2017) Generic strategy LCA and LCC: guidance for LCA and LCC focused on prevention, valorisation and treatment of side flows from the food supply chain.SP Rapport 2017:01, ISBN 978-91-88349-84-2. Accessible at: http://eu-refresh.org/generic-strategy-lca-and-lcc
85. Kaklauskas A (2016) Analysis of the life cycle of a built environment. Nova Science Publishers, Incorporated
86. Guo M, Murphy RJ (2012) LCA data quality: sensitivity and uncertainty analysis. Sci Total Environ 435:230–243
87. Wei W, Larrey-Lassalle P, Faure T et al (2015) How to conduct a proper sensitivity analysis in life cycle assessment: taking into account correlations within LCI data and interactions within the LCA calculation model. Environ Sci Technol 49:377–385

Chapter 28

Design and Analysis of Macroalgae-Based Biorefineries

Alexander Golberg

Abstract

Displacing fossil fuels and their derivatives with renewables and increasing sustainable food and materials production are among the major challenges facing the world in the coming decades. A possible, sustainable direction for addressing these challenges is biorefinery. Plants and microalgae are possible feedstocks; however, concerns over net energy balance, potable water use, environmental hazards, and uncertainty in the processing technologies question their actual potential to meet the anticipated food, feed, materials, and energy challenges in a sustainable way. Alternative sustainable sources for biorefineries are offshore-produced macroalgae. However, implementation of the offshore biorefineries requires detailed analysis of their technological, economic, and environmental performance. In this chapter, the basic principles of marine biorefineries design are shown. The methods to integrate thermodynamic efficiency, investment and environmental aspects are discussed. The performance improvements by development of new cultivation and dewatering methods that fit macroalgae physiology and development of new conversion methods that address macroalgae's unique chemical composition are shown.

Keywords Macroalgae, Seaweed, Offshore cultivation, Marine biorefinery design, Environmental exergonomics, Exergy efficiency, Fermentation, Co-products

1 Introduction

Population growth and increased per-person resource consumption in the era of changing climate will increase the demand for food, chemicals, and fuels. A possible sustainable direction for addressing this challenge is the production of food, chemicals, and fuels from the delicately cultivated biomass. A complex system of biomass cultivation, harvest, and conversion to multiple products is entitled biorefinery. This biomass can come from terrestrial or aquatic biomass. Terrestrial biomass has been traditionally used for food supply for most of the global population. However, concerns over net energy balance, potable water use, environmental hazards, and uncertainty in the processing technologies—mostly the problems with lignin—raise questions regarding the actual potential of terrestrial biomass to meet the anticipated food, chemicals, materials, and energy challenges in a sustainable way [1, 2].

Kristian Spilling (ed.), *Biofuels from Algae*, Methods in Molecular Biology, vol. 2996,
https://doi.org/10.1007/978-1-0716-5031-8_28,

Additional sources of biomass for the biorefinery are algal biomass. Microalgae are promising species because of their high growth rates. However, the real-scale implementation of microalgae systems for bioenergy and commodity chemical production is limited today by costs associated with reactor construction and maintenance, contamination, and energy required for the separation of these single-cell organisms from water [3].

Alternative sources for biorefineries are macroalgae (or seaweeds). Macroalgae have been harvested throughout the world as a food source and as a commodity for the production of hydrocolloids for centuries. However, to date macroalgae still present only a tiny percent of the global biomass supply of ~17×10^6 fresh weight (FW) tons of macroalgae in comparison to 16×10^{11} tons of terrestrial crops, grasses, and forests [4–6]. A large body of evidence from the last decades suggests that offshore-cultivated macroalgae, which contain very little lignin—a major problem molecule in the processing of terrestrial biomass—and do not compete with food crops for arable land or potable water, can provide an alternative source of biomass for sustainable production of food [7–12], materials [13–16], chemicals [17], and fuels [18–20].

2 Marine Biorefinery System

A marine biorefinery adheres to the Merriam-Webster dictionary definition of a *system*: "a set of interacting or interdependent component parts forming a complex/intricate whole, which property is different from the properties of its individual components." A concept of an offshore marine biorefinery, shown in Fig. 1, is a complex system. Systematic design of its components requires:

- Determination of the optimum location
- System design to function in the chosen location
- System integration within the natural environment
- Cultivation platform design-upstream processing
- Conversion processes—downstream processing
- Analysis of social and environmental impacts.

Determination of the optimum marine biorefinery location is predicated on environmental and socio-economic conditions. Environmental conditions are numerous and include the conditions for macroalgae growth and risks. Assessment of offshore locations for cultivation suitability is an expensive and time-consuming task. For the initial assessment, the following model has been recently suggested to assess the biomass growth rate at a specific geographical location [21]:

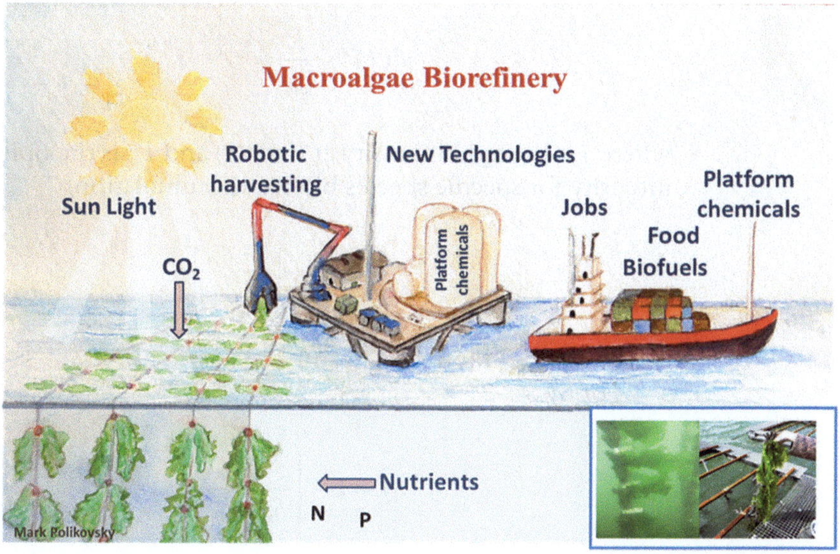

Fig. 1 The concept of off-shore biorefineries for the production of food, platform chemicals, and biofuels. We assume that the cultivation is done by extensive methods with ropes/cages. Insert on the right left shows the example off-shore cultivation of macroalga from *Ulva* genus [88]. Figure adapted from [21] with permission

$$\mu = \mu_{max} \cdot f(I, T, S, N, P) - r_{resp}, \tag{1}$$

where macroalgae biomass growth rate (μ) is calculated based on its maximum possible growth rate (μ_{max}) as a function of light intensity (I), temperature (T), salinity (S), nutrients (N and P for nitrate and phosphate, respectively), and respiration rate (r_{resp}).

In this model, we assume that each of the factors has a separate impact on the biomass growth rate. Therefore, the approximated function for biomass growth rate appears in Eq. (2):

$$\mu = \mu_{max} \cdot f(I) \cdot f(T) \cdot f(S) \cdot f(N) \cdot f(P) - r_{resp}, \tag{2}$$

where μ_{max} (which is a function of the stocking density [22]) is the maximum growth rate, and r_{resp} is the respiration rate (d^{-1}) defined as [23–25]:

$$r_{resp} = r_{resp\_ref}\, \theta^{T - T_{ref}}, \tag{3}$$

where r_{resp_ref} is the maximum respiration rate at reference temperature (T °C), and θ is the empirical factor. The function $f(I, T, S, N, P)$ is defined as follows in Eqs. (4)–(8):

$$f(I)=\frac{I}{I_{opt}}e^{\left(1-\frac{I}{I_{opt}}\right)}, \tag{4}$$

where I is the light intensity at time (t) and I_{opt} is the optimum light intensity for specific species biomass accumulation.

$$f(T)=e^{-a\left(\frac{T-T_{opt}}{T_x-T_{opt}}\right)^2}, \tag{5}$$

where $T_x=T_{min}$ for $T<=T_{opt}$ and $T_x=T_{max}$ for $T>T_{opt}$.

For $S>S_{cr}$

$$f(S)=1-\left(\frac{S-S_{opt}}{S_x-S_{opt}}\right)^m, \tag{6}$$

where $S_x=S_{min}$ and m_{min} for $S<S_{opt}$ and $S_x=S_{max}$ and m_{max} for $S>=S_{opt}$

for $S< S_{cr}$

$$f(S)=\frac{S-S_{min}}{S_{opt}-S_{min}}. \tag{7}$$

For $(N>N_{min})$ and $(P>P_{min})$:

For N:

If l<N:P and N:P<h:

$$\mathrm{f(N,P,C)}=1$$

If N:P<l f(N,P,C)=f(N):

$$f(N)=\frac{N_{int}-N_{int\_min}}{keq+N_{int}-N_{int\_min}}$$

If N:P>hh f(N,P,C)=f(P):

If $P_{int}<P_{int\_max}$:

$$f(P)=\frac{P_{int}}{P_{int\_max}}$$

If $P_{int}>P_{int\_max}$:

$$f(p)=1,$$

where μ_{max}, r_{resp_ref}, T_{ref}, θ, I_{opt}, T_{min}, T_{max}, T_{opt}, S_{cr}, S_{min}, S_{max}, N_{min}, P_{min}, l, h, hh, keq, N_{int_min}, and P_{int_max} are species specific-parameters. An example of the use of this model for Ulva growth can be found in

[21]. Daily production per m^2 of FW biomass is derived by multiplying the calculated growth rate (μ) with biomass density (σ, kg m^{-2}) [22], Eq. (8):

$$BM_{FW} = \mu \cdot \sigma.$$

The example of the model use has been demonstrated on the macroalgae *Ulva* species, to assess its global cultivation potential [21]. The work identified potential provinces where *Ulva* species can be cultivated for the maximum biomass yields offshore (Fig. 2).

System engineering design that will fit the chosen location in the harsh marine environment is a challenge. Attaching seaweeds to ropes, lines, or nets is a traditional way of cultivation since installation and maintenance costs are very low. One of the cultivation methods is done by seedlings directly to the ropes [26]. Another approach is via transplantation: seedlings are grown indoors, then cultured in greenhouse tanks, and finally, the small fronds are transplanted onto ropes in the sea. The ropes should be tensed to reduce diffusion boundary layers surrounding the thalli, thus increasing water exchange for efficient nutrient availability and fast growth [27, 28]. Horizontal and vertical rope systems were also discussed by [26]. A system with concentrically horizontal ropes was tested in [29]. Seabed planting can be cited, as well as a cage system (open-topped cage anchored to the sea floor) for nearshore cultivation [30]. Co-management with other offshore systems, like wind farms, for instance was considered [31–34] and discussed as an environmentally and economically advantageous approach [35].

Optimizing downstream processing of the biorefineries is a complex task, which depends upon the species choice, their chemical composition, available technologies, human resources and economic and environmental impacts. Exergonomics, introduced by Yantovsky for the analysis of energy systems [36], links invested and operational exergy expenditures and allows one to find the optimal exergy efficiency of production systems [36, 37]. Yantovsky also suggested that "for more reliable decision making, the simultaneous optimization of three target functions: exergy, money, and pollution, is needed" [36]. However, the designer should reduce not only pollution from the system. The designer must also consider the multiple complex effects the energy system—especially a large-scale renewable energy system—has on the ecosystem services of the surrounding environment. In the next section, we will examine the use of environmental exergonomcs, a recently developed tool for energy and production system analysis [38], for the design and analysis of offshore marine biorefineries. Environmental exergonomics includes mechanical (technological efficiency of the system), capital efficiency of the system, and environmental efficiency, measured by the eco-exergy (exergy contribution of ecosystems to the biorefinery).

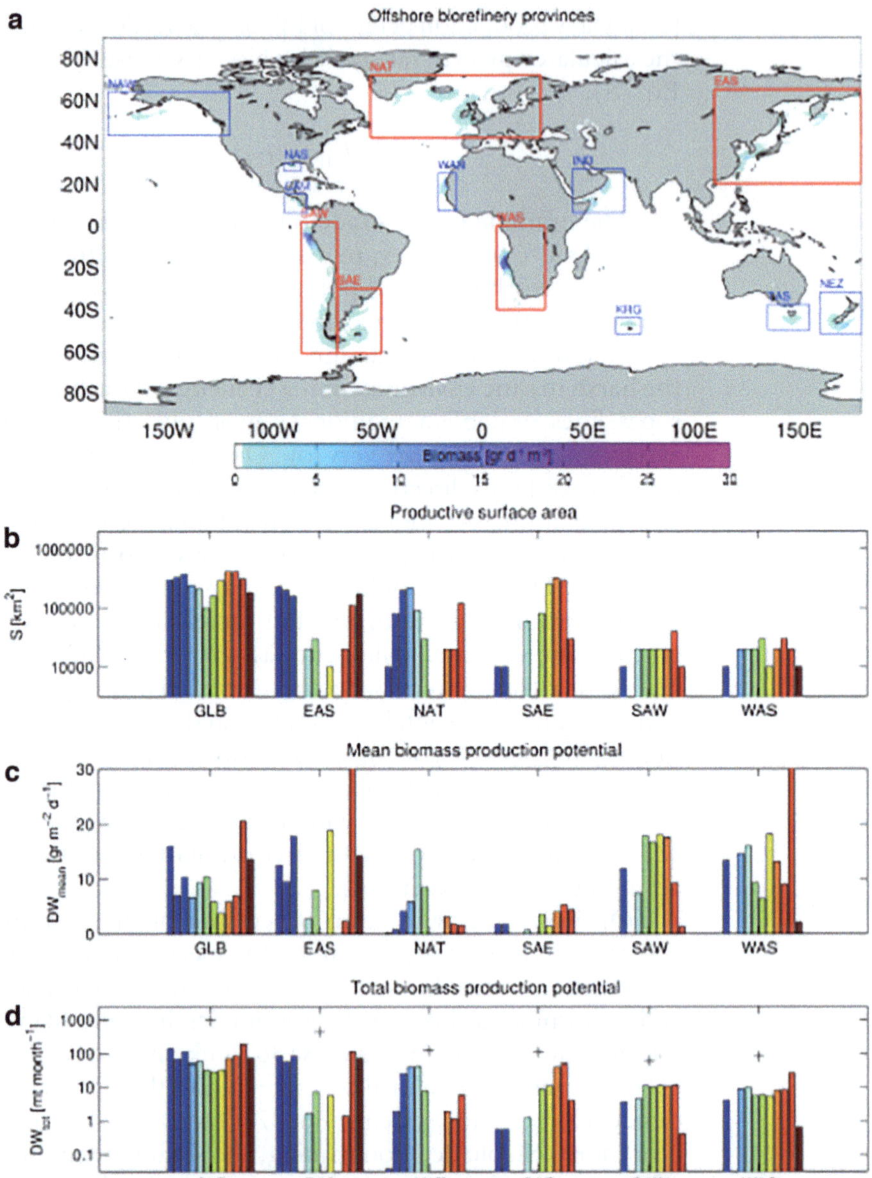

Fig. 2 Regional potential for offshore biorefinery. (**a**) Potential for biomass production at a distance of less than 400 km from land, taking biomass stocking density of 4 kg m^{-2}. Values in the map will change linearly with changes in stocking density. Boxes delineate major offshore biorefinery provinces, with those permitting biomass production at water depth of up to 100 m (defined as near-future deployable biorefinery provinces—NDBP) marked in red, and those permitting biomass production only at deeper waters marked in blue. (**b–d**) Monthly estimates of (**b**) productive surface area; (**c**) mean biomass production potential; and (**d**) total production potential within the 5 NDBP (red boxes and associated abbreviations in panel **a**) and integrated globally (denoted GLB). Colors denote different months of the year. The analysis is performed over locations associated with water depth of 100 m or shallower. The + signs mark annually integrated biomass production potential at each region. Assumed biomass density of 4 kg m^{-2}. Production potential will change linearly with changes in assumed biomass density. Abbrevations: East Asia offshore waters (EAS), North Atlantic (NAT), South America offshore waters—East (SAE), South America offshore waters—West (SAW), West Africa offshore waters—South (WAS). Figure adapted from [88] with permission

3 Environmental Exergonomics for Marine Biorefineries

Consider the marine biorefinery system on the production scale. The system converts solar and mechanical energies into concentrated energy products, such as food, energy and chemicals. If the system performance is measured in the units of exergy, in the most general case (Fig. 3a), the inputs to the process are represented by an exergy stream of solar energy supply (e_{s_e}), mechanical energy supply (e_{m_e}), materials (e_m), capital inflow (e_k), human labor (e_l), and information, represented by eco-exergy (e_{eco}). The outputs are the delivered exergy contained in food (e_f), useful energy, such as biofuels (e_e), and platform chemicals (e_c), exergy rejection to the environment (e_{en}), materials waste (e_w), and eco-exergy information loss (or gain) ($e_{eco\text{-}c}$). As both physical and information exergies are conserved in these systems, the system will experience continuous physical and information exergies losses. The Sankey exergy diagram for the exergy flow in marine biorefineries, which includes the eco-system's eco-exergy losses, is shown in Fig. 3b.

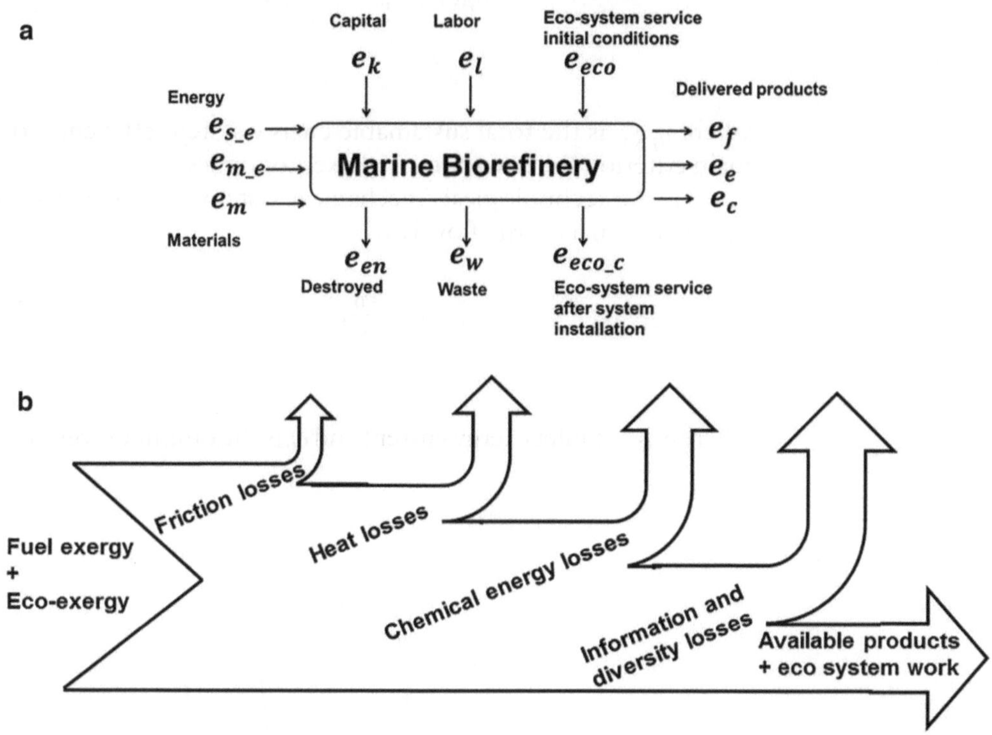

Fig. 3 Exergy flow in marine biorefineires. (**a**) Exergy currents. (**b**) Environmental exergonomics exergy diagram for marine biorefineries

3.1 Marine Biorefinery System Boundary

The biorefinery system includes physical, financial, and environmental components. The physical boundaries for the analyzed energy systems are the boundaries of the physical territory (allocated ocean area and required onshore facilities) where the system is constructed and operated. The capital boundaries include the capital invested in system construction, maintenance, and deconstruction. The environmental boundaries include the ecosystems that are located in the area that is occupied by the cultivation site and processing facility or that is affected by the facility construction. The system also includes the produced products, wastes, jobs, and local eco-systems. The exergy time history of energy unit—including construction exergy current ($\dot{e}_c$), operation exergy current ($\dot{e}_{con}$), and deconstruction exergy current ($\dot{e}_d$), as well as the exergy currents changes in surrounding eco-system ($\dot{e}_{eco}$)—are summarized in Fig. 4.

3.2 Marine Biorefinery System Efficiency

Under the assumption that each of the input and output factors can indeed be described using physical and informational exergy functions, the efficiency of a system using the Environmental Exergonomics method is described in Eq. (9):

$$\eta_{ENV} = \frac{1}{\frac{1}{\eta} + \frac{1}{K} + \frac{1}{E}}, \tag{9}$$

where η_{ENV} is the total sustainable energy system efficiency, or the main criterion for environmental exergonomics;

η is the technological/mechanical system efficiency, based on the operational exergy flow [36]:

$$\eta = \frac{\delta_0}{\delta_i}, \tag{10}$$

where δ_i is the inlet exergy current and δ_0 is the output exergy current.

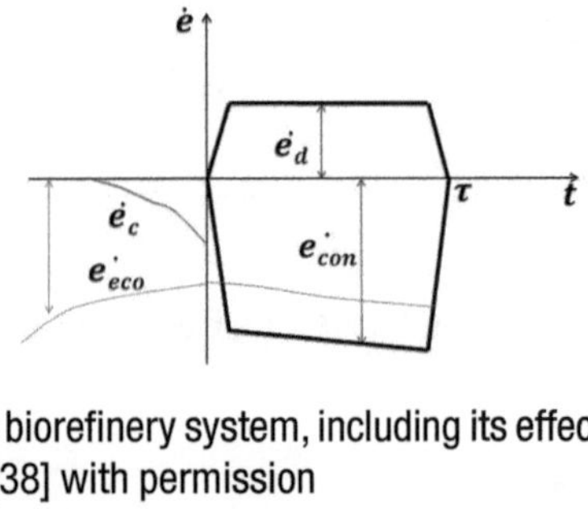

Fig. 4 The history of a biorefinery system, including its effects on the environment. Figure adapted from [38] with permission

K is the net exergy financial coefficient, the ratio of delivered exergy to invested exergy (money) [36]:

$$K = \frac{\frac{de_d}{dt} \cdot \tau}{e_c + e_l}, \tag{11}$$

where e_d is the delivered exergy, e_c is the invested exergy needed for system construction, e_l is the invested labor, and τ is system operation time.

And E is the ecological or ecosystem efficiency of the energy conversion system based on the eco-exergy flow. We will use a ratio of eco-exergy before and after energy system construction and use:

$$E = \frac{\frac{de_d}{dt} \cdot \tau}{e_i}, \tag{12}$$

where e_i is the consumed eco-exergy, which is described by the reduced ability of the ecosystem to perform work:

$$e_i = e_{eco\_0} - e_{eco\_\tau}. \tag{13}$$

The Environmental Exergonomics main criterion function is defined as:

$$Z_{env} = \frac{1}{\eta} + \frac{1}{K} + \frac{1}{E}. \tag{14}$$

Assuming that K and E are independent, for the arbitrary functions K(η) and E(η):

$$\frac{dZ_{env}}{d\eta} = -\frac{1}{\eta^2} + \frac{-\frac{dK}{d\eta}}{K^2} + \frac{-\frac{dE}{d\eta}}{E^2}. \tag{15}$$

For $\frac{dZ_{env}}{d\eta} = 0$:

$$\eta^2 = \frac{K^2 E^2}{-E^2 \frac{dK}{d\eta} - K^2 \frac{dE}{d\eta}}, \tag{16}$$

and thus

$$Z_{min} = \frac{\left(-E^2 \frac{dK}{d\eta} - K^2 \frac{dE}{d\eta}\right)^{1/2} + K + E}{KE}, \tag{17}$$

and

$$\eta_{opt} = \frac{KE}{\left(-E^2\frac{dK}{d\eta} - K^2\frac{dE}{d\eta}\right)^{1/2}}, \tag{18}$$

where η_{opt} is the optimum efficiency of the system. A correlation between K and η, and E and η, are study-specific. If a correlation between the these efficiencies is found, as suggested in the early works of Szargut [39], further functional analysis using a single parameter, for example η, or monitary cost is possible [36].

3.3 Calculation of Exergy Currents

3.3.1 Physical Exergy Currents

Physical exergy is defined as the maximum amount of reversible work that can be produced by bringing the temperature, pressure, velocity, and position within a gravitational field, and by bringing chemical composition to equilibrium with the defined reference state. Equation (19) describes the physical exergy of the system in the most general form [40]:

$$\begin{aligned}\delta = {} & [h - h_0 - T_0(s - s_0)] + \frac{(V - V_2)^2}{2} + g(z - z_0) \\ & + \sum_i (\mu_i c_i - \mu_0 c_0).\end{aligned} \tag{19}$$

The first term of the equation includes the classical thermodynamic properties—enthalpy (h), temperature (T), entropy (s)—known for many substances and mixtures in a wide range of states. The second and third terms are a result of measured position (z) and velocity (V) relative to the reference state, and their exergy and energy contents have the same numerical value as proposed in [41]. The fourth term is the chemical exergy of basic system elements (μ) and is the chemical potential. For all properties, subscript "0" stands for the value of the property at standard conditions.

3.3.2 Capital Exergy Currents

The capital exergy currents can be divided into monetary and labor currents. This subdivision and separation of the labor current from the monetary investment proposed by Sciubba [42, 43] emphasize the important impact of energy systems on workers and society. The detailed analyses of capital exergy currents can be found in [42, 43]. For simplicity, in this work, the capital exergy current is defined as the exergy required to build the unit and the exergy equivalent of working hours invested by the system staff during the system's life time:

$$e_{c+l} = e_c + e_l, \tag{20}$$

where e_c is the exergy required to build the unit, and

$$e_l = K_{labor} \cdot n_{workers} \cdot WH, \tag{21}$$

where K_{labor} is the exergenic equivalent of labor [44], and WH is the work hours in a year.

3.3.3 Eco-exergy Currents

The term eco-exergy was developed in ecology [45]. The concept of eco-exergy was first applied to ecology in the 1970s [46, 47], and the past four decades led to the formulation of the "maximum exergy principle in ecology," which described the formation of biodiverse communities in terms of thermodynamics [48]. Eco-exergy has been used in ecology to express emergent properties of ecosystems arising from self-organization processes in the evolution of their development [49]. Exergy has also been used as an objective function in ecological models to assess the changes and concentrations of various species in the ecosystem under stress [48].

Eco-exergy is a measure of the maximum amount of work that an ecosystem can perform when it is brought into thermodynamic equilibrium with its environment [50], Eq. (22):

$$e_{eco} = RT\sum_{i=0}^{n}\left[\left(C_i \ln\left(\frac{C_i}{C_i^{eq}}\right)\right) + \left(C_i - C_i^{eq}\right)\right], \tag{22}$$

where R is the gas constant, C_i is the concentration of species i in the system, and C_i^{eq} is the concentration of species i in the reference environment. The term eco-exergy evolved from the use of entropy in information theory, where entropyis the average amount of information contained in each message received [51]. In application to ecosystems, the information is coded in the genetic load of the organisms. In early works, Jørgensen et al. [50] proposed the following equation for the calculation of an ecosystem's exergy [48]:

$$e_{eco} = RT\left(\mu_1 - \mu_1^{eq}\right)\left|\sum_{i=1}^{N} C_i - RT\sum_{i=2}^{N}\left(C_i \ln\left(P_{i,a}\right)\right)\right., \tag{23}$$

where $\mu_1 - \mu_0$ is calculated from standard chemical potentials of the organic matter, C_i is the concentration of the species in the environment, and $P_{i,a}$ is the probability of producing the component i at thermodynamic equilibrium. $P_{i,a}$ can be found from the number of permutations among which the characteristic amino acid sequence for the considered species. Since living organisms use 20 different amino acids and each gene determines on average 700 amino acids, P can be evaluated using Eq. (24) [48]:

$$P_{i,a} = 20^{-700G}, \tag{24}$$

where G is the number of genes (for the standard table).

Eco-exergy has been used as an ecological indicator to assess ecological condition and ecosystem health [52]. The most recent definition of eco-exergy is [53]:

$$e_{eco} = f \sum_{i=1}^{n} \left(B_i \beta_i \right), \tag{25}$$

where f is the work energy per unit of biomass [53], which on average is 18.7 kJ gr^{-1}. B_i is the biomass weight of the species, i (gr), and β_i is the weighting factor available in tables in Appendix A [54]. β_i is equal to RTK, where R is the gas constant, T is absolute temperature, and K is Kullback's measure of information based on information embedded in the genes of the species, Eq. (26) [55]. The Kullback's measure of information defines the incremental changes in the system information as a result of the transition from a "reference state (i_o)" to a current one (i) as follows [55]:

$$K = \sum_{i=1}^{n} p_i \ln \frac{p_i}{p_{i0}}. \tag{26}$$

The impacts assessment of the large-scale biorefinery systems on the local and global ecosystems are rarely found in the literature. The installation of large scale biorefinery requires deployment of the large territories of sea. This, in turn, may cause change to local biodiversity and may affect even larger ecosystem services [56]. These novel uses of the sea affect the habitant, food and water availability, and preying strategy in animal species. It can also lead to the introduction of invasive species that decrease the natural biomass biodiversity [57]. The above-mentioned examples of ecological changes in areas with biorefinery installations can affect the biodiversity and thus the exergy of the ecosystem. The change in the eco-exergy in the area in which the biorefinery system is installed can be calculated using Eq. (27):

$$e_i = f \sum_{i=1}^{n} \left(B_i \beta_i \right)_0 - f \sum_{i=1}^{n} \left(B_i \beta_i \right)_\tau, \tag{27}$$

where the first term (subscript "0") stands for the eco-exergy of the ecological system before biorefinery construction, and the second term (subscript "τ") stands for the eco-exergy of the ecological system after the biorfinery deconstruction.

4 Determination of Optimum Scale and Serviced Area for Marine Biorefineries

Previous studies on the cost function of agricultural processing systems [58] and onshore macroalgae for biofuel biorefinery energy efficiency analysis [59] show that feedstock transportation costs limit the size of the biorefinery. Transportation costs limit the maximum possible distance of the cultivation site to the processing facility. However, different from the nearshore facilities, where the costs on biomass transportation from the sea to the onshore processing facility are known (~30% of the macroalgae costs [60]), the real monetary transportation costs from the open ocean offshore biorefineries can be only estimated. A more realistic approach is to estimate the energy expenses required for transportation that will limit the distance of an offshore cultivation site from the processing facility.

The maximum economic distance from the processing facility of an offshore cultivation area is calculated using Eq. (28):

$$D_t = \frac{\varepsilon \sum_{p=1}^{n} \mathrm{E}_p}{2E_t}, \tag{28}$$

where D_t (km) is the maximum economic transportation distance, and ε is the ratio of the energy embedded in the final products that can be used for transportation of the feedstock to keep the process economically viable. Here we assume that the transportation vessel makes only one direction with cargo and is empty on its way back.

To exemplify the estimation of the transportation energy constrains of the off-shore cultivation, following [61], we assumed that the transportation will be done with Aframax ship tanker. The tanker capacity is 100 000 tons and the average fuel consumption (between full and empty cargo) is 25.4 gal km^{-1} (4019.55 MJ km^{-1}) of heavy ship oil. Previous extensive studies in the bioethanol industry showed that for profitability and positive net energy balance, the energetic cost of transportation should be at ~1.8% of the total energy embedded in the final products, distributed equality between biomass transportation and final products distribution [62]. Therefore, we constrained the total energy expenditures (ε) on transportation on 0.9% of the energy embedded in the potential products of the transported macroalgae biomass. The recent work on offshore cultivation analysis of *Ulva* species showed the D_t of the farm varies from 115 to 690km, depending on the moisture content of the macroalgae during transportation.

5 Environmental Risks and Sustainability

Large-scale macroalgae cultivation can be responsible for positive and negative impact on coastal and marine ecosystems [63]. Therefore, the balance is necessary to attained in between food, chemicals and fuel production and its environmental cost [64]. Although scale-up reduces production costs of macroalgae, the offshore cultivation is challenging because of the harsh environment and also could possess risks to the environment. Risk management framework should be developed for each individual case to address these factors.

The overall risk management framework is generally used for the decision-making process and provides more clear idea to make decision about any technological term; includes and comprises the defining the challenge or problem, stakeholders involved, consideration of almost possible concerns, identification of actual risk, review and judgment, and finally the decision. The proposed framework of risk management for offshore macroalgae cultivation is shown

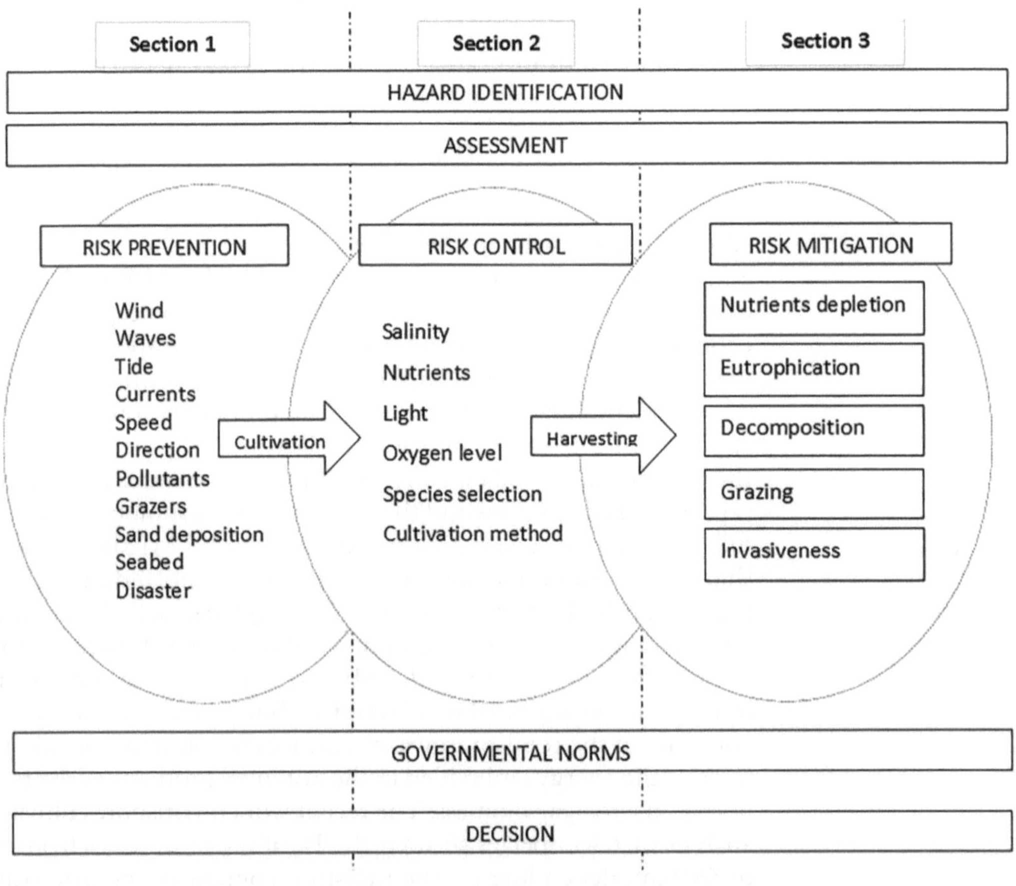

Fig. 5 Entire framework of the risk management for offshore macroalgal cultivation. Figure inspired by [88]

in Fig. 5. The framework is divided into three sections as follows. Section 1 shows the possible risks, which can be prevented before the cultivation or during the cultivation and are related with the requirement of macroalgae cultivation. Section 2 shows the risks that might be controlled in the process production of biomass and harvesting. Section 3 shows the risks that will need to be mitigated, as these are the potential impacts of cultivation of macroalgae on the marine environment.

6 Strategies for Improving the Total Efficiency of Marine Biorefineries

6.1 Smart Mixing Systems to Improve Total Productivity

One of the limiting factors in the macroalgae cultivation in the food, chemicals and fuels value chain is the biomass yields. One of the limitations of the biomass yield is the ability of photosynthetic receptors to capture photons and regenerate, a process known as light/dark reactions [65]. Plants have evolved the ability to harvest almost 100% of the arriving protons; however, the total photosynthetic efficiency is only around 5% due to the low utilization of photons. In the normal illumination, at the out-door environment, the rate of photosynthesis is not limited by a number of photons. Multiple studies have shown that the rate of photosynthesis is limited by multiple physiological processes such as diffusion, plant metabolism, carbon fixating reaction metabolism [66–70]. Studies have shown that equal photosynthetic rates can be achieved by continuous and pulsed light (with specific frequencies) [71]. This property has been widely used in the design of the on-shore photobioreactors, when mixing was used to intensify the total microalgae and macroalgae yields [72, 73]. In this processes, however, mixing is usually used to improve the nutrients diffusion and provide optimum aeration to the plants. In such mixing regimes, most of the energy is wasted on the friction with walls.

A new concept for an offshore macroalgae cultivation system in which an increase of yields per area could be achieved by external mixing, adapted to the natural photon capture: carbon fixing rates ratio was introduced. In this mode, "smart" mixing enables utilization of the volume of the floating off-shore macroalgae reactor, by exposing the macroalgae cells to the solar energy for a short time to capture photons and then to take the algae to the depth for a period of time required for carbon fixation, when a new layer of algae is exposed to the sun. The ultimate goal of this system is to increase the total energy efficiency of the biorefinery to capturing increasing the photon utilization yields per area of installed macroalgae offshore farm.

In the offshore environment, pulsed light effect can be achieved by macroalgae mixing when the thallus (with the characteristic dimension l) is exposed to the light only for a limited time (t_{light}) to capture photons. The rest of the time in the cycle (t_{dark}) the thallus can be in the shadow to complete the carbon fixing reaction and recovering the photosynthesis center absorbent capacity. The schematic representation of the simplest macroalgae mixing system (in 2 dimension) in the off-shore condition is shown in Fig. 6. Different from the on-shore tubular and flat photobioreactors, which can absorb light from different directions, the photon flux in the off-shore conditions comes only from the water surface. The length of the path the single thallus is making under the ideal mixing conditions is:

$$L = v(t_{dark} + t_{light})/l, \tag{29}$$

where L is the path length, l is the effective size of a single thallus, v is the water speed, t_{dark} is the time of light independent reaction (carbon fixation, Photosynthetic System II regeneration), and t_{light} is the maximum time required for receptors to capture maximum amount of photons [74]. The total number of thalli than can be circulated in a path is k, where $k = L/l$. If the total available energy embedded in the single thallus of a macroalgae is E_s, the total energy embedded in the circulating algae, if mixed is $k \cdot E_s$.

The total gain of energy in this system, in other words, the total reduction of exergy during photosyntheses (e_{en_p}) is the difference in the exergy acquired through the increase of the number of macroalgae grown in the same areas with the exergy destroyed in mixing. Equation (30) is a key equation for calculating the total system energy gain in all types of photobioreactors with mixing (in the two-dimensional problem):

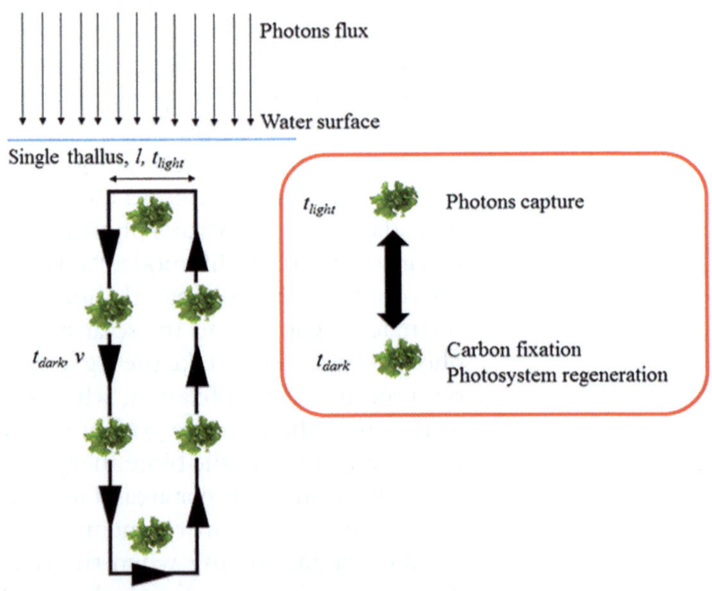

Fig. 6 Macroalgae flow path under ideal mixing conditions. Figure adapted from [89] with permission

$$-\Delta e_{0\_mix} = (k-1)E_s - E_{mix}, \tag{30}$$

where $\Delta e_{0_{mix}}$ is the net exergy gain per characteristic dimension l.

The goal of the next section is to show the example of the optimization calculation when mixing is done by pushing the macroalgae through the water bulk. Alternative mixing systems, such as airlift, which push the whole volume of water with the macroalgae, are possible, and the calculations of E_{mix} for them will be different, though following the same principle.

For the case of mechanical mixing, when the macroalgae biomass is pushed through the bulk of water and assuming that $k>>1$ and $t_{dark}>>t_{light}$, the net exergy gain with mixing is described by Eq. (31):

$$-\Delta e_{0\_\text{mix}} = \frac{vt_{dark}}{l}E_s - c_D\rho v^3 t_{dark}\, l^2 \cdot \frac{vt_{dark}}{l}, \tag{31}$$

where ρ is the density of macroalgae, v is the speed of macroalgae movement in the water, and c_D is a drag coefficient of macroalgae against the water. The drag force on macroalgae is defined through the drag coefficient, $F_D = c_D\rho v^2$, and the required power for mixing is $F_D v$. Equation (32) shows the objective function of the exergy losses during cultivation which should be minimized:

$$-\Delta e_{0\_mix} = \frac{vt_{dark}}{l}E_s - c_D\rho v^4 t_{dark}^2\, l. \tag{32}$$

The parameter under control in this type of mixing system is a macroalgae movement velocity. Thus, for minimization of the energy, we differentiate the exergy losses during cultivation by the macroalgae rotation velocity and derive the optimum mixing velocity:

$$v_{opt} = \sqrt[3]{\frac{E_s}{4c_D\rho t_{dark} l^2}}.$$

And thus, the maximum possible exergy gain (in joules) in this type of mixing system is:

$$-\Delta e_{0\_mix\_max} = \frac{\sqrt[3]{\frac{E_s}{4c_D\rho t_{dark}\, l^2}} \cdot t_{dark} \cdot E_s}{l} - c_D\rho t_{dark}^2\, l^3 \sqrt{\frac{E_s}{4c_D\rho t_{dark}\, l^2}}^{\,4} \tag{33}$$

$$-\Delta e_{0\_mix\_max} = 0.47\frac{E_s^{\frac{4}{3}} t_{dark}^{\frac{2}{3}}}{(c_D\rho)^{\frac{1}{3}} l^{\frac{5}{3}}}.$$

Therefore, the impact of mixing on the total exergy efficiency improvement of the biorefinery with a cultivation territory of a characteristic size D (this area includes $(D/l)^2$ units of floating on the surface macroalgae thalli):

$$\eta_{max\_mix} = 1 - \frac{e_{en} + e_w + e_{eco\_c} - e_{0\_mix\_max} \frac{(D)^2}{l^2}}{e_{s\_e} + e_{m\_e} + e_m + e_k + e_l + e_{eco}}. \quad (34)$$

Thus, using mixing approaches adapted for algae biology, it would be possible to increase the system's yields.

6.2 Computational Modeling of Fermentation Processes

Biomass conversion is a critical step in biorefineries. In the biological conversion pathway, once the set of feedstocks possibly available for fermentation is identified, biorefinery designers need to select the organisms for the fermentation process. Natural feedstock biomass is composed of various different molecules, such as monosugars (glucose, galactose, rhamnose, xylose, etc.), amino acids (valine, histidine, lysine, etc.), fatty acids (myristic, oleic, palmitic, etc.), fibers (cellulose, hemicellulose, lignin, ulvan, etc.), and others. Thus, selection of the fermenting organism is not a trivial issue, since most of the organisms cannot metabolize part of the existing biomass components, leading to a significant amount of residual media and low fermentation efficiency. For example, wild-type *Saccharomyces cerevisiae*, which is a first-choice organism for bioethanol production, poorly utilizes carbohydrates such as *xylose, rhamnose,* and *galactose*. One approach to overcoming this deficiency and thereby improving the bioethanol yields is to genetically modify *S. cerevisiae* to improve sugar uptake mechanisms. Studies in this direction are undertaken for several years, but successful implementation remains an open challenge [75]. Another approach is to induce or increase the required functionality in the organism, leading to broader digestion ability. For example, Bond-Watts et al. [76] proposed different plasmid inserts into *E. coli* to introduce butanol-producing pathways. However, broad digestion ability of some single organisms reduces the total yields of desired products. For example, *E. coli* is less efficient in the production of ethanol from glucose than *S. cerevisiae* [77, 78].

Fermentation by bacterial communities is a natural alternative to genetic modifications of selected organisms. Community members can be selected to naturally digest the broader range of existing biomass compounds and further convert them to desired products. However, fermentation by communities has some serious drawbacks, such as the need for expertise in growing of several organisms, and the understanding of inter-organism interactions and of competition for resources. Mathematical modeling of the community-based fermentation process is also further complicated, since the natural inter-organism interactions are still poorly understood [79] (relatively to intra-organism metabolism) and since mapping of metabolites between models representing individual organisms is not trivial [80, 81]. There are several approaches for mathematical modeling of community behavior. For example, OptCom [79] is a methodology that proposes a computational framework to describe different intra-species interactions grown together. This framework aims to describe trade-offs between individual vs. community-level fitness criteria. Similarly, cFBA [82] is a method that integrates inter-species interactions to achieve the maximal growth rate of the entire bacterial community, while SUMEX [83] performs the same task by maximizing the total molar output exchange minus input exchange of metabolites. Another approach [84] tests the fermentation efficiency of co-culturing species that require different media compounds.

One of the possible approaches to allow for better and predictable control of the inter-organism relations in communities stems from the idea of serial fermentation. In this approach, the possible benefits of inter-organism interactions, as demonstrated by the above methods, are traded off for increased control. Briefly, this means constructing a process in which each organism is grown separately, and then residual media together with resulted grown and decomposed biomass are transferred to the next organism in the chain. This approach has an advantage in the context of broader biomass components utilization allowing for tighter process control. In addition, serial fermentation removes the demand for community growing expertise (only single organism growing expertise is required), increases the process flexibility (the modification of one fermenting organism will have low impact on the inter-organism interactions), and simplifies the mathematical process modeling. The simplification of the mathematical modeling is crucial in the fermentation design stage, since estimation of expected system efficiency prior to its implementation by simulation will reduce the number of experiments required for process optimization and will provide a new tool to support efficient process design.

BioLEGO [85] web service is one example of a computational modeling framework that enables modeling and evaluation of the expected performance of single- and two-step fermentation processes. It implements a flexible modular modeling approach enabling smooth generation of different fermentation configurations consisting of independent encapsulated modules, representing individual organisms. Each such module contains an encapsulated metabolic model of an individual organism constructed by experts based on experiments, literature, and automatic reconstruction tools, like MIRAGE [86]. In addition, the BioLEGO web service enables assessment of possible product amounts that can be achieved from specific biomass, mix of biomasses or general media according to various fermentation configurations. Moreover, it provides recommendations for media content improvement according to biomass component gradient analysis of estimated production rates.

To demonstrate the usability of the BioLEGO approach, the efficiency of a two-step fermentation process for a 2:1 mix of *U. lactuca* and *K. alvarezii* algal biomasses by *E. coli* and wild-type *S. cerevisiae* was evaluated. Both possible directions, i.e., *E. coli* followed by yeast and vice versa, were tested. In addition, the efficiency of two-step fermentation process to the single-step fermentation by each organism separately and by the organism co-culture were compared. The input setup was the choice of two default media—"*U. lactuca*" and "*K. alvarezii*" and of both desired organisms. In addition, one needs to provide the ratio of the selected biomasses and check all the desired fermentation configurations (Fig. 7a).

The simulation results (Fig. 7b) predict that two-step fermentation starting with WT yeast should produce 168.4–170.8 g ethanol per 1 kg of media mix, which has a clear advantage over all other scenarios, predicting ethanol production yields of 142.4–148.1 g of ethanol for WT *S. cerevisiae* alone or around 90–100 g of ethanol for scenarios with *E.coli* as the first fermenting organism. Note that the co-culturing setup estimation yields a prediction of 90.4–97.7 g of ethanol per 1 kg of media mix. The example above clearly demonstrates the potential advantage of two-step fermentation processes over the single-step in this selected biomass composition.

a

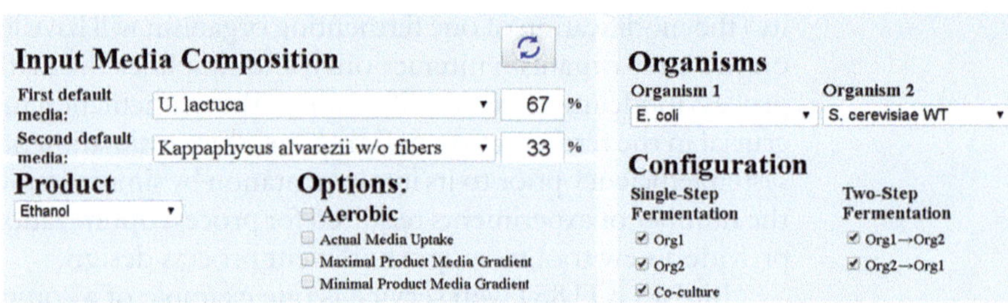

b

Configuration	*E. coli* growth [h^{-1}]	*S. cerevisiae* WT growth [h^{-1}]	Min Ethanol		Max Ethanol		Details
			Production [g/Kg]	Carbon Utilization [%]	Production [g/Kg]	Carbon Utilization [%]	
S. cerevisiae WT	---	0.0469	142.3676	20.7%	148.0627	21.6%	>>>
E. coli	0.1016	---	95.5066	13.9%	95.5066	13.9%	>>>
E. coli with *S. cerevisiae* WT	0.1081	0	90.3788	13.2%	97.6970	14.2%	>>>
S. cerevisiae WT ⇒ *E. coli*	0.0501	0.0469	168.3556	24.5%	170.8098	24.9%	>>>
E. coli ⇒ *S. cerevisiae* WT	0.1016	0.0001	95.5066	13.9%	99.8183	14.5%	>>>

Fig. 7 BioLEGO example. We did the evaluation of the efficiency of a two-step fermentation process for a 2:1 mix of *U. lactuca* together with *K. alvarezii* biomasses by *E. coli* and WT *S. cerevisiae*. (**a**) Simulation input configuration; (**b**) Simulation output summary. Figure adapted from [85] with permission

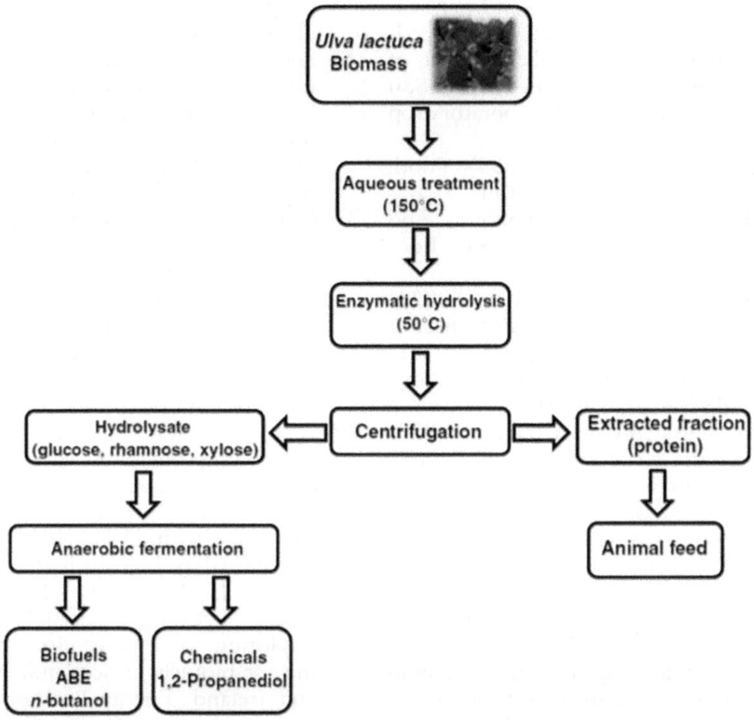

Fig. 8 Example of a multiproduct macroalgae *Ulva* based biorefinery. Figure adapted from [87] with permission

6.3 Co-production of Proteins, Platform Chemicals, and Biofuels

One of the important strategies to increase the exergy efficiency of marine biorefineries is to volatilize the whole biomass. As macroalgae biomass is composed of carbohydrates, proteins, fibers, lipids, multiple secondary metabolites, and inorganic molecules, technologies for separation of these chemicals will generate biorefineries with multiple product streams. A recent work showed the possibility to generate protein and sugar mix as two separate fractions [87]. In a cascade biorefinery approach, these materials were further used for animal feed and fermentation of acetone, ethanol, butanol, and 1,2-propane diol, all coming from the same initial material (Fig. 8). Producing multiple products reduces the biomass waste and thus increases exergy delivered to users in products.

References

1. Bentsen NS, Felby C (2012) Biomass for energy in the European Union – a review of bioenergy resource assessments. Biotechnol Biofuels 5:25. https://doi.org/10.1186/1754-6834-5-25
2. Gerbens-Leenes W, Hoekstra AY, van der Meer TH (2009) The water footprint of bioenergy. Proc Natl Acad Sci U S A 106:10219–10223. https://doi.org/10.1073/pnas.0812619106
3. Greenwell HC, Laurens LML, Shields RJ et al (2010) Placing microalgae on the biofuels priority list: a review of the technological challenges. J R Soc Interface 7:703–726. https://doi.org/10.1098/rsif.2009.0322

4. Roesijadi G, Jones SBB, Snowden-Swan LJ, Zhu Y (2010) Macroalgae as a biomass feedstock: a preliminary analysis. Department of Energy under contract DE-AC05-76RL01830 by Pacific Northwest National Laboratory, pp 1–50
5. Pimentel M, Pimentel MH (2008) Food, energy, and society. CRC, Boca Raton
6. Pimentel D (2012) Global economic and environmental aspects of biofuels. CRC, Boca Raton
7. Hofmann LC, Strauss S, Shpigel M et al (2024) The green seaweed Ulva: tomorrow's "wheat of the sea" in foods, feeds, nutrition, and biomaterials. Crit Rev Food Sci Nutr. https://doi.org/10.1080/10408398.2024.2370489
8. Polikovsky M, Gillis A, Steinbruch E et al (2020) Biorefinery for the co-production of protein, hydrochar and additional co-products from a green seaweed Ulva sp. with subcritical water hydrolysis. Energy Convers Manag https://doi.org/10.1016/j.enconman.2020.113380
9. Kazir M, Abuhassira Y, Robin A et al (2019) Extraction of proteins from two marine macroalgae, Ulva sp. and Gracilaria sp., for food application, and evaluating digestibility, amino acid composition and antioxidant properties of the protein concentrates. Food Hydrocoll 87:194–203. https://doi.org/10.1016/j.foodhyd.2018.07.047
10. Polikovsky M, Fernand F, Sack M et al (2016) Towards marine biorefineries: selective proteins extractions from marine macroalgae Ulva with pulsed electric fields. Innov Food Sci Emerg Technol 37:194–200. https://doi.org/10.1016/j.ifset.2016.03.013
11. Steinbruch E, Kashyap M, Chemodanov A et al (2024) Continuous pulsed electric field processing for intensification of aqueous extraction of protein from fresh green seaweed Ulva sp. biomass. Food Hydrocoll 157:110477. https://doi.org/10.1016/J.FOODHYD.2024.110477
12. Prabhu MS, Levkov K, Livney YD et al (2019) High-voltage pulsed electric field preprocessing enhances extraction of starch, proteins, and ash from marine macroalgae Ulva ohnoi. ACS Sustain Chem Eng 7:17453–17463. https://doi.org/10.1021/acssuschemeng.9b04669
13. Ghosh S, Coons J, Yeager C et al (2022) Halophyte biorefinery for polyhydroxyalkanoates production from Ulva sp. hydrolysate with Haloferax mediterranei in pneumatically agitated bioreactors and ultrasound harvesting. Bioresour Technol 344:125964. https://doi.org/10.1016/J.BIORTECH.2021.125964
14. Gnaim R, Polikovsky M, Unis R et al (2021) Marine bacteria associated with the green seaweed Ulva sp. for the production of polyhydroxyalkanoates. Bioresour Technol 328:124815. https://doi.org/10.1016/j.biortech.2021.124815
15. Steinbruch E, Drabik D, Epstein M et al (2020) Hydrothermal processing of a green seaweed Ulva sp. for the production of monosaccharides, polyhydroxyalkanoates, and hydrochar. Bioresour Technol 318:124263. https://doi.org/10.1016/j.biortech.2020.124263
16. Ghosh S, Greiserman S, Chemodanov A et al (2021) Polyhydroxyalkanoates and biochar from green macroalgal Ulva sp. biomass subcritical hydrolysates: process optimization and a priori economic and greenhouse emissions break-even analysis. Sci Total Environ 770:145281. https://doi.org/10.1016/j.scitotenv.2021.145281
17. van der Wal H, Sperber BLHMHM, Houweling-Tan B et al (2013) Production of acetone, butanol, and ethanol from biomass of the green seaweed Ulva lactuca. Bioresour Technol 128:431–437. https://doi.org/10.1016/j.biortech.2012.10.094
18. Rostan J, Billing SL, Doran J, Hughes A (2022) Creating a social license to operate? Exploring social perceptions of seaweed farming for biofuels in Scotland, Northern Ireland and Ireland. Energy Res Soc Sci 87:102478. https://doi.org/10.1016/j.erss.2021.102478
19. Herrmann C, FitzGerald J, O'Shea R et al (2015) Ensiling of seaweed for a seaweed biofuel industry. Bioresour Technol 196:301–313. https://doi.org/10.1016/j.biortech.2015.07.098
20. del Río PG, Gomes-Dias JS, Rocha CMR et al (2020) Recent trends on seaweed fractionation for liquid biofuels production. Bioresour Technol 299:122613. https://doi.org/10.1016/j.biortech.2019.122613
21. Lehahn Y, Ingle KN, Golberg A (2016) Global potential of offshore and shallow waters macroalgal biorefineries to provide for food, chemicals and energy: feasibility and sustainability. Algal Res 17:150–160. https://doi.org/10.1016/j.algal.2016.03.031
22. Nikolaisen L, Jensen PD, Bech KS et al (2008) Energy production from marine biomass (Ulva lactuca). PSO Project No. 2008-1-0050, pp 1–72
23. Martins I, Marques JC (2002) A model for the growth of opportunistic macroalgae (Enteromorpha sp.) in tidal estuaries. Estuar Coast Shelf Sci 55:247–257. https://doi.org/10.1006/ecss.2001.0900
24. Lee CS, Ang P (1991) A simple model for seaweed growth and optimal harvesting strategy. Ecol Model 55:67–74
25. Seip KL (1980) A computational model for growth and harvesting of the marine alga Ascophyllum nodosum. Ecol Model 8:189–199. https://doi.org/10.1016/0304-3800(80)90037-X

26. Peteiro C, Freire Ó (2012) Outplanting time and methodologies related to mariculture of the edible kelp Undaria pinnatifida in the Atlantic coast of Spain. J Appl Phycol 24:1361–1372. https://doi.org/10.1007/s10811-012-9788-2
27. Neushul M, Benson J, Harger BWW, Charters AC (1992) Macroalgal farming in the sea: water motion and nitrate uptake. J Appl Phycol 4:255–265. https://doi.org/10.1007/BF02161211
28. Hurd CL (2000) Water motion, marine macroalgal physiology, and production. J Phycol 36:453–472.https://doi.org/10.1046/j.1529-8817.2000.99139.x
29. Buck BH, Buchholz CM (2004) The offshore-ring: a new system design for the open ocean aquaculture of macroalgae. J Appl Phycol 16:355–368. https://doi.org/10.1023/B:JAPH.0000047947.96231.ea
30. Hanisak M (1987) Cultivation of Gracilaria and other macroalgae in Florida for energy production. Dev Aquacult Fish Sci 191:218
31. Buck BH, Krause G, Michler-Cieluch T et al (2008) Meeting the quest for spatial efficiency: progress and prospects of extensive aquaculture within offshore wind farms. Helgol Mar Res 62:269–281. https://doi.org/10.1007/s10152-008-0115-x
32. Jung Ka, Lim S-RR, Kim Y, Park JM (2013) Potentials of macroalgae as feedstocks for biorefinery. Bioresour Technol 135:182–190. https://doi.org/10.1016/j.biortech.2012.10.025
33. Michler-Cieluch T, Krause G, Buck BH (2009) Marine aquaculture within offshore wind farms: social aspects of multiple-use planning. GAIA 18:158–162
34. Michler-Cieluch T, Krause G, Buck BH (2009) Reflections on integrating operation and maintenance activities of offshore wind farms and mariculture. Ocean Coast Manag 52:57–68. https://doi.org/10.1016/j.ocecoaman.2008.09.008
35. Reith JH, Curversa PWM, Kamermans P et al (2005) Bio-offshore Grootschalige teelt van zeewieren in combinatie met offshore windparken in de Noordzee. pp 1–137
36. Yantovski E (2000) Exergonomics in education. Energy 25:1021–1031. https://doi.org/10.1016/S0360-5442(00)00027-X
37. Yantovsky EI (1989) Non-equilibrium thermodynamics in thermal engineering. Energy 14:393–396. https://doi.org/10.1016/0360-5442(89)90134-5
38. Golberg A (2015) Environmental exergonomics for sustainable design and analysis of energy systems. Energy. https://doi.org/10.1016/j.energy.2015.05.053
39. Szargut J (1971) Anwendung der Exergie zur angenaherten wirtschaftlichen Optimierung. Brennst Wärme Kraft 23:516–519
40. Simpson AP, Edwards CF (2011) An exergy-based framework for evaluating environmental impact. Energy 36:1442–1459. https://doi.org/10.1016/j.energy.2011.01.025
41. Hermann WA (2006) Quantifying global exergy resources. Energy 31:1349–1366. https://doi.org/10.1016/j.energy.2005.09.006
42. Sciubba E (2003) Extended exergy accounting applied to energy recovery from waste: the concept of total recycling. Energy 28:1315–1334. https://doi.org/10.1016/S0360-5442(03)00111-7
43. Sciubba E (2012) A thermodynamically correct treatment of externalities with an exergy-based numeraire. Sustainability 4:933–957. https://doi.org/10.3390/su4050933
44. Dai J, Chen B, Sciubba E (2014) Ecological accounting based on extended exergy: a sustainability perspective. Environ Sci Technol 48:9826–33. https://doi.org/10.1021/es404191v
45. Jørgensen SE (2007) An integrated ecosystem theory. Ann Eur Acad Sci 2006–2007:19–33
46. Jørgensen SE, Mejer H (1977) Ecological buffer capacity. Ecol Model 3:39–61
47. Jørgensen SE (1990) Ecosystem theory, ecological buffer capacity, uncertainty and complexity. Ecol Model 52:125–133
48. Xu FL, Dawson RW, Tao S et al (2002) System-level responses of lake ecosystems to chemical stresses using exergy and structural exergy as ecological indicators. Chemosphere 46:173–185. https://doi.org/10.1016/S0045-6535(01)00127-8
49. Zhang J, Gurkan Z, Jørgensen SE (2010) Application of eco-exergy for assessment of ecosystem health and development of structurally dynamic models. Ecol Model 221:693–702
50. Jørgensen SE (2007) An integrated ecosystem theory. Ann Eur Acad Sci 2006–2007:19–33
51. Harte J, Newman EA (2014) Maximum information entropy: a foundation for ecological theory. Trends Ecol Evolut 29:384–389. https://doi.org/10.1016/j.tree.2014.04.009
52. Dalsgaard JPT, Lightfoot C, Christensen V (1995) Towards quantification of ecological sustainability in farming systems analysis. Ecol Eng 4:181–189
53. Jørgensen SE (2015) New method to calculate the work energy of information and organisms. Ecol Model 295:18–20. https://doi.org/10.1016/j.ecolmodel.2014.09.001
54. Jørgensen SE, Ladegaard N, Debeljak M, Marques JC (2005) Calculations of exergy for organisms. Ecol Model 185:165–175. https://doi.org/10.1016/j.ecolmodel.2004.11.020

55. Svirezhev YM, Steinborn WH (2001) Exergy of solar radiation: information approach. Ecol Model 145:101–110. https://doi.org/10.1016/S0304-3800(01)00409-4
56. Hernandez RR, Easter SB, Murphy-Mariscal ML et al (2014) Environmental impacts of utility-scale solar energy. Renew Sustain Energy Rev 29:766–779
57. Crowl TA, Crist TO, Parmenter RR et al (2008) The spread of invasive species and infectious disease as drivers of ecosystem change. Front Ecol Environ 6:238–246. https://doi.org/10.1890/070151
58. French B (1960) Some considerations in estimating assembly cost functions for agricultural processing operations. J Farm Econ 42:767–778
59. Golberg A, Vitkin E, Linshiz G et al (2014) Proposed design of distributed macroalgal biorefineries: thermodynamics, bioconversion technology, and sustainability implications for developing economies. Biofuels Bioprod Biorefin 8:67–82. https://doi.org/10.1002/bbb.1438
60. Valderrama D, Cai J, Hishamunda N (2013) Social and economic dimensions of carrageenan seaweed farming. FAO fisheries and aquaculture technical paper 530. https://icsfarchives.net/16509/
61. Lenstra WJ, van Hal JW, Reith JH (2011) Ocean seaweed biomass. For large scale biofuel production. Paper presented at the Ocean Seaweed Biomass Conference, Bremerhaven, Germany, 5–7 September 2011
62. USDA (2008) Energy balance of the corn-ethanol industry http://www.usda.gov/oce/reports/energy/2008Ethanol_June_final.pdf
63. Hughes AD, Kelly MS, Black KD, Stanley MS (2012) Biogas from macroalgae: is it time to revisit the idea? Biotechnol Biofuels 5:86. https://doi.org/10.1186/1754-6834-5-86
64. Wei N, Quarterman J, Jin Y-S (2013) Marine macroalgae: an untapped resource for producing fuels and chemicals. Trends Biotechnol 31:70–77. https://doi.org/10.1016/j.tibtech.2012.10.009
65. Kok B, Forbush B, McGloin M (1970) Cooperation of charges in photosynthetic O_2 evolution-I. A linear four step mechanism. Photochem Photobiol 11:457–475
66. Stitt M, Schulze D (1994) Does Rubisco control the rate of photosynthesis and plant growth? An exercise in molecular ecophysiology. Plant Cell Environ 17:465–487. https://doi.org/10.1111/j.1365-3040.1994.tb00144.x
67. Marcus Y, Altman-Gueta H, Wolff Y, Gurevitz M (2011) Rubisco mutagenesis provides new insight into limitations on photosynthesis and growth in Synechocystis PCC6803. J Exp Bot 62:4173–4182. https://doi.org/10.1093/jxb/err116
68. Paul MJ (2001) Sink regulation of photosynthesis. J Exp Bot 52:1383–1400. https://doi.org/10.1093/jexbot/52.360.1383
69. Stitt M (1986) Limitation of photosynthesis by carbon metabolism : I. Evidence for excess electron transport capacity in leaves carrying out photosynthesis in saturating light and CO(2). Plant Physiol 81:1115–1122
70. Farquhar GD, Sharkey TD (1982) Stomatal conductance and photosynthesis. Annu Rev Plant Physiol 33:317–345. https://doi.org/10.1146/annurev.pp.33.060182.001533
71. Tennessen DJ, Bula RJ, Sharkey TD (1995) Efficiency of photosynthesis in continuous and pulsed light emitting diode irradiation. Photosynth Res 44:261–269. https://doi.org/10.1007/BF00048599
72. Bruhn A, Dahl J, Nielsen HB et al (2011) Bioenergy potential of Ulva lactuca: biomass yield, methane production and combustion. Bioresour Technol 102:2595–2604. https://doi.org/10.1016/j.biortech.2010.10.010
73. Sforza E, Simionato D, Giacometti GM, et al (2012) Adjusted light and dark cycles can optimize photosynthetic efficiency in algae growing in photobioreactors. PLoS One 7:e38975. https://doi.org/10.1371/journal.pone.0038975
74. Carvalho AP, Silva SO, Baptista JM, Malcata FX (2011) Light requirements in microalgal photobioreactors: an overview of biophotonic aspects. Appl Microbiol Biotechnol 89:1275–1288. https://doi.org/10.1007/s00253-010-3047-8
75. van Maris AJA, Abbott DA, Bellissimi E et al (2006) Alcoholic fermentation of carbon sources in biomass hydrolysates by Saccharomyces cerevisiae: current status. Antonie van Leeuwenhoek 90:391–418
76. Bond-Watts BB, Bellerose RJ, Chang MCY (2011) Enzyme mechanism as a kinetic control element for designing synthetic biofuel pathways. Nat Chem Biol 7:222–227. https://doi.org/10.1038/nchembio.537
77. Kim Y, Ingram LO, Shanmugam KT (2007) Construction of an Escherichia coli K-12 mutant for homoethanologenic fermentation of glucose or xylose without foreign genes. Appl Environ Microbiol 73:1766–1771. https://doi.org/10.1128/AEM.02456-06
78. Talebnia F, Niklasson C, Taherzadeh MJ (2005) Ethanol production from glucose and dilute-acid hydrolyzates by encapsulated S. cerevisiae. Biotechnol Bioeng 90:345–353. https://doi.org/10.1002/bit.20432
79. Zomorrodi AR, Maranas CD (2012) OptCom: a multi-level optimization framework for the metabolic modeling and analysis of microbial communities. PLoS Comput Biol 8:e1002363. https://doi.org/10.1371/journal.pcbi.1002363

80. Durinck S, Spellman PT, Birney E, Huber W (2009) Mapping identifiers for the integration of genomic datasets with the R/Bioconductor package biomaRt. Nat Protoc 4:1184–1191. https://doi.org/10.1038/nprot.2009.97
81. van Iersel MP, Pico AR, Kelder T et al (2010) The BridgeDb framework: standardized access to gene, protein and metabolite identifier mapping services. BMC Bioinformat 11:5. https://doi.org/10.1186/1471-2105-11-5
82. Khandelwal RA, Olivier BG, Röling WFM et al (2013) Community flux balance analysis for microbial consortia at balanced growth. PLoS One https://doi.org/10.1371/journal.pone.0064567
83. Zarecki R, Oberhardt MA, Yizhak K et al (2014) Maximal sum of metabolic exchange fluxes outperforms biomass yield as a predictor of growth rate of microorganisms. PLoS One https://doi.org/10.1371/journal.pone.0098372
84. Hanly TJ, Henson MA (2011) Dynamic flux balance modeling of microbial co-cultures for efficient batch fermentation of glucose and xylose mixtures. Biotechnol Bioeng 108:376–385. https://doi.org/10.1002/bit.22954
85. Vitkin E, Golberg A, Yakhini Z (2015) BioLEGO — a web-based application for biorefinery design and evaluation of serial biomass fermentation. Techology 3:1–10. https://doi.org/10.1142/S2339547815400038
86. Vitkin E, Shlomi T (2012) MIRAGE: a functional genomics-based approach for metabolic network model reconstruction and its application to cyanobacteria networks. Genome Biol 13, R111. https://doi.org/10.1186/gb-2012-13-11-r111
87. Bikker P, van Krimpen MM, van Wikselaar P et al (2016) Biorefinery of the green seaweed Ulva lactuca to produce animal feed, chemicals and biofuels. J Appl Phycol 28:3511–3525. https://doi.org/10.1007/s10811-016-0842-3
88. van der Burg S, Stulver M, Veenstra F et al (2013) A triple P review of the feasibility of sustainable offshore seaweed production in the North Sea. LEI Report 13-077 ISBN 978-90-8615-652-8
89. Golberg A, Liberzon A (2015) Modeling of smart mixing regimes to improve marine biorefinery productivity and energy efficiency. Algal Res 11:28–32. https://doi.org/10.1016/j.algal.2015.05.021

Check for updates

Chapter 29

Metabolic Engineering of Microalgae for Biofuel Production

Mohammad Pooya Naghshbandi, Meisam Tabatabaei, Mortaza Aghbashlo, Muhammad Nauman Aftab and Irfana Iqbal

Abstract

Microalgae are considered promising cell factories to produce various types of biofuels, including bioethanol, biodiesel, and biohydrogen, by using carbon dioxide and sunlight. Despite the unique advantages of these microorganisms, the commercialization of microalgal biofuels has been hindered by unfavorable economic factors. Metabolic engineering is among the most promising strategies put forth to overcome this challenge. In this chapter, metabolic pathways involved in lipid and hydrogen production by microalgae are reviewed and discussed. Moreover, metabolic and genetic engineering approaches investigated for improving the rate of lipid production (as a feedstock for biodiesel production) and biohydrogen synthesis are presented. Finally, genetic engineering tools and approaches employed for engineering microalgal metabolic pathways are described in detail. A thorough step-by-step protocol for reconstructing the metabolic pathways of various microorganisms, including microalgae, is also presented.

Keywords Microalgae, Biodiesel, Biohydrogen, Genetic engineering, Metabolic engineering

1 Introduction

Microalgae are photosynthetic cell factories and have been the center of research activities aimed at producing a wide array of bioproducts, including biofuels [1]. Despite the fact that the economic feasibility of algal fuels has been seriously questioned recently [2, 3], the unique features of these organisms—such as fast growth rate, high lipid and/or carbohydrate content, ability to grow in unconventional water resources, and capability for photobiological production of biohydrogen—remain motivating [4, 5]. In addition, it is generally believed that algal fuels could be regarded as a solution to growing environmental concerns, i.e., global warming and climate change, mainly caused by the widespread combustion of fossil fuels. It has been estimated that the combustion of biofuels of microalgal origin could potentially mitigate CO_2, SO_2, and particulate matter emissions by 78%, 98%, and 50%, respectively [6, 7]. It should also be mentioned that microalgae offer another environmental benefit due

Kristian Spilling (ed.), *Biofuels from Algae*, Methods in Molecular Biology, vol. 2996,
https://doi.org/10.1007/978-1-0716-5031-8_29,

to their capability to fix large amounts of atmospheric CO_2 (i.e., 183 tonnes of CO_2 per 100 tonnes of microalgal biomass) [8]. Therefore, it can be deduced that biofuel production utilizing microalgae has multidimensional advantages over the use of conventional energy crops. Some of these advantageous features are presented herein as follows [9, 10]:

1. Limited land requirements due to the high growth rate of microalgae
2. Lower water requirements
3. No need for herbicides or pesticides
4. Higher efficiency in CO_2 sequestration (greater tolerance to high CO_2 content in gas streams)
5. Greater feedstock sustainability, as microalgae can be harvested in batches nearly year-round

Despite the above-mentioned advantages, the commercialization of algal fuels has been hindered by their comparatively higher production costs than their counterparts produced from conventional energy crops, i.e., first-generation biofuels. These high production costs can be attributed to a number of technical barriers, as outlined below, which are yet to be fully addressed by the research community [4, 11]:

1. High cost of downstream processing
2. Highly variable large-scale outdoor cultivation conditions and the resulting inconsistency in algal biomass production
3. Contamination by other microorganisms (wild algae and bacteria) under large-scale production conditions
4. Limited light penetration in dense microalgal cultivation systems
5. Expensive oil/carbohydrate extraction techniques
6. Low biomass recovery rates
7. Potential growth inhibition caused by a combination of high dissolved oxygen and intense sunlight.

A number of strategies are outlined below that could be considered to partially address some of the above-mentioned technical barriers [12, 13]:

1. Designing advanced photobioreactors with features such as high light penetration efficiency
2. Implementing a biorefinery approach aimed at generating multiple products to enhance the economic viability of the entire system
3. Selecting and developing high-yield, oxygen-tolerant, oil-rich microalgae species through genetic and metabolic engineering methods

4. Improving high- and/or low-temperature tolerance in oil-rich microalgae using genetic and metabolic engineering techniques
5. Developing low-cost technologies for biomass harvesting, drying, and oil extraction.

As highlighted in recent research, the selection, modification, and cultivation methods of microalgae are critical for optimizing biofuel production. Notably, the use of highly adaptable strains that can uptake high concentrations of inorganic carbon sources and exhibit stress resistance is essential for large-scale production. Furthermore, due to the controllable cultivation environment they offer, the use of photobioreactors as the primary approach is increasing, which is a key factor in improving production. Adjusting photobioreactor performance and parameters to enhance lipid accumulation is also critically important [14].

Among the above-mentioned strategies, metabolic engineering combined with genetic engineering can be regarded as highly promising, as the host metabolic system can be extensively manipulated to achieve several desirable attributes simultaneously. Examples of such attributes include improved production yield, reduced formation of unfavorable by-products, the ability to utilize a wide spectrum of carbon sources, and enhanced resistance to environmental conditions (such as high/low temperatures, broad pH ranges, and toxic substances) [15].

The recent advancements in synthetic biology, omics technologies, and gene editing are powerful tools that have reduced the risks associated with genetically modified microalgae while also improving the economic viability of biofuel production. Optimizing metabolic flux and/or the overexpression or deletion of genes encoding specific enzymes—or regulating their expression rates—are the main methods used to manipulate the biochemical pathways of microalgae through metabolic engineering. It should be highlighted that it has also become technically feasible to introduce entire metabolic pathways from one microorganism into a superior host organism through synthetic biology [6, 16, 17].

Overall, such efforts aim to reconstruct or re-engineer metabolic pathways to develop algal cells with favorable characteristics [18]. To ensure the successful construction of an effective biochemical pathway and the generation of desired products, selecting an appropriate algal host and targeting the right pathway are critical. In addition, achieving stable integration of heterologous DNA fragments into the algal host is essential. The complex nature of metabolic pathways—due to the involvement of many genes and enzymes—and the occurrence of unfavorable feedback responses, such as feedback inhibition, have made the construction or reconstruction of pathways capable of efficient biofuel production a challenging task [6, 19, 20].

2 Metabolic Engineering for Biodiesel Production by Microalgae

Like canola and sunflower, microalgae can also produce triglycerides. The oil content of microalgae can reach as high as 30–70% of their dry biomass under suitable conditions, which has been a major source of interest for biodiesel production [17, 21, 22]. Metabolic engineering is the primary approach for producing economically viable biodiesel from microalgae by augmenting lipid accumulation—a critical factor influencing the overall feasibility of algal-based biofuels. However, the main challenge lies in optimizing lipid yield without compromising cellular growth, as highlighted in a recent review [23].

The lipid metabolism pathway is crucial for fuel extraction from microalgae and is considered highly complex and tightly regulated. This pathway also varies across different microalgae species due to metabolic divergence among phylogenetic groups [1, 18]. Moreover, an elaborate regulatory network of nodes and internodes further complicates the task of manipulating this pathway [1, 18]. Therefore, accurate identification of specific pathways within a given phylogenetic group of algae is essential for improving lipid production yield through metabolic and genetic engineering [24, 25].

The main steps of lipid biogenesis in microalgae are: (a) the carboxylation of acetyl-CoA to malonyl-CoA by acetyl-CoA carboxylase (ACCase), which is regarded as the first committed step in fatty acid biosynthesis; (b) fatty acid production through a series of reactions catalyzed by type II fatty acid synthase (FAS); and finally, (c) triacylglycerol (TAG) formation [26, 27]. Figure 1 shows the metabolic pathway for lipid production in microalgae, with the rate-limiting steps highlighted by asterisks.

As mentioned earlier, the FAS complex and ACCase are considered key conserved (rate-limiting) enzymes in the lipid biosynthesis pathway that occurs in the chloroplast (Fig. 1) [28, 29]. In light of this, the majority of efforts have focused on the overexpression of the gene encoding ACCase to enhance fatty acid production [30]. On the other hand, attempts have also been made to suppress enzymes involved in fatty acid oxidation by knocking out or silencing their respective genes—namely, acyl-CoA dehydrogenase, acyl-CoA synthase, acyl-CoA oxidase, and carnitine acyltransferase I [31, 32].

The diatom *Cyclotella cryptica* was the first alga genetically engineered to improve fatty acid production through overexpression of the ACCase gene. However, despite a two- to threefold increase in ACCase activity, fatty acid accumulation remained unchanged [33]. Subsequent attempts to increase lipid content in various microalgal systems by overexpressing ACCase were also somewhat disappointing [34, 35]. These results may suggest that ACCase activity is not as critical as initially believed—or, in other words, the reaction catalyzed by this enzyme might not be a true rate-limiting step in the microalgal fatty acid production pathway. Such findings underscore the need

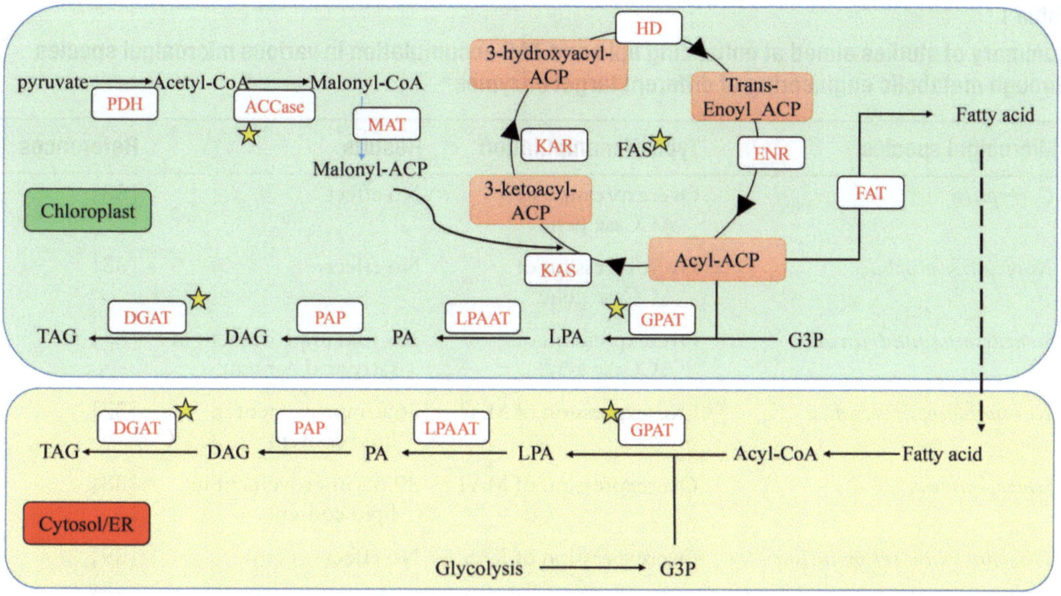

Fig. 1 Overview of lipid biosynthesis pathways in microalgae. Enzymes are shown in boxes with red annotations, while metabolites and pathways are shown in black. Yellow asterisks indicate the rate-limiting enzymes/steps. Free fatty acids are produced in the chloroplast, while TAGs may be synthesized either in the endoplasmic reticulum via the Kennedy pathway or in the chloroplast. Abbreviations: ACCase, acetyl-CoA carboxylase; ACP, acyl carrier protein; CoA, coenzyme A; DAGAT, diacylglycerol acyltransferase; ENR, enoyl-ACP reductase; FAT, fatty acyl-ACP thioesterase; GPAT, glycerol-3-phosphate acyltransferase; HD, 3-hydroxyacyl-ACP dehydratase; KAR, 3-ketoacyl-ACP reductase; KAS, 3-ketoacyl-ACP synthase; LPAAT, lysophosphatidic acid acyltransferase; LPAT, lysophosphatidylcholine acyltransferase; MAT, malonyl-CoA: ACP transacylase; PDH, pyruvate dehydrogenase; TAG, triacylglycerols

for further research to more thoroughly investigate the various roles and regulatory aspects of this enzyme.

Malonyl-CoA ACP transacylase (MAT) has also been recognized as a promising enzyme to target for manipulation of the fatty acid synthesis pathway. This enzyme catalyzes the formation of malonyl-ACP by transferring a malonyl group from malonyl-CoA to the acyl carrier protein (ACP) [36]. Some successful attempts have been made so far, through which the lipid content of different microalgal species was improved by overexpressing MAT (Table 1).

Following the formation of malonyl-ACP, this complex enters a series of reactions—collectively known as the FAS cycle—that lead to the production of fatty acids. One of the targeted enzymes in the FAS complex for achieving higher lipid production is β-ketoacyl-ACP synthase (KAS), which is the first enzyme in the FAS cycle [26]. However, overexpression of KAS in *P. tricornutum* did not result in any increase in lipid production [39]. It is also worth noting that, contrary to expectations, KAS overexpression in plants led to a 5—10% decrease in TAG content [58]. Thus, it can be deduced that

Table 1
Summary of studies aimed at enhancing lipid and TAG accumulation in various microalgal species through metabolic engineering of different target enzymes

Microalgal species	Type of manipulation	Results	References
C. cryptica	Overexpression of ACCase gene	No effect	[33]
Navicula saprophila	Overexpression of ACCase gene	No effect	[33]
Scenedesmus quadricauda	Overexpression of ACCase gene	1.6-fold improvement in fatty acid content	[35]
Nannochloropsis oceanica	Overexpression of MAT	36% improvement in lipid content	[37]
Schizochytrium	Overexpression of MAT	39.6% improvement in lipid content	[38]
Phaeodactylum tricornutum	Overexpression of KAS	No effect	[39]
Phaeodactylum triornuum	Overexpression of LPAAT	Improvement in lipid accumulation	[40]
Chlamydomonas reinhardtii	Overexpression of LPAAT	20% improvement in lipid content	[41]
P. triornuum	Expression of GPAT under the Pt211 promoter	2.6-fold improvement in lipid content	[42]
P. triornuum	Expression of GPAT	2-fold improvement in lipid content	[43]
C. reinhardtii	Overexpression of GPAT	Above 50% improvement in TAG content	[44]
P. tricornutum	Overexpression of DGAT1	35% improvement in neutral lipid content	[45]
N. oceanica	Overexpression of DGAT1	47% improvement in TAG content	[46]
C. reinhardtii	Overexpression of DGAT2	9-fold improvement in TAG content	[47]
C. reinhardtii	Overexpression of DGAT2	29-fold improvement in TAG content	[48]
Scenedesmus obliquus	Overexpression of the DGAT2 from *C. reinhardtii*	84.9% improvement in lipid content	[49]
N. oceanica	Overexpression of DGAT2	78% improvement in TAG content	[50]
N. oceanica	Overexpression of DGAT2	69% improvement in neutral lipid content	[51]

Table 1
continued

Microalgal species	Type of manipulation	Results	References
Chlorella minutissima	Overexpression of G3PDH, GPAT, LPAAT, PAP, and DGAT	2-fold improvement in neutral lipid content	[52]
Nannochloropsis salina	Recombinant expression of DGAT1	74.9% improvement in TAG content	[53]
C. reinhardtii	Overexpression of DGAT2	Increase in mRNA level (7–29.1 times) No effect on lipid accumulation	[54]
P. tricornutum	Overexpression of malic enzyme	2.5-fold improvement in total lipid content	[33]
C. minutissima	Overexpression of GPAT, LPAAT, and DGAT	2-fold improvement in TAG content	[52]
P. tricornutum	Overexpression of DGAT2	35% improvement in TAG content	[45]
C. reinhardtii	Overexpression of DGAT2-1; DGAT2-5	20% and 44% improvement in lipid content, respectively	[55]
Phaeodactyllum tricornutum	Silencing of PDK	82% improvement in neutral lipids	[56]
Thalassiosira pseudonana	Deletion of multifunctional lipase	3.3-fold improvement in total lipids	[57]

KAS may not be a suitable candidate gene for microalgal metabolic engineering aimed at improving lipid production. In fact, in addition to KAS, other subunits of the FAS complex have also been found to be challenging targets for enhancing lipid accumulation in microalgae [59]. Accordingly, future research should more thoroughly investigate the roles and mechanisms of action of these enzymes within the broader cellular network.

Apart from the fatty acid production pathway, manipulation of genes involved in the TAG production pathway (Fig. 1) has also been employed to increase lipid content in various microalgal species. TAGs are, in fact, the major storage lipids in microalgae. Two main pathways have been proposed for TAG formation: (1) acylation of diacylglycerol (DAG) *via* the acyl-CoA-dependent pathway (Kennedy pathway) and (2) acylation of DAG *via* the acyl-CoA-independent pathway (also known as the recycling of membrane lipids into TAGs). A significant number of studies have focused on improving the metabolic flux through the TAG production pathway *via* metabolic engineering—specifically, by targeting and modifying the rate-limiting steps involved [17, 26].

In the first step of the TAG production cycle, glycerol-3-phosphate is acylated into lysophosphatidate (LPA) by GPAT. Being the initial reaction in the pathway, coupled with the low activity of the catalyzing enzyme, makes this step a significant rate-limiting point. Subsequently, LPA is converted into phosphatidate by lysophosphatidic acid acyltransferase (LPAAT). Phosphatidate is then dephosphorylated to generate DAG [60, 61]. A summary of studies aimed at increasing TAG formation in various microalgal species through overexpression of GPAT or LPAAT genes is presented in Table 1.

In the final step of the TAG production pathway, TAG is synthesized by transferring an acyl group from acyl-CoA to DAG in a reaction catalyzed by diacylglycerol acyltransferase (DGAT). DGAT is also considered a rate-limiting enzyme in the Kennedy pathway, as a deficiency in TAG production has been associated with DAG accumulation [62]. Two types of DGAT are directly involved in TAG formation in microalgae: DGAT1, located in the chloroplast, and DGAT2, which functions in the Kennedy pathway and is located in the endoplasmic reticulum membrane [63].

Although overexpression of DGAT1 in *P. tricornutum* and *N. salina* has been shown to increase TAG production [45, 64], DGAT2 has been identified as a more promising enzyme. Specifically, overexpression of DGAT2 in *C. reinhardtii* and its expression in *S. obliquus* reportedly led to dramatic increases in TAG production [47, 49]. Several other studies have also reported the positive effects of DGAT2 overexpression on TAG production in microalgae (Table 1). Therefore, DGAT can be regarded as one of the most favorable targets for improving TAG production in microalgae.

Despite the favorable outcomes achieved through the overexpression or deletion of various genes in the lipid biosynthesis metabolic pathway, different strains have occasionally been reported to respond differently to the manipulation of the same gene(s). For example, DGAT overexpression did not lead to increased TAG production in *C. reinhardtii*, whereas, in contrast, it resulted in up to a 35% increase in TAG accumulation in *P. tricornutum* [45, 54]. Therefore, obtaining more in-depth information about microalgal metabolic pathways—through tools such as whole-genome sequencing—is essential for enhancing or reconstructing the respective pathways. Gaining such insights could contribute to the development of more economically viable solutions to overcome the challenges associated with microalgal biofuel production [65].

In addition to the fatty acid and TAG biosynthesis pathways and their associated genes, manipulation of certain genes in other biochemical pathways has also been reported to improve microalgal lipid content. The mitochondrial PDH complex catalyzes the transformation of pyruvate into acetyl-CoA *via* an oxidative decarboxylation reaction. As a result, increased PDH complex activity leads to greater availability of acetyl-CoA, the substrate for ACCase, thereby enhancing fatty acid production. However, the PDH complex is deactivated through phosphorylation by pyruvate dehydrogenase kinase (PDK). Therefore, suppressing PDK expression can enhance PDH complex activity and subsequently increase acetyl-CoA production.

Based on this principle, Ma et al. [56] used an antisense PDK construct in *P. tricornutum* and successfully demonstrated an increase in total neutral lipid content without altering the lipid profile of the algal cells. Furthermore, pyruvate and NADH are essential sources of reducing power for lipogenesis. The malic enzyme catalyzes the decarboxylation of malate to pyruvate, producing NADH and carbon dioxide in the process. Taking this into account, Xue et al. [66] demonstrated that total lipid accumulation in *P. tricornutum* could be increased through the overexpression of the endogenous malic enzyme gene.

3 Metabolic Engineering for Biohydrogen Production by Microalgae

Biohydrogen is a carbon-free fuel, the combustion of which results in the generation of water. Hydrogen is also the primary fuel used for electricity production in fuel cells. Microalgae produce hydrogen through at least three known pathways, which have been well reviewed in our previous report (Fig. 2) [17]. Briefly, these pathways can be generally categorized as light-dependent (biophotolysis) or light-independent (dark fermentation). The light-dependent route can be further divided into direct and indirect pathways. In the direct pathway (also known as the PSII-dependent pathway), H_2 is derived from electrons and protons generated by water splitting at photosystem II (PSII). In contrast, the indirect pathway primarily relies on electrons and protons obtained from the degradation of intracellular carbon compounds. In microalgae, these carbon compounds—such as starch—are produced through the fixation of CO_2 using light energy *via* photosynthesis.

The light-independent pathway involves anaerobic fermentation of high-molecular weight organic substrates into soluble metabolic products such as ethanol, formate, and others. In addition to these metabolites, pyruvate oxidation by the enzyme pyruvate–ferredoxin oxidoreductase leads to the formation of acetyl-CoA and CO_2. This reaction is also associated with the reduction of ferredoxin, through which electrons are transferred to hydrogenase for the catalytic production of H_2 [67, 69]. Hydrogenase is, in fact, the key enzyme throughout the hydrogen production process in microalgae. However, its active site is highly sensitive to oxygen, necessitating a bi-phasic production strategy for hydrogen generation under photosynthetic conditions [70]. It should be noted that hydrogenase activity can be externally induced by adding DCMU, a PSII electron transport chain inhibitor [71, 72].

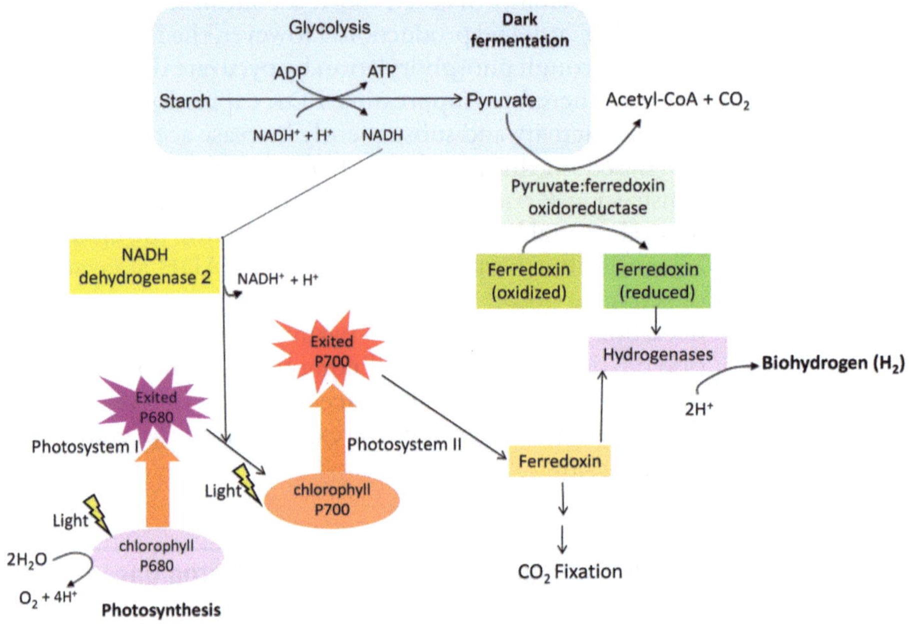

Fig. 2 Biohydrogen production pathways in microalgae. In the alga *C. reinhardtii*, at least three pathways have been identified for hydrogen production. Two of these pathways are light-dependent: (1) in the direct pathway, the PSII supplies electrons to hydrogenases *via* the photosynthetic electron transport chain, driving hydrogen production; and (2) in the indirect pathway, hydrogen is produced without the involvement of PSII, through a process dependent on photosystem I (PSI). In the third pathway, dark fermentation, glycolysis of starch leads to hydrogen production. Abbreviations: ADP, adenosine diphosphate; ATP, adenosine triphosphate; NADH, nicotinamide adenine dinucleotide (reduced form); NAD^+, nicotinamide adenine dinucleotide (oxidized form). (Reproduced with permission from Majidian et al. [17]. © 2018 Elsevier. License No.: 5995140808669)

To address the issue of hydrogenase's oxygen sensitivity, microalgae should first be grown photosynthetically to accumulate biomass, after which hydrogen production should be initiated under anoxic conditions. Achieving anoxic conditions can be facilitated by inhibiting PSII through the application of herbicides or sulfur deprivation [73, 74]. Several genetic engineering approaches have also been proposed to enhance hydrogen production in microalgae. One such strategy involves inducing anaerobic conditions, which can suppress the expression of the D2 protein—an essential component of PSII—leading to a temporary period of anaerobiosis and, consequently, increased hydrogen production. Since the expression of the D2 protein depends on a copper-responsive nuclear transgene, the addition of copper may negatively impact the production rate of this protein [75].

Other strategies include increasing hydrogenase activity and its resistance to oxygen, preventing electron diversion to competing pathways in order to boost electron flux toward the hydrogen production pathway, and enhancing the availability of electron sources

[76, 77]. For instance, Kruse et al. [78] reported a 5- to 13-fold increase in hydrogen production in an engineered *C. reinhardtii* strain (compared to the wild-type) by blocking cyclic electron transfer around Photosystem I, thereby increasing electron flow toward hydrogenases.

Furthermore, advanced metabolic engineering techniques, such as genome editing, are being employed to significantly enhance biohydrogen production in microalgae. Gene manipulation of hydrogenase and the blocking of competing metabolic processes have been shown to improve enzyme activity in both light- and dark-dependent hydrogen production pathways. In addition, combinatorial approaches that have proven effective in *E. coli* are now being explored for application in microalgae. Advanced engineering strategies aim to overcome thermodynamic barriers, enable heterologous protein expression, improve substrate conversion, and increase electron flux toward proton reduction. Dark fermentation is particularly favored for industrial applications, where deleting competing pathways—such as nitrate assimilation in *Synechocystis* sp.—has been shown to dramatically increase hydrogen production.

Moreover, overexpression of hexose transporters, such as Chlorella HUP1, or the introduction of human glucose transporters (GLUT1) into microalgae like *C. reinhardtii* has been demonstrated to improve carbohydrate uptake and, consequently, enhance hydrogen production. Metabolic models, such as AlgaGEM for *C. reinhardtii*, serve as valuable tools for predicting the effects of genetic modifications on hydrogen production. These models assist in identifying and targeting specific pathways for optimization—for example, by preventing cyclic electron flow to enhance hydrogen production through dark fermentation [79].

4 Genetic Engineering of Microalgae

Transformation in microalgae can occur in the chloroplast, nuclear, or mitochondrial genomes [80]. Although most enzymes are encoded in the nuclear genome, it should be noted that some important enzymes are transported to the chloroplast to perform their functions [81, 82]. *C. reinhardtii* has been the primary subject of genetic engineering efforts in microalgae over the past decades. The main reasons for introducing *C. reinhardtii* as a model microalga for genetic and metabolic engineering studies include (a) its well-characterized genome, (b) ease of cultivation and manipulation, and (c) its long-standing use in research [17, 83, 84].

One of the main drawbacks of nuclear transformation is the typically low expression rates of the genes of interest [1]. Additionally, another major obstacle is the presence of efficient gene-silencing microRNA systems in the nuclear genomes of some algae, which can inhibit or reverse artificial genetic modifications [85, 86].

In contrast, recombinant gene expression in the chloroplast generally results in higher protein accumulation rates [70, 87].

It should be emphasized that the success of metabolic engineering is largely dependent on advancements in genetic engineering tools such as genome sequencing, transformation techniques, gene targeting, and selection methods. These tools have significantly facilitated the design of new pathways or the modification of existing ones to enhance biofuel production in microalgae. Omics technologies also represent promising tools for achieving this goal [88, 92].

In recent years, CRISPR-Cas systems have become increasingly valuable, enabling precise and multiplexed gene editing in a variety of microalgal species. This powerful tool is accelerating the development of efficient strains for biofuel production [93]. Furthermore, the integration of multi-omics data with advanced computational modeling has deepened our understanding of microalgal metabolic networks and improved the predictive capabilities of metabolic engineering strategies. Single-cell omics technologies and advancements in synthetic biology have further empowered engineering strategies by enabling precise control over cultivation conditions and facilitating the construction of complex metabolic pathways, respectively [94, 96].

Among the available tools, transformation and selection methods are critical steps in achieving specific gene modifications in microalgae. The main techniques for microalgal transformation include (1) electroporation, (2) agitation and shaking with glass beads, (3) Agrobacterium-mediated transformation, and (4) particle bombardment (biolistics) [97, 98].

5 Methods and Protocols of Metabolic Engineering

The following protocol (in three steps) is commonly used for the metabolic engineering of various microorganisms, including microalgae [99]:

Step 1: Primary Laboratory Experiments

1. Selection of host strain and determination/formulation of nutrient requirements: As mentioned earlier, the expression and/or silencing of the same genes in different microalgal species may lead to varying results. Therefore, it is essential to select the appropriate species/strain as well as an optimal culture medium.
2. Collection of information on host features: Most wild-type strains exhibit low production yields, limited productivity, or poor resistance to environmental stressors. Hence, it is important to gather comprehensive data on the physiological and metabolic characteristics of the selected host.

3. Random mutation and selection: These experiments introduce unwanted, unidentified, and random genetic alterations, which may still contribute to strain improvement through selection.
4. Targeted metabolic engineering: Specific genetic modifications are performed to alter metabolic pathways based on prior analyses and objectives.

Step 2: Systems Metabolic Engineering

1. High-throughput and genome-scale analysis
2. Integration of omics data
3. Modeling, simulation, and network analysis
4. Prediction of engineering targets
5. Analysis of physiological performance

Step 3: Fermentation and Downstream Processing

1. Medium and process optimization
2. Product purification and recovery
3. Evaluation of product yield and productivity

In general, the highest workload and major bottlenecks during a metabolic engineering project occur in Steps 2–3, where metabolic network reconstruction is performed. This step is critical for understanding and predicting cellular behavior and guiding rational engineering strategies. The four main steps for metabolic network reconstruction are well detailed in the protocol proposed by Thiele and Palsson [100] and are summarized in Table 2.

Table 2
Protocol for metabolic pathway reconstruction in various microorganisms, including microalgae

Steps	Notes and commands
Stage 1: Creating a draft reconstruction	
Step 1: Obtain genome annotation (by sequencing centers and the National Center for Biotechnology Information (NCBI))	Genome position
	Coding region
	Strand
	Locus name
	Alias
	Gene function
	Protein classification

continued

Table 2
continued

Steps	Notes and commands
Step 2: Identify candidate metabolic functions(Use any way to collect candidate metabolic functions)	Search for E.C. numbers and metabolic terms (i.e., dehydrogenase and kinase)
Step 3: Obtain candidate metabolic reactions for these functions	Use KEGG, Brenda
Step 4: Assemble draft reconstruction	Collect all obtained data in a spreadsheet about each gene and its potential reactions.
Step 5: Collect experimental data	Literature review of all experiments
Stage 2: Manual reconstruction refinement	
Step 6: Determine and verify substrate and cofactor usage	Use KEGG, Brenda
Step 7: Obtain a neutral formula for each metabolite in the reaction	Use KEGG, Brenda, PubChem
Step 8: Determine the charged formula for each metabolite in the reaction	Use Pipeline Pilot and pKa DB software
Step 9: Calculate reaction stoichiometry	Make equal every element and the charge on each side of the equation
Step 10: Determine reaction directionality	Irreversible reactions: (1) Phosphate transfer from ATP to an accepter (exception: the ATP synthetase occurs in reverse); (2) Reactions involving quinones are generally irreversible
Step 11: Add information for gene and reaction localization (difficult to obtain)	- use algorithms: PSORT and PASUB **(Critical Step)**
Step 12: Add subsystem information to the reaction	Great help for the debugging Use the KEGG text toolbox.
Step 13: Verify gene-protein-reaction (GPR) association	**(Critical Step)**
Step 14: Add metabolite identifier	Associate each metabolite with at least one of the following identifiers: ChEBI, Kegg, and PubChem.
Step 15: Determine and add confidence score	–

continued

Table 2
continued

Steps	Notes and commands
Step 16: Flag reactions for which information from other organisms was used	–
Step 17: Add references and notes based on experimental information	Allows other users of the reconstruction to retrace the evidence easily
Step 18: Repeat Steps 6 to 17 for all genes identified in the draft reconstruction	–
Step 19: Add spontaneous reactions to the reconstruction	Use biochemical literature and databases
Step 20: Add extracellular and periplasmic transport reactions to the reconstruction	Based on experimental data
	A transport reaction should exist
Step 21: Add exchange reactions to the reconstruction	–
Step 22: Add intracellular transport reactions to the reconstruction	Only for multi-compartment reconstructions
Step 23: Draw a metabolic map (optional)	Very useful
Determine biomass composition	
Step 24: Determine the chemical composition of the cell, i.e., protein, RNA, DNA, lipids, Cofactor content	Needs experimental data or primary literature
Step 25: Determine the amino acid content	Either experimentally or by estimation using the CMR database
Step 26: Use the molar percentage and molecular weight of each amino acid to calculate the weight per mol protein	–
Step 27: Determine the nucleotide content either experimentally (option A) or by estimation (option B)	A. Determination of nucleotide content experimentally
	B. Estimation of nucleotide composition from genome information

continued

Table 2
continued

Steps	Notes and commands
Step 28: Calculate the fractional distribution of each nucleotide to the biomass composition by repeating Step 26	–
Step 29: Determine the lipid content	From fatty acids and phospholipids
Step 30: Determine the content of the soluble pool (polyamines, vitamins, and cofactors)	The soluble pool contains, for example, spermidine, coenzyme A, and folic acid
Step 31: Determine the ion content	Calculation of the molar fraction of the ions
Step 32: Determine growth-associated maintenance (GAM)	Use experimental data
Step 33: Compile and add biomass reaction to the reconstruction	All precursors are assembled in one single reaction: the biomass reaction **(Critical Step)**
Step 34: Add non-growth associated ATP maintenance reaction (NGAM)	Add the following reaction to the reconstruction reaction list: 1 ATP + 1 $H2O \rightarrow$ 1 ADP + 1 Pi + 1 H^+
Step 35: Add demand reactions to the reconstruction	For compounds that are known to be produced by the organism
Step 36: Add sink reactions to the reconstruction	**(Critical Step)**
Step 37: Determine growth medium requirements	Use experimental data
Stage 3: Conversion from reconstruction to mathematical model	
Step 38: Initialize the COBRA Toolbox	Install MATLAB, the required Toolboxes Enter this command: initCobraToolbox
Step 39: Load reconstruction into Matlab	Use:model = xls2model(RxnFileName,MetFileName)
Step 40: Verify the S matrix. Use	Use: spy(matrix) should be repeated when reactions are added to the reconstruction to ensure that they are connected to the network
Step 41: Set objective function	Use COBRA Toolbox: model = changeObjective(model, rxnNameList, objectiveCoeff) **(Critical Step)**

continued

Table 2
continued

Steps	Notes and commands
Step 42: Set simulation constraints	Use:model = changeRxnBounds(model, rxnNameList,value,boundType)
Stage 4: Network evaluation = "Debugging mode"	
Step 43: Check for stoichiometrically unbalanced reactions	Use:[UnbalancedRxns] = CheckmMassChargeBalance(model,RxnList)
Step 44: Evaluate stoichiometrically unbalanced reactions	Solve the errors
Step 45: Identify metabolic dead-ends	Use: [Gaps] = AnalyzeGaps(model)
Step 46: Identify candidate reactions to fill gaps	– Use primary literature and genome annotation: find candidate genes and reactions to fill the gap
	– Use KEGG maps and biochemical textbooks: identify the metabolic "environment" of the dead-end metabolite
Step 47: Add gap reactions to the reconstruction	**(Critical Step)**
Step 48: Add notes and references to dead-end metabolites	**(Critical Step)**
Step 49: Add missing exchange reactions to the model	Exchange reactions need to be added to the reconstruction. Repeat Step 21
Step 50: Set exchange constraints for a simulation condition	Use:model = changeRxnBounds(model,rxnNameList,value,boundType)
Test for stoichiometrically balanced cycles or Type III pathways (optional)	
Step 51: Test for Type III pathways	Use:TestForTypeIIIPathways(model,ListExch)
Step 52: Analyze output if Type III pathways are found	–
Step 53: Identify Type III pathways	–
Step 54: Analyze the directionality of each reaction participating in a Type III pathway	–

continued

Table 2
continued

Steps	Notes and commands
Step 55: Analyze if any reaction participating in a Type III pathway may be falsely included in the reconstruction by reviewing the supporting evidence	–
Step 56: If none of the reactions or reaction directions can be corrected based on experimental or thermodynamic information, you can try to limit the directionality of the loop reactions iteratively	–
Step 57: Adjust directionality for all reactions identified in Steps 54 and 55, note the change and reasons	–
Step 58: After eliminating a reaction direction or a deletion of a reaction, repeat the Type III pathway analysis	**(Critical Step)**
Step 59: Re-compute the gap list	Use: [Gaps] = AnalyzeGaps(model)
Test if biomass precursors can be produced in a standard medium (set in Step 42)	
Step 60: Obtain the list of biomass components	Use:[BiomassComponent, BiomassFraction] = PrintBiomass(model,BiomassNumber)
Step 61: Add demand function for each biomass precursor	Use:[modelNew,rxnNames] = addDemandReaction(model,metaboliteNameList)
Step 62: Change the objective function to the demand function	Use:modelNew = changeObjective function(modelNew, rxnName)
Step 63: Maximize ("max") for new objective function	Use:FBAsolution = optimizeCbModel (modelNew,'max')

continued

Table 2
continued

Steps	Notes and commands
Step 64: Identify reactions that are mainly responsible for synthesizing the biomass component	–
Step 65: For each of these reactions, follow the wire diagram given in Figure 14	–
Step 66: Test if biomass precursors can be produced in other growth media	–
Test if the model can produce known secretion products	
Step 67: Collect a list of known secretion products and medium conditions	–
Step 68: Set the constraints to the desired medium condition	Use:model = changeRxnBounds(model,rxnNameList,value,boundType)
Step 69: Change the objective function to the exchange reaction of your secretion product	Use:modelNew = AddRatioReaction(model, ListOfRxns, RatioCoeff)
Step 70: Maximize ("max") for the new objective function	Use:FBAsolution = optimizeCBModel(model,'max')
Test if the model can produce a certain ratio of two secretion products	
Step 71: Set the constraints to the desired medium condition	Use:model = changeRxnBounds(model,rxnNameList,value,boundType)
Step 72: Verify that both by-products can be produced independently	–
Step 73: Add a row to the S matrix	Use:modelNew = AddRatioReaction(model, ListOfRxns, RatioCoeff)
Step 74: Change the objective function to the exchange reaction of one of your secretion products	Use:model = changeObjective(model, rxnNameList, objectiveCoeff)

continued

Table 2
continued

Steps	Notes and commands
Step 75: Maximize for the new objective function	FBAsolution = optimizeCBModel (modelNew,'max')
Check for blocked reactions	
Step 76: Change simulation conditions to a rich medium or open all exchange reactions	Use:model = changeRxnBounds(model,rxnNameList,value,boundType)
Step 77: Run analysis for blocked reactions	Use:BlockedReactions = FindBlockedReaction(model)
Step 78: Connect the reaction to the remaining network (optional)	–
Compute single gene deletion phenotypes	
Step 79: Compute single gene deletion phenotypes	Use:[grRation,grRateKO,grRateWT] = singleGeneDeletion (model, method, geneList)
Step 80: Compare with experimental data	–
Step 81: Set simulation condition	–
Step 82: Use single reaction deletion to identify candidate reactions that enable the model's capability despite known incapability	Use: [grRation,grRateKO,grRateWT,hasEffect,delRxns,fluxSolution] = singleGeneDeletion (model, method, geneList)
Test if the model can predict the correct growth rate or other quantitative properties	
Step 83: Compare predicted physiological properties with known properties	–
Test if the model can grow fast enough	
Step 84: Optimize for biomass reaction in different medium conditions and compare with experimental data	–
Step 85: Test if any of the medium components are growth-limiting	Use:model = changeRxnBounds(model,rxnNameList,value,boundType)

continued

Table 2
continued

Steps	Notes and commands
Step 86: Maximize biomass	–
Step 87: Determine reduced cost associated with network reactions when optimizing for the objective function	Use:FBAsolution = optimizeCbModel (model,osenseStr,primalOnlyFlag))
Test if the model grows too fast	
Step 88: Optimize for biomass reaction in different medium conditions and compare with experimental data	–
Step 89: Verify that the model constraints are set as intended	Use:PrintConstraints(model,minlnf, Maxlnf)
Perform one or more of the following tests to identify possible errors in the network	
Step 90: Verify that all fractions and precursors in the biomass reaction are consistent with current knowledge	–
Step 91: Identify shuttling reactions	–
Step 92: Re-investigate the thermodynamic information associated with the network reaction	–
Step 93: Use single reaction deletion	Use:[grRatio,grRateKO,grRateWT] = singleRxnDeletion(model,method,rxnList)
Step 94: Reduced cost	Use:FBAsolution = optimizeCbModel (model,osenseStr,primalOnlyFlag)
Data assembly and Dissemination	
Step 95: Print Matlab model content	Use:eriteCBmodel(model,format, FileName) where format is xls **(Critical Step)**
Step 96: Add gap information to the reconstruction output	–

References

1. Gimpel JA, Henríquez V, Mayfield SP (2015) In metabolic engineering of eukaryotic microalgae: potential and challenges come with great diversity. Front Microbiol 6:1376. https://doi.org/10.3389/fmicb.2015.01376
2. Stephens E, Ross IL, King Z et al (2010) An economic and technical evaluation of microalgal biofuels. Nat Biotechnol 28:126–128. https://doi.org/10.1038/nbt0210-126
3. Peng K, Li J, Jiao K et al (2018) The bioeconomy of microalgal biofuels. In: Energy from microalgae. Springer, Cham, pp 157–169. https://doi.org/10.1007/978-3-319-69093-3_7
4. Cheng JJ, Timilsina GR (2011) Status and barriers of advanced biofuel technologies: a review. Renew Energy 36:3541–3549. https://doi.org/10.1016/j.renene.2011.04.031
5. Singh J, Gu S (2010) Commercialization potential of microalgae for biofuels production. Renew Sustain Energy Rev 14:2596–2610. https://doi.org/10.1016/j.rser.2010.06.014
6. Banerjee C, Dubey KK, Shukla P (2016) Metabolic engineering of microalgal based biofuel production: prospects and challenges. Front Microbiol 7:432. https://doi.org/10.3389/fmicb.2016.00432
7. Brown LM, Zeiler KG (1993) Aquatic biomass and carbon dioxide trapping. Energy Convers Manag 34:1005–1013. https://doi.org/10.1016/0196-8904(93)90048-F
8. Chisti Y (2008) Biodiesel from microalgae beats bioethanol. Trends Biotechnol 26:126–131. https://doi.org/10.1016/j.tibtech.2007.12.002
9. Schenk PM, Thomas-Hall SR, Stephens E et al (2008) Second generation biofuels: high-efficiency microalgae for biodiesel production. Bioenergy Res 1:20–43. https://doi.org/10.1007/s12155-008-9008-8
10. Greenwell HC, Laurens LML, Shields RJ et al (2010) Placing microalgae on the biofuels priority list: a review of the technological challenges. J R Soc Interface 7:703–726. https://doi.org/10.1098/rsif.2009.0322
11. Pienkos PT, Darzins AL (2009) The promise and challenges of microalgal-derived biofuels. Biofuels Bioprod Biorefin 3:431–440. https://doi.org/10.1002/bbb.159
12. De Bhowmick G, Koduru L, Sen R (2015) Metabolic pathway engineering towards enhancing microalgal lipid biosynthesis for biofuel application—a review. Renew Sustain Energy Rev 50:1239–1253. https://doi.org/10.1016/j.rser.2015.04.131
13. Ho S-H, Ye X, Hasunuma T et al (2014) Perspectives on engineering strategies for improving biofuel production from microalgae—a critical review. Biotechnol Adv 32:1448–1459. https://doi.org/10.1016/j.biotechadv.2014.09.002
14. Wang M, Ye X, Bi H et al (2024) Microalgae biofuels: illuminating the path to a sustainable future amidst challenges and opportunities. Biotechnol Biofuels Bioprod 17:10. https://doi.org/10.1186/s13068-024-02461-0
15. Zhang F, Rodriguez S, Keasling JD (2011) Metabolic engineering of microbial pathways for advanced biofuels production. Curr Opin Biotechnol 22:775–783. https://doi.org/10.1016/j.copbio.2011.04.024
16. Liu D, Evans T, Zhang F (2015) Applications and advances of metabolite biosensors for metabolic engineering. Metab Eng 31:35–43. https://doi.org/10.1016/j.ymben.2015.06.008
17. Majidian P, Tabatabaei M, Zeinolabedini M et al (2018) Metabolic engineering of microorganisms for biofuel production. Renew Sustain Energy Rev 82:3863–3885. https://doi.org/10.1016/j.rser.2017.10.085
18. Jagadevan S, Banerjee A, Banerjee C et al (2018) Recent developments in synthetic biology and metabolic engineering in microalgae towards biofuel production. Biotechnol Biofuels 11:1–21. https://doi.org/10.1186/s13068-018-1181-1
19. Costa JAV, De Morais MG (2011) The role of biochemical engineering in the production of biofuels from microalgae. Bioresour Technol 102:2–9. https://doi.org/10.1016/j.biortech.2010.06.014
20. Zorrilla López U, Masip Vilà G, Arjó Pont G et al (2013) Engineering metabolic pathways in plants by multigene transformation. https://doi.org/10.1387/ijdb.130162pc
21. Chisti Y (2007) Biodiesel from microalgae. Biotechnol Adv 25:294–306. https://doi.org/10.1016/j.biotechadv.2007.02.001
22. Gaurav N, Sivasankari S, Kiran GS et al (2017) Utilization of bioresources for sustainable biofuels: a review. Renew Sustain Energy Rev 73:205–214. https://doi.org/10.1016/j.rser.2017.01.070
23. Khoo KS, Ahmad I, Chew KW et al (2023) Enhanced microalgal lipid production for biofuel using different strategies including genetic modification of microalgae: a review. Prog Energy Combust Sci 96:101071. https://doi.org/10.1016/j.pecs.2023.101071
24. Bellou S, Baeshen MN, Elazzazy AM et al (2014) Microalgal lipids biochemistry and biotechnological perspectives. Biotechnol Adv 32:1476–1493. https://doi.org/10.1016/j.biotechadv.2014.10.003

25. Yu W-L, Ansari W, Schoepp NG et al (2011) Modifications of the metabolic pathways of lipid and triacylglycerol production in microalgae. Microb Cell Fact 10:1–11. https://doi.org/10.1186/1475-2859-10-91
26. Radakovits R, Jinkerson RE, Darzins A et al (2010) Genetic engineering of algae for enhanced biofuel production. Eukaryot Cell 9:486–501. https://doi.org/10.1128/ec.00364-09
27. Shokravi Z, Shokravi H, Chyuan OH et al (2020) Improving 'lipid productivity' in microalgae by bilateral enhancement of biomass and lipid contents: a review. Sustainability 12:9083. https://doi.org/10.3390/su12219083
28. Post-Beittenmiller D, Jaworski JG, Ohlrogge JB (1991) In vivo pools of free and acylated acyl carrier proteins in spinach. J Biol Chem 266:1858–1865. https://doi.org/10.1016/S0021-9258(18)52372-3
29. Post-Beittenmiller D, Roughan G, Ohlrogge JB (1992) Regulation of plant fatty acid biosynthesis: analysis of acyl-coenzyme A and acyl-acyl carrier protein substrate pools in spinach and pea chloroplasts. Plant Physiol 100:923–930. https://doi.org/10.1104/pp.100.2.923
30. Mulgund A (2022) Increasing lipid accumulation in microalgae through environmental manipulation, metabolic and genetic engineering: a review in the energy NEXUS framework. Energy Nexus 5:100054. https://doi.org/10.1016/j.nexus.2022.100054
31. Chang W-C, Zheng H-Q, Chen C-NN (2016) Comparative transcriptome analysis reveals a potential photosynthate partitioning mechanism between lipid and starch biosynthetic pathways in green microalgae. Algal Res 16:54–62. https://doi.org/10.1016/j.algal.2016.03.007
32. Liang M-H, Jiang J-G (2013) Advancing oleaginous microorganisms to produce lipid via metabolic engineering technology. Prog Lipid Res 52:395–408. https://doi.org/10.1016/j.plipres.2013.05.002
33. Dunahay TG, Jarvis EE, Dais SS et al (1996) Manipulation of microalgal lipid production using genetic engineering. Appl Biochem Biotechnol 57:223–231. https://doi.org/10.1007/BF02941703
34. Sheehan J, Dunahay T, Benemann J et al (1998) A look back at the US Department of Energy's aquatic species program: biodiesel from algae. Natl Renew Energy Lab 328:1–294. NREL/TP-580-24190
35. Gomma AE, Lee S-K, Sun SM et al (2015) Improvement in oil production by increasing malonyl-CoA and glycerol-3-phosphate pools in *Scenedesmus quadricauda*. Indian J Microbiol 55:447–455. https://doi.org/10.1007/s12088-015-0546-4
36. Zhang X, Agrawal A, San K (2012) Improving fatty acid production in *Escherichia coli* through the overexpression of malonyl coA-acyl carrier protein transacylase. Biotechnol Prog 28:60–65. https://doi.org/10.1002/btpr.716
37. Chen J, Liu W, Hu D et al (2017) Identification of a malonyl CoA-acyl carrier protein transacylase and its regulatory role in fatty acid biosynthesis in oleaginous microalga *Nannochloropsis oceanica*. Biotechnol Appl Biochem 64:620–626. https://doi.org/10.1002/bab.1531
38. Li Z, Meng T, Ling X et al (2018) Overexpression of malonyl-CoA: ACP transacylase in *Schizochytrium* sp. to improve polyunsaturated fatty acid production. J Agric Food Chem 66:5382–5391. https://doi.org/10.1021/acs.jafc.8b01026
39. Fan Y, Yuan C, Jin Y et al (2018) Characterization of 3-ketoacyl-CoA synthase in a nervonic acid producing oleaginous microalgae *Mychonastes afer*. Algal Res 31:225–231. https://doi.org/10.1016/j.algal.2018.02.017
40. Balamurugan S, Wang X, Wang H-L et al (2017) Occurrence of plastidial triacylglycerol synthesis and the potential regulatory role of AGPAT in the model diatom *Phaeodactylum tricornutum*. Biotechnol Biofuels 10:1–14. https://doi.org/10.1186/s13068-017-0786-0
41. Yamaoka Y, Achard D, Jang S et al (2016) Identification of a *Chlamydomonas plastidial* 2-lysophosphatidic acid acyltransferase and its use to engineer microalgae with increased oil content. Plant Biotechnol J 14:2158–2167. https://doi.org/10.1111/pbi.12572
42. Zou L-G, Chen J-W, Zheng D-L et al (2018) High-efficiency promoter-driven coordinated regulation of multiple metabolic nodes elevates lipid accumulation in the model microalga *Phaeodactylum tricornutum*. Microb Cell Fact 17:1–8. https://doi.org/10.1186/s12934-018-0906-y
43. Niu Y-F, Wang X, Hu D-X et al (2016) Molecular characterization of a glycerol-3-phosphate acyltransferase reveals key features essential for triacylglycerol production in *Phaeodactylum tricornutum*. Biotechnol Biofuels 9:1–11. https://doi.org/10.1186/s13068-016-0478-1
44. Iskandarov U, Sitnik S, Shtaida N et al (2016) Cloning and characterization of a GPAT-like gene from the microalga *Lobosphaera incisa* (Trebouxiophyceae): overexpression in *Chlamydomonas reinhardtii* enhances TAG production. J Appl Phycol

28:907–919. https://doi.org/10.1007/s10811-015-0634-1

45. Niu Y-F, Zhang M-H, Li D-W et al (2013) Improvement of neutral lipid and polyunsaturated fatty acid biosynthesis by overexpressing a type 2 diacylglycerol acyltransferase in marine diatom *Phaeodactylum tricornutum*. Mar Drugs 11:4558–4569. https://doi.org/10.3390/md11114558
46. Wei H, Shi Y, Ma X et al (2017) A type-I diacylglycerol acyltransferase modulates triacylglycerol biosynthesis and fatty acid composition in the oleaginous microalga *Nannochloropsis oceanica*. Biotechnol Biofuels 10:1–18. https://doi.org/10.1186/s13068-017-0858-1
47. Hung C-H, Ho M-Y, Kanehara K, Nakamura Y (2013) Functional study of diacylglycerol acyltransferase type 2 family in *Chlamydomonas reinhardtii*. FEBS Lett 587:2364–2370. https://doi.org/10.1016/j.febslet.2013.06.002
48. Iwai M, Ikeda K, Shimojima M, Ohta H (2014) Enhancement of extraplastidic oil synthesis in *Chlamydomonas reinhardtii* using a type-2 diacylglycerol acyltransferase with a phosphorus starvation–inducible promoter. Plant Biotechnol J 12:808–819. https://doi.org/10.1111/pbi.12210
49. Chen C, Kao A, Tsai Z et al (2016) Expression of type 2 diacylglycerol acyltransferase gene DGTT1 from enhances lipid production in *Scenedesmus obliquus*. Biotechnol J 11:336–344. https://doi.org/10.1002/biot.201500272
50. Zienkiewicz K, Zienkiewicz A, Poliner E et al (2017) *Nannochloropsis*, a rich source of diacylglycerol acyltransferases for engineering of triacylglycerol content in different hosts. Biotechnol Biofuels 10:1–20. https://doi.org/10.1186/s13068-016-0686-8
51. Li D-W, Cen S-Y, Liu Y-H et al (2016) A type 2 diacylglycerol acyltransferase accelerates the triacylglycerol biosynthesis in heterokont oleaginous microalga *Nannochloropsis oceanica*. J Biotechnol 229:65–71. https://doi.org/10.1016/j.jbiotec.2016.05.005
52. Hsieh H-J, Su C-H, Chien L-J (2012) Accumulation of lipid production in *Chlorella minutissima* by triacylglycerol biosynthesis-related genes cloned from *Saccharomyces cerevisiae* and *Yarrowia lipolytica*. J Microbiol 50:526–534. https://doi.org/10.1007/s12275-012-2041-5
53. Beacham TA, Ali ST (2016) Growth dependent silencing and resetting of DGA1 transgene in *Nannochloropsis salina*. Algal Res 14:65–71. https://doi.org/10.1016/j.algal.2016.01.005
54. La Russa M, Bogen C, Uhmeyer A et al (2012) Functional analysis of three type-2 DGAT homologue genes for triacylglycerol production in the green microalga *Chlamydomonas reinhardtii*. J Biotechnol 162:13–20. https://doi.org/10.1016/j.jbiotec.2012.04.006
55. Deng X-D, Gu B, Li Y-J et al (2012) The roles of acyl-CoA: diacylglycerol acyltransferase 2 genes in the biosynthesis of triacylglycerols by the green algae *Chlamydomonas reinhardtii*. Mol Plant 5:945–947. https://doi.org/10.1093/mp/sss040
56. Ma Y-H, Wang X, Niu Y-F et al (2014) Antisense knockdown of pyruvate dehydrogenase kinase promotes the neutral lipid accumulation in the diatom *Phaeodactylum tricornutum*. Microb Cell Fact 13:1–9. https://doi.org/10.1186/s12934-014-0100-9
57. Trentacoste EM, Shrestha RP, Smith SR et al (2013) Metabolic engineering of lipid catabolism increases microalgal lipid accumulation without compromising growth. Proc Natl Acad Sci 110:19748–19753. https://doi.org/10.1073/pnas.1309299110
58. Dehesh K, Tai H, Edwards P et al (2001) Overexpression of 3-ketoacyl-acyl-carrier protein synthase IIIs in plants reduces the rate of lipid synthesis. Plant Physiol 125:1103–1114. https://doi.org/10.1104/pp.125.2.1103
59. Sun X-M, Ren L-J, Zhao Q-Y et al (2019) Enhancement of lipid accumulation in microalgae by metabolic engineering. Biochim Biophys Acta Mol Cell Biol Lipids 1864:552–566. https://doi.org/10.1016/j.bbalip.2018.10.004
60. Klok AJ, Lamers PP, Martens DE et al (2014) Edible oils from microalgae: insights in TAG accumulation. Trends Biotechnol 32:521–528. https://doi.org/10.1016/j.tibtech.2014.07.004
61. Shan D, Li J, Wu L et al (2010) GPAT3 and GPAT4 are regulated by insulin-stimulated phosphorylation and play distinct roles in adipogenesis. J Lipid Res 51:1971–1981. https://doi.org/10.1194/jlr.M006304
62. Oelkers P, Cromley D, Padamsee M et al (2002) The DGA1 gene determines a second triglyceride synthetic pathway in yeast. J Biol Chem 277:8877–8881. https://doi.org/10.1074/jbc.M111646200
63. Cases S, Smith SJ, Zheng Y-W et al (1998) Identification of a gene encoding an acyl CoA: diacylglycerol acyltransferase, a key enzyme in triacylglycerol synthesis. Proc Natl Acad Sci 95:13018–13023. https://doi.org/10.1073/pnas.95.22.13018
64. Li J, Han D, Wang D et al (2014) Choreography of transcriptomes and lipidomes of *Nannochloropsis* reveals the mechanisms of oil synthesis in microalgae. Plant Cell 26:1645–1665. https://doi.org/10.1105/tpc.113.121418

65. Boyle NR, Morgan JA (2009) Flux balance analysis of primary metabolism in *Chlamydomonas reinhardtii*. BMC Syst Biol 3:1–14. https://doi.org/10.1186/1752-0509-3-4
66. Xue J, Niu Y-F, Huang T et al (2015) Genetic improvement of the microalga *Phaeodactylum tricornutum* for boosting neutral lipid accumulation. Metab Eng 27:1–9. https://doi.org/10.1016/j.ymben.2014.10.006
67. Eroglu E, Melis A (2016) Microalgal hydrogen production research. Int J Hydrogen Energy 41:12772–12798. https://doi.org/10.1016/j.ijhydene.2016.05.115
68. Guan Y, Deng M, Yu X et al (2004) Two-stage photo-biological production of hydrogen by marine green alga *Platymonas subcordiformis*. Biochem Eng J 19:69–73. https://doi.org/10.1016/j.bej.2003.10.006
69. Melis A, Melnicki MR (2006) Integrated biological hydrogen production. Int J Hydrogen Energy 31:1563–1573. https://doi.org/10.1016/j.ijhydene.2006.06.038
70. Johanningmeier U, Fischer D (2010) Perspective for the use of genetic transformants in order to enhance the synthesis of the desired metabolites: engineering chloroplasts of microalgae for the production of bioactive compounds. Bio-Farms Nutraceuticals Funct Food Saf Contr Biosens 2010:144–151. https://doi.org/10.1007/978-1-4419-7347-4_11
71. Gimpel JA, Specht EA, Georgianna DR et al (2013) Advances in microalgae engineering and synthetic biology applications for biofuel production. Curr Opin Chem Biol 17:489–495. https://doi.org/10.1016/j.cbpa.2013.03.038
72. Shuba ES, Kifle D (2018) Microalgae to biofuels: 'Promising' alternative and renewable energy, review. Renew Sustain Energy Rev 81:743–755. https://doi.org/10.1016/j.rser.2017.08.042
73. Khetkorn W, Rastogi RP, Incharoensakdi A et al (2017) Microalgal hydrogen production–a review. Bioresour Technol 243:1194–1206. https://doi.org/10.1016/j.biortech.2017.07.085
74. Beer LL, Boyd ES, Peters JW, Posewitz MC (2009) Engineering algae for biohydrogen and biofuel production. Curr Opin Biotechnol 20:264–271. https://doi.org/10.1016/j.copbio.2009.06.002
75. Surzycki R, Cournac L, Peltier G et al (2007) Potential for hydrogen production with inducible chloroplast gene expression in *Chlamydomonas*. Proc Natl Acad Sci 104:17548–17553. https://doi.org/10.1073/pnas.0704205104
76. Esquível MG, Amaro HM, Pinto TS et al (2011) Efficient H_2 production via *Chlamydomonas reinhardtii*. Trends Biotechnol 29:595–600. https://doi.org/10.1016/j.tibtech.2011.06.008
77. Dubini A, Ghirardi ML (2015) Engineering photosynthetic organisms for the production of biohydrogen. Photosynth Res 123:241–253. https://doi.org/10.1007/s11120-014-9991-x
78. Kruse O, Rupprecht J, Bader KP, Thomas-Hall S, Schenk PM, Finazzi G, Hankamer B (2005) Improved photobiological H2 production in engineered green algal cells. J Biol Chem, 280(40), 34170–34177. https://doi.org/10.1074/jbc.M503840200
79. Zhang J, Xue D, Wang C et al (2023) Genetic engineering for biohydrogen production from microalgae. iScience 26. https://doi.org/10.1016/j.isci.2023.107255
80. Specht E, Miyake-Stoner S, Mayfield S (2010) Micro-algae come of age as a platform for recombinant protein production. Biotechnol Lett 32:1373–1383. https://doi.org/10.1007/s10529-010-0326-5
81. Heydarizadeh P, Poirier I, Loizeau D x(2013) Plastids of marine phytoplankton produce bioactive pigments and lipids. Mar Drugs 11:3425–3471. https://doi.org/10.3390/md11093425
82. Terashima M, Specht M, Hippler M (2011) The chloroplast proteome: a survey from the *Chlamydomonas reinhardtii* perspective with a focus on distinctive features. Curr Genet 57:151–168. https://doi.org/10.1007/s00294-011-0339-1
83. Rasala BA, Mayfield SP (2015) Photosynthetic biomanufacturing in green algae; production of recombinant proteins for industrial, nutritional, and medical uses. Photosynth Res 123:227–239. https://doi.org/10.1007/s11120-014-9994-7
84. Coragliotti AT, Beligni MV, Franklin SE, Mayfield SP (2011) Molecular factors affecting the accumulation of recombinant proteins in the *Chlamydomonas reinhardtii* chloroplast. Mol Biotechnol 48:60–75. https://doi.org/10.1007/s12033-010-9348-4
85. Molnar A, Bassett A, Thuenemann E et al (2009) Highly specific gene silencing by artificial microRNAs in the unicellular alga *Chlamydomonas reinhardtii*. Plant J 58:165–174.https://doi.org/10.1111/j.1365-313X.2008.03767.x
86. Cerutti H, Ma X, Msanne J, Repas T (2011) RNA-mediated silencing in algae: biological roles and tools for analysis of gene function. Eukaryot Cell 10:1164–1172. https://doi.org/10.1128/ec.05106-11
87. Purton S, Szaub JB, Wannathong T et al (2013) Genetic engineering of algal chloroplasts: progress and prospects. Russ J Plant Physiol 60:491–499. https://doi.org/10.1134/S1021443713040146

88. Anand V, Singh PK, Banerjee C, Shukla P (2017) Proteomic approaches in microalgae: perspectives and applications. 3 Biotech 7:1–10. https://doi.org/10.1007/s13205-017-0831-5
89. Banerjee C, Singh PK, Shukla P (2016) Microalgal bioengineering for sustainable energy development: recent transgenesis and metabolic engineering strategies. Biotechnol J 11:303–314. https://doi.org/10.1002/biot.201500284
90. Kasai Y, Oshima K, Ikeda F et al (2015) Construction of a self-cloning system in the unicellular green alga *Pseudochoricystis ellipsoidea*. Biotechnol Biofuels 8:1–12. https://doi.org/10.1186/s13068-015-0277-0
91. Shin SE, Lim JM, Koh HG et al (2016) CRISPR/Cas9-induced knockout and knock-in mutations in *Chlamydomonas reinhardtii*. Sci Rep 6:27810. https://doi.org/10.1038/srep27810
92. Sizova I, Greiner A, Awasthi M et al (2013) Nuclear gene targeting in *Chlamydomonas* using engineered zinc-finger nucleases. Plant J 73:873–882. https://doi.org/10.1111/tpj.12066
93. Arshad S, Qadir ML, Hussain N et al (2025) Advances in CRISPR/Cas9 technology: shaping the future of photosynthetic microorganisms for biofuel production. Funct Plant Biol 52. https://doi.org/10.1071/FP24255
94. Naduthodi MIS, Claassens NJ, D'Adamo S et al (2021) Synthetic biology approaches to enhance microalgal productivity. Trends Biotechnol 39:1019–1036. https://doi.org/10.1016/j.tibtech.2020.12.010
95. Stavridou E, Karapetsi L, Nteve GM et al (2024) Landscape of microalgae omics and metabolic engineering research for strain improvement: an overview. Aquaculture 740803. https://doi.org/10.1016/j.aquaculture.2024.740803
96. Abdullah M, Ali Z, Yasin MT et al (2024) Advancements in sustainable production of biofuel by microalgae: recent insights and future directions. Environ Res 119902. https://doi.org/10.1016/j.envres.2024.119902
97. Andrianantoandro E, Basu S, Karig DK, Weiss R (2006) Synthetic biology: new engineering rules for an emerging discipline. Mol Syst Biol 2:28–2006. https://doi.org/10.1038/msb4100073
98. Bashir KMI, Kim M-S, Stahl U, Cho M-G (2016) Microalgae engineering toolbox: selectable and screenable markers. Biotechnol Bioprocess Eng 21:224–235. https://doi.org/10.1007/s12257-015-0386-4
99. Jang Y-S, Park JM, Choi S et al (2012) Engineering of microorganisms for the production of biofuels and perspectives based on systems metabolic engineering approaches. Biotechnol Adv 30:989–1000. https://doi.org/10.1016/j.biotechadv.2011.08.015
100. Thiele I, Palsson BØ (2010) A protocol for generating a high-quality genome-scale metabolic reconstruction. Nat Protoc 5:93–121. https://doi.org/10.1038/nprot.2009.203

MIX
Papier aus verantwortungsvollen Quellen
Paper from responsible sources
FSC® C105338

FSC
www.fsc.org

If you have any concerns about our products,
you can contact us on
ProductSafety@springernature.com

In case Publisher is established outside the EU,
the EU authorized representative is:
Springer Nature Customer Service Center GmbH
Europaplatz 3, 69115 Heidelberg, Germany

Printed by Libri Plureos GmbH
in Hamburg, Germany